Organic
Substances
and Sediments
in Water

Volume 1

Humics and Soils

Robert A. Baker

Editor

LEWIS PUBLISHERS

Library of Congress Cataloging-in-Publication Data
Organic substances and sediments in water / Robert A. Baker, Editor.

    p.   cm.

    Papers from two symposia held at the American Chemical Society Meeting in Boston, Apr. 22–27, 1990.

    Includes bibliographical references and index.

    Contents: v. 1. Humics and soils — v. 2. Processes and analytical — v. 3. Biological.

    1. Organic water pollutants—Congresses. 2. Sediments, Suspended–Congresses.   3. Water chemistry—Congresses.   4. Biochemistry--Congresses. I. Baker, Robert Andrew, 1925- . II. American Chemical Society. Meeting (1990 : Boston, Mass.)
TD427.O70753   1991
628.1′68--dc20   91–7855
ISBN 0–87371–342–7 (v. 1)

LEWIS PUBLISHERS, INC.
121 South Main Street, Chelsea, Michigan 48118

PRINTED IN THE UNITED STATES OF AMERICA

To
Peggy Baker
for her continued understanding and encouragement

Robert A. Baker is affiliated with the National Research Program, Water Resources Division of the U.S. Geological Survey. He holds a BChE from North Carolina State University, MChE and MS from Villanova University, and a D.Sc. from the Graduate School of Public Health, University of Pittsburgh. Dr. Baker's professional career has involved research, consultation, and management related to environmental science and engineering problems. He is active in professional societies and has authored over 70 books, patents, and papers.

Preface

Water resources managers, regulators, and researchers require definitive information that describes the highly correlated, interdisciplinary factors that influence fate and transport of water contaminants. Not unexpectedly, evolving questions stay ahead of advances in scientific and engineering developments. One of the most important and significant aspects currently being intensely investigated is the role of particulates and sediments in contaminant behavior. This three-volume compilation documents the proceedings of a symposium dedicated to the subject of organic substances and sediments in water. Stress was placed on the organic substances because so many of the anthropogenic contaminants which pose potential problems at all trophic levels are organic in nature.

The symposium program from which the proceedings derive included critical reviews which describe the state-of-the-science, and often identify major needs. This should be especially valuable to the reader, regardless of individual interest. As in any symposium proceedings, topics are treated with varying depth. However, coverage over the interdisciplinary subject is reasonably complete.

The first volume delves into the roles of humic substances and soils-sediments in the sorption and mobility of contaminants. Both regimes are introduced by comprehensive review papers, and both reviews are followed by papers that treat specific topics in depth.

The second volume combines papers that summarize various processes involved in contaminant fate and transport as well as analytical developments. The processes section has been divided into aquatic particle-organic chemical interaction (characterization and contaminant geochemistry); fate and transport; and interfacial and organic-inorganic processes. The processes and analytical sections present theoretical as well as case study developments.

The third volume is devoted to biological processes. It begins with a state-of-the-science summary which incorporates references to the other papers deriving from the symposium. The papers are divided under subheadings: integrating chemistry and toxicology of sediment-water interactions; uptake and accumulation (bioavailability and bioaccumulation); biodegradation (aerobic dechlorinations and co-metabolism).

This compilation extends over the broad interdisciplinary subject of organic substances and sediments in water. It should prove valuable to experienced scientists as well as those making initial inquiries.

Acknowledgments

An American Chemical Society symposium on the subject "Contaminants and Sediments" was held in Honolulu, April 1–6, 1979. A two-volume publication of the same title was published by Ann Arbor Science in 1980. These publications have frequently been cited in the literature. Several years ago colleagues suggested that the writer consider organization of another symposium to foster technology transfer and to update the proceedings of the previous state-of-the-science summary. This led to the symposium "Organic Substances and Sediments in Water" held at the American Chemical Society Meeting in Boston, April 22–27, 1990. The symposium emphasized organic substances and the complex processes effecting their fate and transport, particularly as these occur at the interface of suspended and fixed surfaces. Interdisciplinary contributions were solicited and development of topical sessions shared with recognized experts. These were: V. D. Adams, Tennessee Technical University; D. Armstrong, University of Wisconsin; S.A. Boyd, Michigan State University; C.T. Chiou, U.S. Geological Survey; B. Dempsey, Pennsylvania State University; B.J. Eadie, Great Lakes Environmental Research Laboratories; S.J. Eisenreich, University of Minnesota; P.F. Landrum, Great Lakes Environmental Research Laboratory; J. Leenheer, U.S. Geological Survey; R.L. Malcolm, U.S. Geological Survey; J.F. McCarthy, Oak Ridge National Laboratory; A.V. Palumbo, Oak Ridge National Laboratory; and A. Stone, Johns Hopkins University. Their dedication and cooperation was of the finest from onset through final manuscript peer review.

In addition to North American participants, scientists and engineers from other continents contributed. Five invited European scientists were: Jacques Buffle, University of Geneva, Switzerland; Hans Borén, Linköping University, Sweden; Egil Gjessing, Norwegian Institute for Water Research, Oslo, Norway; Jussi Kukkonen, University of Joensuu, Finland; and Paolo Sequi, Istituto D. Chimica Agraria, Bologna, Italy. Their perceptions and comments were as valuable as their technical contributions. A grant from the U.S. Environmental Protection Agency provided travel support for the invited speakers. Louis Swaby, Office of Exploratory Research, Washington, DC, and Wayne Garrison, Environmental Research Laboratory, Athens, Georgia provided program development assistance and liaison.

Chemical sciences are often an integral aspect of scientific and engineering processes perceived as nonchemical in nature. To improve knowledge of such situations and to facilitate communication among interdisciplinary contributions, the American Chemical Society, through its Committee on Science, has established a Pedagogical Symposium program. These tutorial symposia typically offer overview and research presentations by acknowledged experts in related fields. A competitive proposal to conduct a pedagogical symposium on the same subject as the research symposium was awarded by the Committee on Science. The tutorial was held on April 24, 1990. The lecturers were: E.J. Bouwer, Johns Hopkins University; D.M.D. Toro, Manhattan College; J.W. Farrington, University of Massachusetts; I. Knight, University of Maryland; J.R. Pratt, Pennsylvania State University; R.E. Speece, Vanderbilt University; and J.A. Symons, University of Houston. Their presentations dramatically demonstrated the interdependence of various scientific and engineering processes as well as the benefits of interdisciplinary technology transfer. Drs. Pratt, Speece, and Knight contributed papers to these proceedings.

Financial support for the pedagogical symposium was from the Committee on Science and from the Environmental Chemistry Division of the American Chemical Society. The research symposium and the pedagogical symposium were held under the auspices of the Environmental Chemistry Division. Encouragement and support of the officers and members of these organizational units is gratefully acknowledged.

The endeavor would have been of no avail without the contribution of the scientists and engineers whose manuscripts are contained in these proceedings. The editor appreciates their willingness to share knowledge.

Contents

PART I
HUMIC AND OTHER SUBSTANCES

1. Organic Substance Structures That Facilitate Contaminant
    Transport and Transformations in Aquatic Sediments,
    J.A. Leenheer 3

2. The Importance of Humic Substance-Mineral Particle
    Complexes in the Modeling of Contaminant Transport in
    Sediment-Water Systems, *R.L. Wershaw* 23

3. Composition of Humin in Stream Sediment and Peat,
    James A. Rice and Patrick MacCarthy....................... 35

4. Particulate and Colloidal Organic Material in Pueblo Reservoir,
    Colorado: Influence of Autochthonous Source on Chemical
    Composition, *James F. Ranville, Richard A. Harnish, and
    Diane M. McKnight* 47

5. Evidence for the Diffusion of Aquatic Fulvic Acid from the
    Sediments of Lake Fryxell, Antarctica, *George Aiken,
    Diane McKnight, Robert Wershaw, and Laurence Miller* 75

6. Change in Properties of Humic Substances by H_2SO_4
    Acidification, *Egil T. Gjessing, Harry Efraimsen,
    Magne Grande, Torsten Källqvist, and Gunnhild Riise* 89

7. Polynuclear Aromatic Hydrocarbon Binding to Natural Organic
    Matter: A Comparison of Natural Organic Matter Fractions,
    *Gary L. Amy, Martha H. Conklin, Houmao Liu, and
    Christopher Cawein* 99

8. Calorimetric Acid-Base Titrations of Fulvic Acid,
    Mike Machesky 111

9. The Transport and Composition of Humic Substances in
    Estuaries, *Lewis E. Fox* 129

10.   The Hydrolysis of Suwannee River Fulvic Acid,
      Ronald C. Antweiler.   163

PART II
SORPTION INTERACTIONS WITH SOILS, SEDIMENTS, AND DISSOLVED ORGANIC MATTER

11.   Immobilization of Organic Contaminants by Organo-Clays:
      Application to Soil Restoration and Hazardous Waste
      Containment, *Stephen A. Boyd, William F. Jaynes, and
      Brenda S. Ross*   181

12.   Effects of Surfactants on the Mobility of Nonpolar Organic
      Contaminants in Porous Media, *James A. Smith, David M.
      Tuck, Peter R. Jaffé, and Robert T. Mueller*   201

13.   The Effects of Pore-Water Colloids on the Transport of
      Hydrophobic Organic Compounds from Bed Sediments, *G.J.
      Thoma, A. C. Koulermos, K. T. Valsaraj, D. D. Reible, and
      L. J. Thibodeaux*   231

14.   A Thermodynamic Partition Model for Binding of Nonpolar
      Organic Compounds by Organic Colloids and Implications for
      Their Sorption to Soils and Sediment, *Yu-Ping Chin,
      Walter J. Weber, Jr., and Cary T. Chiou*   251

15.   Applicability of Linear Partitioning Relationships for Sorption
      of Organic Vapors onto Soil and Soil Minerals, *S. K. Ong,
      S. R. Lindner, and L. W. Lion*   275

16.   Competitive Effects in the Sorption of Nonpolar Organic
      Compounds by Soils, *Joseph J. Pignatello*   291

PART III
BIODEGRADATION OF ORGANIC CONTAMINANTS IN SOILS AND SEDIMENTS

17.   Biodegradation of PCBs by Aerobic Microorganisms,
      Peter Adriaens, Chi-Min Huang, and Dennis D. Focht   311

18.   Microbial Oxidation of Natural and Anthropogenic Aromatic
      Compounds Coupled to Fe(III) Reduction, *Debra J. Lonergan
      and Derek R. Lovley*   327

19.   Occurrence and Speciation of Naturally Produced
      Organohalogens in Soil and Water, *A. Grimvall, H. Borén, and
      G. Asplund*   339

20. Organic Fertilizers and Humification in Soil, *Paolo Sequi, Claudio Ciavatta, and Livia Vittori Antisari* 351

List of Authors... 369

Index 373

Organic Substances and Sediments in Water

Volume 1

Humics and Soils

PART I

HUMIC AND OTHER SUBSTANCES

Organic Substance Structures That Facilitate Contaminant Transport and Transformations in Aquatic Sediments

J. A. Leenheer

INTRODUCTION

Organic substances in aquatic sediments are known to be an important phase for contaminant partitioning and transport, but detailed mechanistic understanding of organic substance interactions with various contaminants is generally lacking because of the chemical and physical complexity of organic substances in various aquatic environments. Organic substances in aquatic sediment are usually different than soil organic matter because of autochthonous organic inputs to the sediment; therefore, studies of contaminant-binding mechanisms to soil organic matter might not be directly applicable.[1] Studies of contaminant binding with suspended sediments have been hindered by difficulties associated with dewatering requirements needed to obtain sufficient sample quantities, and by different operational definitions of particulate, colloidal, and dissolved organic fractions.

The objectives of this chapter are to provide an overview of potential organic substance structures and mechanisms that bind contaminants to aquatic sediments, to describe the assessment research being performed on the nature of organic substances in aquatic sediments, and to discuss various physical and chemical processes that transform organic substances in sediment and alter contaminant binding. Research needs and new approaches to address these research needs will also be presented.

CONTAMINANT-BINDING MECHANISMS

Most aquatic sediments consist of a clay matrix coated with aluminum, iron, and manganese sesquioxides; viable organic biota, such as bacteria and phytoplankton; and nonviable organic substances. A diagrammatic representation of sediment by Jenne[2] is shown in Figure 1.1. Nonviable organic coatings are

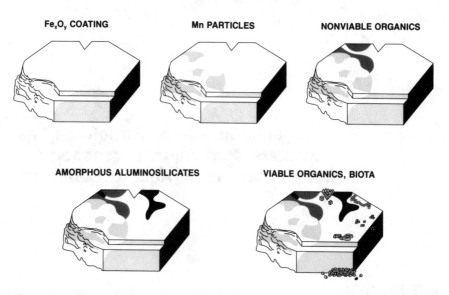

Figure 1.1. Diagrammatic representation of the important organic and inorganic coatings on the surface of a clay mineral particle. From Jenne.[2]

hypothesized to consist of a heterogeneous mixture of proteins, carbohydrates, lipids, humic substances, and aromatic geopolymers, such as coal or charcoal, bound to a clay matrix, as depicted in Figure 1.2A.

Nonionic, nonpolar organic contaminants have been found to partition by a solubility mechanism into organic matter coatings.[3,4] A review of literature describing contaminant-partitioning interactions has been compiled by Chiou.[5] The partitioning of the contaminant 2,3,7,8-tetrachlorodibenzodioxane into the lipid tripalmitin is illustrated in Figure 1.2B. Aromatic carbon adsorbents are solids onto which nonionic, nonpolar contaminants, such as a PCB isomer, adsorb as illustrated in Figure 1.2C. The presence of natural organic substances has been found to diminish the adsorptive capacity and retard the adsorptive kinetics for nonionic, nonpolar contaminant interactions with activated carbon.[6]

Nonionic organic contaminants with aromatic pi electrons might form charge-transfer complexes (pi bonds) with electron donating or accepting moieties in organic matter. This reaction is illustrated in Figure 1.2D by the pi complex between 2,3,5,6-tetrachloro-p-benzoquinone (chloranil) and an aromatic moiety in a humic acid molecule. Charge-transfer interactions are evidenced by changes in the electronic spectra of contaminants, such as chloranil.[7]

Moderately polar, organic base contaminants might be adsorbed through a combination of hydrogen bonding, cation exchange, and nonpolar partitioning interactions.[8] In Figure 1.2E, atrazine interacts with the nonpolar lipid portion of the organic coating by nonpolar partitioning interactions, and with a carboxyl group of a dicarboxylic acid through hydrogen bonding. Diquat, a divalent cation of the bipyridylium class of herbicides, interacts with adjacent carboxylate groups on a humic acid molecule by an electrostatic cation-exchange interaction, as shown in Figure 1.2F.

Both polar and nonpolar organic contaminants can form a variety of bound conjugates with organic matter through biochemical processes.[1] A glycoside conjugate of chloropropham, a pesticide of the carbamate class, is illustrated in Figure 1.2E.[9] Bound conjugates are difficult to cleave from the organic substrata for contaminant-residue analysis.

A variety of metals, ranging from the major cations to trace metals in water, interact with organic coatings on sediment. Proteins containing the amino acid cysteine that bind trace metals such as copper, cadmium, mercury, silver, and zinc are known as metallothioniens,[10] and are produced by organisms to detoxify trace metals. A mercury cysteine complex is shown in Figure 1.2G.

Humic substances are known for the ability to complex a variety of trace metals; a review of chemical structures and functional groups in humic substances that complex trace metals is given by Stevenson.[11] Schnitzer[12] and Gamble et al.[13] concluded that functional group arrangements similar to salicylate and phthalate groups were responsible for much of trace-metal binding in humic substances. More recently, substituted malonate groups have been found to be important in certain fulvic acids from soil[14] and water.[15] Copper complexed to a malonic acid group in a sediment-bound fulvic acid structure is illustrated in Figure 1.2H.

NATURE OF ORGANIC SUBSTANCES IN AQUATIC SEDIMENTS

Clay/Organic Matter Bonding Mechanisms

The nature of organic coatings on sediments is partly determined by the surface mineralogy and bonding mechanisms of organic coatings to inorganic substrates. Stevenson presents a comprehensive summary of clay/humic substance bonding mechanisms, which he divides into primary interactions, secondary interactions, and indirect interactions.[16]

Primary interactions are polar interactions involving both electrostatic and covalent bonding. Cation bridging of carboxylate groups on the protein side chains through divalent calcium and magnesium with the negatively charged clay mineral surface is illustrated in Figure 1.2. Direct cation-exchange interactions are possible through protonated amino groups in basic amino acids. Anion-exchange interactions are possible between organic carboxylate groups and protonated basic sesquioxide coatings containing mainly aluminum, iron, and manganese. Ligand exchange is a mechanism that exchanges an inorganic hydroxyl group in a sesquioxide coating for an organic carboxyl group; infrared spectral evidence indicates that both oxygens in the carboxyl group interact with the metal through a bidentate binding mechanism.[17] Terminal carboxyl

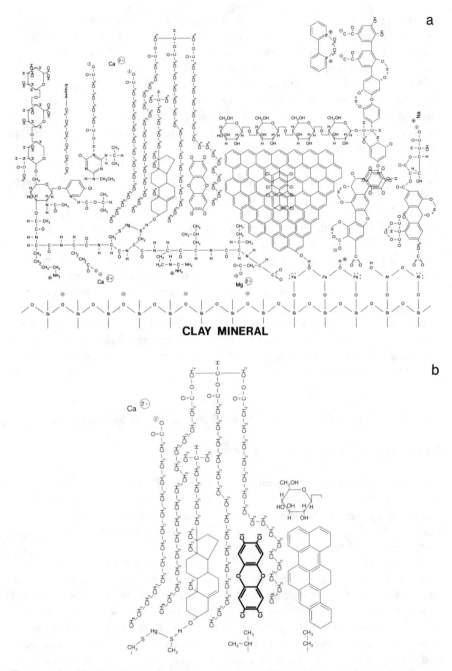

Figure 1.2. *(A)* Hypothetical chemical structures of a nonviable organic coating and associated contaminants on the surface of a clay mineral particle. Chemical structures are shown that bind *(B)* 2,3,7,8-tetrachlorodibenzodioxane, *(C)* PCB isomer, *(D)* chloranil, *(E)* atrazine and chloropropham, *(F)* diquat, *(G)* mercuric ion, and *(H)* cupric ion.

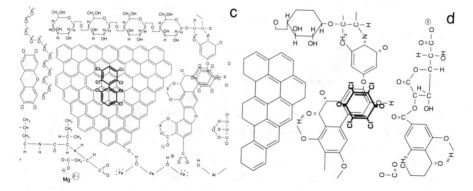

c

d

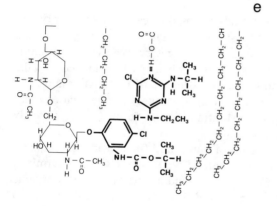

e

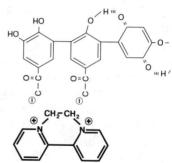

f

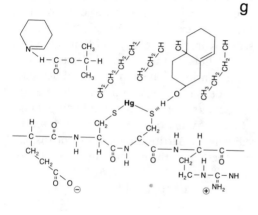

g

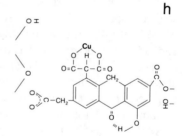

h

groups in the protein, humic acid, and fulvic acid structures are ligand exchanged to the sesquioxide coating on the clay mineral in Figure 1.2.

Secondary interactions include hydrogen bonding and physical (or Van der Waals) sorption between nonpolar surfaces. Hydrogen bonding might occur directly between polar organic hydroxyl, carboxyl, amide, and amino functional groups, and oxide and hydroxide surfaces on clay minerals, or they might occur indirectly through water bridging between polar functional groups and hydrated ions. Hydrogen bonding also solvates and desorbs polar organic coatings in aqueous systems. Physical sorption between nonpolar surfaces is illustrated in Figure 1.2 by the interface between the lipid, tripalmitin, and the nonpolar portions of the amino acids, leucine and isoleucine.

The major indirect interaction is the entropy gain that results from the displacement of water from the clay mineral surfaces as a result of primary and secondary interactions. Larger entropy gains result from adsorption of large organic macromolecules with multifunctional groups on mineral surfaces than from the adsorption of monofunctional, low–molecular weight, organic molecules.

Proteins have many polar functional groups that can interact by primary interactions with clay minerals as illustrated in Figure 1.2. Scharpenseel and Kruse found that amino acids form strong bridges between clay and humic acids; the organic coatings that resulted could only be partly removed with 0.5 M NaOH extraction.[18] They also found that addition of Fe^{3+}, Al^{3+}, and Ca^{2+} to mixtures of clay and humic acid resulted in strongly bound coatings that they attributed to cation bridging. Montmorillonite formed more strongly bound organic coatings than did kaolinite.

Ligand exchange of carboxyl groups of humic and fulvic acid with iron and aluminum hydrous oxides is well documented.[19-21] Davis and Gloor determined that colloidal alumina from a lake in Switzerland adsorbed dissolved organic matter in the 1000–3000 dalton size range.[22] Natural organic acids bound to hydrous oxide coatings can be extracted by aqueous solutions of complexing agents, such as pyrophosphate, oxalate, and citrate, which dissolve the hydrous oxide coating.[23,24]

Evidence for secondary interactions (hydrogen bonding and Van der Waals forces) that bind organic coatings to inorganic sediments and to other organic coatings is based on studies of organic solvent extraction. Organic solvents and solvent mixtures with water that disrupt hydrogen bonding and Van der Waals forces are the best extracting agents for lipids and moderately polar, humic acid components of soils and sediments. Various solvent mixtures of acetone, N,N-dimethylforamide, dimethylsulfoxide, sulfolane, and pyridine with water and acid are moderately effective extractants of humic and lipid components of organic coatings.[25,26] Rice and MacCarthy discovered that lipids could be efficiently extracted from stream sediment by partitioning a sediment suspension between water and methylisobutyl ketone as the pH and ionic strength of the aqueous phase are decreased.[27]

A bilayer membrane model recently has been proposed for organic coatings on sediment.[28] In the proposed model, polar functionalities that are responsible for primary interactions with mineral surfaces and that are solvated with water are found on the exterior surface of the membrane, and nonpolar hydrocarbon structures, of which lipids are a part, are in the unsolvated interior portion of the membrane. The hypothetical chemical structures of the organic coating in Figure 1.2 are organized based on the bilayer membrane model.

Dependence of Organic Matter Composition on Sediment Particle Size

The general inverse relation of organic matter content with the particle size of aquatic sediments has long been known.[29] The obvious causes of this relation are that a greater proportion of the volume of a particle is occupied by a surface coating as the particle size decreases, and smaller particles, such as clay minerals, frequently have a greater reactivity to form organic coatings than do larger particles, such as silicate sands. The change in the nature of organic coatings as a function of particle size, however, is not well understood. To obtain a better understanding of the nature of organic coatings as a function of sediment particle size, the author conducted a number of elemental and spectrometric studies on suspended-sediment size fractions isolated from the lower Mississippi River and major tributaries during a sampling cruise in May–June 1988.

The procedures used to representatively sample the suspended sediment and to fractionate the sediment into various size fractions are described by Leenheer et al.[30] The size range of the colloid fractions were from 0.005 μm (cutoff of ultrafiltration membrane) to 0.2 μm (cutoff for clay minerals in continuous-flow centrifuge). The mineral colloid fraction was isolated by centrifugation of the concentrate obtained by tangential-flow ultrafiltration, and the organic colloid fraction was obtained by dialysis of the supernatant solution obtained after centrifugation.

The elemental analysis part of the study was to determine organic carbon and nitrogen contents of the various size fractions, and to use the atomic C/N ratios to distinguish allochthonous from autochthonous inputs, as described by Pocklington and Tan.[31] The results for organic elemental composition of suspended-sediment fractions from the Mississippi River are presented in Table 1.1. The organic carbon content systematically increases as the particle size decreases for each site listed in Table 1.1. An unexpected result, however, was finding a minimum in the C/N ratio for the silt-size fraction. A companion study determined that the silt fraction, after ultrasonic dispersion, was largely aggregated colloids in the submicrometer size range, and that the mineral colloid fraction was not significantly different in size from the silt fraction after dispersion.[32] The range of C/N ratios (7.2 to 9.5) is only slightly larger than the Redfield C/N ratio of 6.9 for phytoplankton, which might indicate a large percentage of autochthonous carbon in the suspended silt fraction; however, soil organic nitrogen also is enriched in the clay fraction, possibly as a

Table 1.1. Organic-Carbon and Nitrogen Contents of Size Fractions of Suspended Sediment from the Mississippi River and Major Tributaries, May–June, 1988

Sediment Size Fraction and Analysis	Illinois R. at Meredosia, Illinois	Missouri R. at Herman, Missouri	Mississippi R. at St. Louis, Missouri	Ohio R. at Olmsted, Illinois	Mississippi R. at Hickman, Kentucky	Mississippi R. at Fulton, Tennessee	Mississippi R. at Helena, Arkansas	Yazoo R. above (mile 10) Vicksburg, Mississippi	Mississippi R. below Vicksburg, Mississippi
Sand									
Organic carbon (%)	NA	NA	0.28	NA	NA	0.86	NA	NA	0.63
Nitrogen (%)	NA	NA	0.06	NA	NA	0.07	NA	NA	0.06
Carbon/nitrogen			5.5			14.3			12.3
Silt									
Organic carbon (%)	8.67	3.52	5.21	3.67	3.28	3.25	3.71	3.29	3.80
Nitrogen (%)	1.31	0.50	0.64	0.54	0.47	0.49	0.54	0.53	0.53
Carbon/nitrogen	7.7	8.21	9.5	7.9	8.1	7.7	8.0	7.2	8.4
Mineral colloid									
Organic carbon (%)	20.00	10.70	14.91	18.98	14.45	12.64	12.13	12.48	10.98
Nitrogen (%)	2.60	1.05	1.59	2.36	1.77	1.59	1.47	1.16	1.49
Carbon/nitrogen	9.0	11.90	10.9	9.4	9.5	9.3	9.6	12.5	8.6
Organic colloid									
Organic carbon (%)	45.51	NA	NA	NA	40.64	NA	42.89	NA	25.16
Nitrogen (%)	3.77	NA	NA	NA	3.66	NA	3.59	NA	1.87
Carbon/nitrogen	14.1				13.0		13.9		15.7
Fulvic acid									
Organic carbon (%)	49.94	NA	NA	NA	50.74	NA	51.33	NA	50.32
Nitrogen (%)	2.90	NA	NA	NA	2.99	NA	2.88	NA	3.00
Carbon/nitrogen	20.1				19.8		20.8		19.6

result of aggregating into silt-size particles. Proteins and peptides have C/N ratios in the 3 to 4 range; therefore, the suspended silts may contain 25 to 50% proteinaceous carbon, and these proteins probably are important in forming organic coatings and in causing colloid aggregation.

The silt fractions contained 88% or more of the total suspended-sediment mass for the data in Table 1.1; therefore, the data for the silt fractions approximates the whole sediment data. The low C/N ratio for the silt fractions is surprising, considering the fact that C/N ratios of soils in general range from 9 to 30. Malcolm and Durum also found the C/N ratio of suspended sediments in six rivers with diverse characteristics to be in the 7 to 9 range; they concluded that riverine sediments are significant accumulators of nitrogen.[33] The two colloid fractions in Table 1.1 have substantially greater levels of organic carbon, but higher C/N ratios than the silt fraction.

The organic hydrogen content and protein content of the sediment-size fractions were assessed by infrared spectrometry. The samples were treated with HCl prior to analysis to destroy the carbonate minerals, and the residue was incorporated into KBr pellets for analysis. The organic hydrogen content is evidenced by the C-H stretching band near 2940 cm^{-1}, and protein content is measured by the secondary amide II band (a $C = N$ stretching band) near 1540 cm^{-1}. The infrared spectra of the three sediment-size fractions plus fulvic acid are shown in Figure 1.3. The intensity of the amide II band consistently decreases relative to the C-H band, as the size fraction decreases until the amide II band is absent in the fulvic acid. This trend of decreasing protein content with decreasing particle size substantiates the interpretation of the C/N ratio data in Table 1.1.

The substantial organic carbon contents of the organic colloid fraction (Table 1.1), and the characteristics of its infrared spectrum in the 1200–1000 cm^{-1} region where the C-O stretch of carbohydrates is observed, suggest that the organic colloid can be described as a glycoprotein. This composition is similar to estuarine organic colloids isolated and characterized by Means and Wijayaratne.[34] These free glycoproteins are likely in chemical equilibrium with bound glycoproteins on clay surfaces and might be common components of both freshwater and estuarine systems with large primary and secondary productivity.

Dependence of Organic Matter Composition on Sedimentary Environment

Environmental variables that might affect aquatic sediment composition of organic matter are salinity changes as riverine sediments are transported into estuaries; dissolved-oxygen changes as sediments become buried; seasonal changes that affect primary and secondary productivity, temperature regimes, and allochthonous inputs, such as leaf fall and snowmelt; hydrologic variables, such as river stage and the placement of lakes and reservoirs within a river system; and various anthropogenic inputs such as hydrocarbons in con-

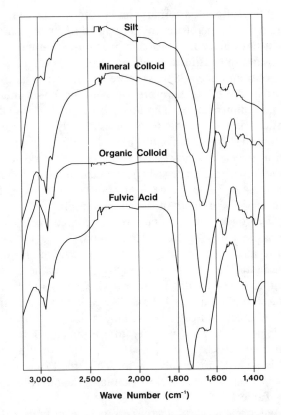

Figure 1.3.   Infrared spectra of suspended silt, mineral colloid, organic colloid, and fulvic acid isolated from the Mississippi River during June 1988. The suspended-silt and mineral colloid fractions were obtained below Hickman, Kentucky, and the organic colloid and fulvic acid fractions were obtained at Fulton, Tennessee.

taminated environments. In this chapter the composition and fate of organic material in sediments of rivers, lakes, estuaries, and oceans will be discussed.

An overview of the nature and amount of particulate organic carbon in 12 major world rivers is presented by Degens and Ittekkot.[35] They estimated that 5–30% of the particulate organic carbon consisted of labile sugars and amino acids that are readily decomposed, and that 70% of riverine particulate organic carbon undergoes decomposition after it enters the coastal marine environment. From analysis of specific amino acid distributions in riverine particulate organic matter in major world rivers, Ittekkot et al. concluded that soil organic matter was the major source of proteinaceous particulate components.[36] Hedges et al. reported lignin content, elemental abundances, and stable carbon isotope compositions for coarse ($> 63 \mu$m) and fine ($< 63 \mu$m) suspended sediments in the lower Amazon River.[37] They concluded that the fine particulate fraction was nitrogen rich and "consists of refractory, lignin-bearing soil humic material that is strongly associated with dispersed fine-grained minerals." Particulate organic matter in the coarse size fraction was composed of recently formed leaf debris and some wood.

Organic compositional data of a number of lake sediments are summarized by Ishiwatari.[38] Lipid content ranged between 3.8 and 7.7%; proteins, between 13 and 29%; carbohydrates, between 3 and 20%; humic acids, between 8 and 24%; and fulvic acid, between 6 and 23%. In contrast to sediments in rivers, organic matter in lake sediments is mainly autochthonous in origin. Characterization of the humic acid fraction from two lake sediments by ^{1}H nuclear magnetic resonance (NMR) spectrometry found that 40–50% of the carbon was in the form of aliphatic, alicyclic structures; no aromatic protons indicative of lignin residues were detected.[39] The Maillard reaction between carbohydrates and proteins produces melanoidin condensation products, which are thought to be a major source of humic substances in lake sediments.[40]

Compositional studies of marine sediments historically have emphasized bed sediments rather than suspended sediments. Particulate organic carbon comprises only a small component (< 10%) of total organic matter in the water column except in zones of high productivity, or in coastal areas of high turbulence or high terrestrial input.[41] The source of marine organic particulates is dominated by primary and secondary production, as in lakes; presumably, its composition is similar.

Vandenbroucke et al. discussed three stages of alteration of biochemicals from autochthonous inputs into humic substances and kerogen in the marine environment.[42] The first stage, which occurs in the water column, involves polycondensation reactions between reactive biopolymers, such as carbohydrates and proteins to form mainly insoluble humin-type polymers. The second stage is the oxidative degradation of humin to form humic and fulvic acids. In the last stage, humic substances are incorporated into anaerobic bottom sediments, oxygen-containing functional groups are lost, and organic matter evolves into kerogen of ancient bottom sediments.

Estuaries are the most complex of all aquatic environments with regard to the source and composition of aquatic sediments. Organic substances in estuarine environments can be regarded as a mixture of terrestrial and marine end members.[43] Depositional processes in estuaries include particulate flocculation and deposition with increasing salinity.[44] Field studies of organic particulate interactions, however, are difficult because of the complex hydrology that affects depositional processes in estuaries.[45]

Estuaries can be regarded as mixing chambers of terrestrial and marine organic substances where various reactions occur as a result of this mixing process. Terrestrial humic acid inputs to estuaries are lost from solution either because of aggregation and sedimentation[46] or biodegradation.[47] Eisma found that mucopolysaccharides that bind microflocs of fluvial origin are mobilized into the dissolved-carbon phase at low salinity in the riverine end of estuaries.[48] The estuarine zone of carbohydrate mobilization corresponded to deflocculation of particulate inputs by rivers. The particulates reflocculate at greater salinities in the seawater reaches of estuaries. Mayer noted that phenolic substances of fluvial origin mix with proteins and amino sugars produced in high productivity regions of estuaries,[43] and mixed phenol, protein, and sugar condensation products, as described by Martin et al.,[49] can result.

Estuaries generally have greater primary productivity than either the riverine or marine end members. Therefore, the composition of estuarine particulates generally is dominated more by the organic materials produced within the system than by materials of fluvial or marine origin. A study of accumulation rates of vascular plant-derived organic material in the lagoonal sediments of Cape Lookout Bight, North Carolina, found that vascular plant-derived carbon accounted for a maximum of 23 ± 17% of the total organic carbon buried in this sediment over the past decade.[50] Studies of mixing reactions of various constituents within estuaries are complicated by the dominance of autochthonous particulates, and the quantitative effect of mixing reactions on organic particulate composition is small.

TRANSFORMATIONS OF ORGANIC SUBSTANCES IN AQUATIC SEDIMENTS

Many types of organic matter are not stable when associated with aquatic sediments. Linear aliphatic and simple aromatic structures, simple sugars, and peptides are readily biodegraded; branched and cyclic aliphatic and substituted aromatic structures are more resistant to biological degradation.[51] Hatcher et al. found that biologically labile carbohydrates, proteins, and lipids were progressively lost from various sediment cores with increasing depth, and the refractory macromolecular organic substances that remained could be regarded as the precursors of kerogen.[52]

Abiotic processes that transform organic substances and associated contaminants in aquatic sediments include photolysis, hydrolysis, pH change, and various salt effects (salting out, salting in, and ion exchange). Organic acids coordinated to iron and manganese oxyhydroxide surfaces on sediment have been found to reduce iron and manganese when irradiated with sunlight.[53,54] These studies have focused more on the mechanisms of metal photoreduction than on organic acid oxidation. The quantitative significance and structural changes of humic and fulvic acid photooxidation catalyzed by metal oxyhydroxides in sediments are not well understood. There also is a possibility of direct photolysis of organic substances on sediments in a manner similar to photolysis of dissolved organic matter,[55] but the attenuation of light by sediment likely minimizes the effect.

Organic structures in sediments that can be altered by hydrolysis include amides, esters, and glycoside linkages of sugars. Amides and glycosides are not readily susceptible to abiotic hydrolysis under ambient conditions in water, and their hydrolysis is generally accomplished through enzymatic processes. Certain esters, however, are subject to abiotic hydrolysis in water under ambient conditions. Phosphate and sulfate esters with organic matter are hydrolyzed under acidic conditions.[56] Many plants contain hydrolyzable tannins that have many phenolic acids esterified to carbohydrates.[57] The half-lives of ester linkages of hydrolyzable tannins in water can vary from hours to centuries depending on the type of ester and its hydrolysis mechanism, the hydroxide ion concentration, and the nature of the electrical double layer near the ester that regulates hydroxide ion concentration.[58]

When aquatic sediments are introduced into acidic waters, they can floccu- late if the pH is below the zero point of charge of the clay colloids in the sediment.[59] This situation frequently occurs in surface waters acidified by mine drainage. A natural occurrence of pH-induced flocculation is illustrated at the confluence of the Rio Branco and Rio Negro in northern Brazil, where kaolin- itic sediments and their associated organic coatings are flocculated by the acidic (pH 4–5) black waters to form the unique river islands of the Anavilhanas Archipelago.[60]

Increasing salt concentrations in riverine sediments as they are deposited into estuaries have complex effects on organic substances and their associated contaminants. Aqueous solutions of increasing ionic strength suppress the electrical double layer and salt out dissolved organic substances and colloids that have significant nonpolar properties. The "salting out" reaction is illus- trated in reaction 1:

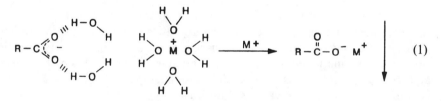

$$ \tag{1} $$

Nonpolar contaminants, such as DDT, show increased partitioning into humic acids as electrolyte salt concentrations are increased.[61]

A characteristic of proteins is that their solubility usually increases with salinity gradients encountered in estuaries.[62] This "salting in" reaction is illus- trated in reaction 2:

$$ \tag{2} $$

Means and Wijayaratne found that the binding of atrizine and linuron with glycoprotein colloids isolated from Chesapeake Bay decreased with increasing salinity.[63] They hypothesized that the decreased interaction might be due to changing properties of the sorbent, such as changes in molecular conformation.

Lastly, ion-exchange reactions occur at exchange sites on riverine sediment particles as they enter saltwater. An exchange of the divalent calcium ion, which aggregates sediment colloids, for the monovalent sodium ion, which disperses sediment colloids and organic coatings, was found to occur in the Mullica River estuary in New Jersey at low salinities.[64] At greater salinities, magnesium ion saturation of the sediment exchange sites increased. An ion-exchange reaction affecting bonding of organic coatings on a sediment surface is illustrated in reaction 3:

$$\underset{\text{Clay}^{\ominus}}{\boxed{}}\!\!\!\text{—}\,Ca^{2+}\text{—}O\text{—}\overset{\displaystyle O}{\overset{\|}{C}}\text{—}R + Na^{+} \longrightarrow \underset{\text{Clay}^{\ominus}}{\boxed{}} + Na^{+}\text{—}\overset{\ominus}{O}\text{—}\overset{\displaystyle O}{\overset{\|}{C}}\text{—}R \qquad (3)$$
$$+\ Ca^{2+}$$

A review of the effects of increasing salinity on the complexation capacity of various trace metals for organic colloids in estuaries noted increases and decreases in measured binding constants, depending on the origin of the colloid (riverine or marine) and on the size of the colloid.[43] The multiple effects of increasing salinity (salting out, salting in, and ion exchange) on sediment composition and contaminant binding make it difficult to predict and model contaminant-sediment interactions in estuaries.

RESEARCH NEEDS

A primary research need is interdisciplinary research studies that integrate the disciplines of chemistry, geology, hydrology, and microbiology to define complex relations of contaminant transport and transformations by organic substances in aquatic sediments. An even broader view provided by biologists and ecologists would be useful to understand the impact of contaminant interactions with sediments on various life-forms.

Improved analytical probes also are needed to determine equilibrium and kinetic relations of contaminant associations with organic substances in aquatic sediments. These probes might exist for a few contaminants such as copper, where specific ion electrodes are used, and polynuclear aromatic hydrocarbons, where fluorescence-quenching measurements are used, but for the majority of contaminants, no specific analytical probes exist apart from complex extraction and analysis procedures. Possibly immunoassay methods,[65] such as those developed for the triazine herbicides, can be developed and applied to determine contaminant associations with aquatic sediments.

Better tools are needed to characterize organic substance structures in aquatic sediments that bind contaminants. NMR spectrometry has been used to elucidate structure in aquatic humic and fulvic acids,[66] but Figure 1.4 illustrates that only poorly resolved structural information can be obtained by ^{13}C NMR spectrometry of whole sediment compared to isolates of dissolved organic substances. Perhaps the quality of ^{13}C NMR spectra of aquatic sediments could be improved by de-ashing the sediment with aqueous HCl/HF, as has been done with soil humin fractions.[67]

Various thermal-degradation methods coupled to gas chromatography and

MINERAL COLLOID
Mississippi River at St. Louis

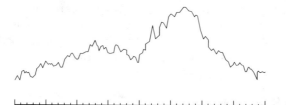

FULVIC ACID
Mississippi River at St. Louis

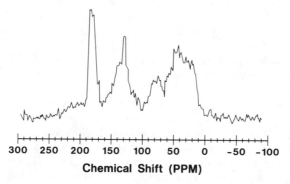

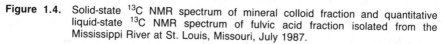

Chemical Shift (PPM)

Figure 1.4. Solid-state ^{13}C NMR spectrum of mineral colloid fraction and quantitative liquid-state ^{13}C NMR spectrum of fulvic acid fraction isolated from the Mississippi River at St. Louis, Missouri, July 1987.

mass spectrometry have been used to characterize natural organic matter in soils, sediments, and various geologic materials.[68,69] These promising techniques have the potential to correlate contaminant-binding properties of various aquatic sediments with the thermograms that provide organic compound class information.

Lastly, there is a need to develop selective extraction and isolation methods for the various organic components in aquatic sediments. As was mentioned at the beginning of this chapter, aquatic sediments are different from soil, and methods developed for the extraction of soil organic matter often do not work efficiently when applied to sediments. Extraction methods for natural organic substances that are more efficient than current extraction methods also might yield benefits in the extraction of various contaminants from aquatic sediments.

ACKNOWLEDGMENTS

The solid-state ^{13}C NMR spectrum shown in Figure 1.4 was obtained from the Colorado State University Regional NMR Center, which is funded by National Science Foundation Grant No. CHE-8616437. The liquid-state spectrum of fulvic acid (also in Figure 1.4) was obtained from Robert L. Wershaw of the U.S. Geological Survey, in Denver, Colorado.

REFERENCES

1. Kaufman, D. D., G. G. Still, G. D. Paulson, and S. K. Bandal, Eds. *Bound and Conjugated Pesticide Residues,* ACS Symposium Series 29 (Washington, DC: American Chemical Society, 1976).

2. Jenne, E. A. "Trace Element Sorption by Sediments and Soil Sites and Processes," in *Symposium on Molybdenum in the Environment,* W. A. Chappel and D. K. Peterson, Eds. (New York: Marcel Dekker, 1977), pp. 425–511.

3. Chiou, C. T., L. J. Peters, and V. H. Freed. "A Physical Concept of Soil-Water Equilibria for Nonionic Organic Compounds," *Science* 206:831–832 (1979).

4. Karickhoff, S. W., D. S. Brown, and T. A. Scott. "Sorption of Hydrophobic Pollutants on Natural Sediments," *Water Res.* 13:241–248 (1979).

5. Chiou, C. T. "Roles of Organic Matter, Minerals, and Moisture in Sorption of Nonionic Compounds in Pesticides by Soil," in *Humic Substances in Soil and Crop Science, Selected Readings,* P. MacCarthy, C. E. Clapp, R. L. Malcolm, and P. R. Bloom, Eds. (Madison, WI: American Society of Agronomy, 1990).

6. Zimmer, G., H.-J. Brauch, and H. Sontheimer. "Activated-Carbon Adsorption of Organic Pollutants," in *Aquatic Humic Substances: Influence on Fate and Treatment of Pollutants,* Advances in Chemistry Series 219, I. H. Suffet and P. MacCarthy, Eds. (Washington, DC: American Chemical Society, 1989), pp. 549–578.

7. Melcer, M. E., M. S. Zalewski, J. P. Hassett, and M. A. Brisk. "Charge-Transfer Interaction Between Dissolved Humic Materials and Chloranil," in *Aquatic Humic Substances: Influence on Fate and Treatment of Pollutants,* Advances in Chemistry Series 219, I. H. Suffet and P. MacCarthy, Eds. (Washington, DC: American Chemical Society, 1989), pp. 173–184.

8. Stevenson, F. J. "Organic Matter Reactions Involving Pesticides in Soil," in *Bound and Conjugated Pesticide Residues,* ACS Symposium Series 29, D. D. Kaufman, G. G. Still, G. D. Paulson, and S. K. Bandal, Eds. (Washington, DC: American Chemical Society, 1976), pp. 180–207.

9. Frear, D. S. "Pesticides Conjugates-Glycosides," in *Bound and Conjugated Pesticide Residues,* ACS Symposium Series 29, D. D. Kaufman, G. G. Still, G. D. Paulson, and S. K. Bandal, Eds. (Washington, DC: American Chemical Society, 1976), pp. 35–54.

10. Baum, R. M. "Research on Chemistry of Heavy Metals in Biology Forges Ahead," *Chem. Eng. News* 68(2):20–23 (1990).

11. Stevenson, F. J. *Humus Chemistry: Genesis, Composition, Reactions* (New York: John Wiley and Sons, 1982), pp. 337–354.

12. Schnitzer, M. "Reactions Between Organic Matter and Inorganic Soil Constituents," in *Transactions of the 9th International Congress on Soil Science,* J. W. Holmes, Ed. (New York: Elsevier, 1969), pp. 1635–1644.

13. Gamble D. S., M. Schnitzer, and I. Hoffman. "Cu^{2+}-Fulvic Acid Chelation Equilibrium in 0.1 M KCl at 25.00 C," *Can. J. Chem.* 48:3197–3204 (1970).
14. Gregor, J. E., and H. K. J. Powell. "Protonation Reactions of Fulvic Acids," *Soil Sci.* 39:243–252 (1988).
15. Leenheer, J. A., R. L. Wershaw, and M. M. Reddy. "Detection and Quantification of Substituted Malonic Acid Functional Groups in Fulvic Acid by ^{1}H-and ^{13}C-NMR Spectrometry," paper presented at the 197th meeting of the American Chemical Society, Miami Beach, FL, September 13, 1989.
16. Stevenson, F. J. *Humus Chemistry: Genesis, Composition, Reactions* (New York: John Wiley and Sons, 1982), pp. 374–402.
17. Kung, K.-H., and M. B. McBride. "Adsorption of Para-Substituted Benzoates on Iron Oxides," *Soil Sci. Soc. Amer. J.* 53(6):1673–1678 (1989).
18. Scharpenseel, H. W., and E. Kruse. "Amino Acids in Clay-Humic Acid Complex Formation," in *Isotopes and Radiation in Soil-Plant Relationships Including Forestry* (Vienna: International Atomic Energy Agency, 1972), pp. 193–210.
19. Parfitt, R. L., A. R. Fraser, and V. C. Farmer. "Adsorption on Hydrous Oxides. III. Fulvic Acid and Humic Acid on Goethite, Gibbsite, and Imogolite," *Soil Sci.* 28:289–296 (1977).
20. Stumm, W., R. Kummert, and L. Sigg. "A Ligand Exchange Model for the Adsorption of Inorganic and Organic Ligands at Hydrous Oxide Interfaces," *Croat. Chem. Acta* 53(2):291–312 (1980).
21. Davis, J. A. "Adsorption of Natural Dissolved Organic Matter at the Oxide/Water Interface," *Geochim. Cosmochim. Acta* 46:2381–2393 (1982).
22. Davis, J. A., and R. Gloor. "Adsorption of Dissolved Organics in Lake Water by Aluminum Oxide: Effect of Molecular Weight," *Environ. Sci. Technol.* 15(10):1223–1229 (1981).
23. Bremner, J. M., and H. Lees. "The Extraction of Organic Matter from Soil by Neutral Reagents," *J. Agri. Sci.* 39:274–279 (1949).
24. Gregor, J. E., and H. K. J. Powell. "Acid Pyrophosphate Extraction of Soil Fulvic Acids," *Soil Sci.* 37:577–585 (1986).
25. Porter, L. K. "Factors Affecting the Solubility and Possible Fractionation of Organic Colloids Extracted from Soil and Leonardite with an Acetone-H_2O-HCl Solvent," *J. Agr. Food Chem.* 15(5):807–811 (1967).
26. Hayes, M. H. B. "Extraction of Humic Substances from Soil," in *Humic Substances in Soil, Sediment, and Water: Geochemistry, Isolation, and Characterization,* G. R. Aiken, D. M. McKnight, R. L. Wershaw, and P. MacCarthy, Eds. (New York: John Wiley and Sons, 1985), pp. 329–362.
27. Rice, J. A., and P. MacCarthy. "Characterization of a Stream Sediment Humin," in *Aquatic Humic Substances: Influence on Fate and Treatment of Pollutants,* Advances in Chemistry Series 219, I. H. Suffet and P. MacCarthy, Eds. (Washington, DC: American Chemical Society, 1989), pp. 41–54.
28. Wershaw, R. L. "A New Model for Humic Materials and Their Interactions with Hydrophobic Organic Chemicals in Soil-Water or Sediment-Water Systems," *J. Cont. Hydrol.* 1:29–45 (1986).
29. Trask, P. D. "Organic Content of Recent Marine Sediments," in *Recent Marine Sediments, A Symposium,* P. D. Trask, Ed. (Tulsa, OK: American Association of Petroleum Geologists, 1939), pp. 428–453.
30. Leenheer, J. A., R. H. Meade, H. E. Taylor, and W. E. Pereira. "Sampling, Fractionation, and Dewatering of Suspended Sediment from the Mississippi River

for Geochemical and Trace-Contaminant Sampling" in *U.S. Geological Survey Toxic Substances Hydrology Program—Proceedings of the Technical Meeting, Phoenix, Arizona, September 26–30, 1988,* G. E. Mallard and S. E. Ragone, Eds., U.S. Geological Survey Water-Resources Inves. Rept. 88–4220 (Reston, VA: U.S. Geological Survey, 1989), pp. 501–512.

31. Pocklington, R., and F. Tan. "Organic Carbon Transport in the St. Lawrence River," in *Transport of Carbon and Minerals in Major World Rivers, Part 2,* E. T. Degens, S. Kempe, and H. Soliman, Eds. (Hamburg: Geol.-Paläont. Inst. Univ. Hamburg Heft 55, 1983), pp. 243–252.

32. Rees, T. F., J. A. Leenheer, and J. F. Ranville. "Use of a Single-Bowl Continuous-Flow Centrifuge for Dewatering Suspended Sediments: Effect on Sediment Physical and Chemical Characteristics," *J. Hydro. Proc.,* in press.

33. Malcolm, R. L., and W. H. Durum. "Organic Carbon and Nitrogen Concentrations and Annual Organic Carbon Load of Six Selected Rivers of the United States," U.S. Geological Water-Supply Paper 1817-F (Reston, VA: U.S. Geological Survey, 1976).

34. Means, J. C., and R. Wijayaratne. "Chemical Characterization of Estuarine Colloidal Organic Matter: Implications for Adsorptive Processes," *Bull. Marine Sci.* 35(3):449–461 (1984).

35. Degens, E. T., and V. Ittekkot. "Particulate Organic Carbon: An Overview," in *Transport of Carbon and Minerals in Major World Rivers, Part 3,* E. T. Degens, S. Kempe, and R. Herrera, Eds. (Hamburg: Geol.-Paläont Inst. Univ. Hamburg Heft 58, 1985), pp. 7–28.

36. Ittekkot, V., O. Martins, and R. Seifert. "Nitrogenous Organic Matter Transported by Major World Rivers," in *Transport of Carbon and Minerals in Major World Rivers, Part 2,* E. T. Degens, S. Kempe, and H. Soliman, Eds. (Hamburg: Geol.-Paläont. Inst. Univ. Hamburg Heft 55, 1983), pp. 119–128.

37. Hedges, J. I., W. A. Clark, P. D. Quay, J. E. Richey, A. H. Devol, and U. M. Santos. "Compositions and Fluxes of Particulate Organic Material in the Amazon River," *Limnol. Oceanogr.* 31(4):717–738 (1986).

38. Ishiwatari, R. "Geochemistry of Humic Substances in Lake Sediments," in *Humic Substances in Soil, Sediment, and Water: Geochemistry, Isolation, and Characterization,* G. R. Aiken, D. M. McKnight, R. L. Wershaw, and P. MacCarthy, Eds. (New York: John Wiley and Sons, 1985), pp. 147–180.

39. Ishiwatari, R. "Chemical Characterization of Fractionated Humic Acids from Lake and Marine Sediments," *Chem. Geol.* 12:113–126 (1973).

40. Hoering, T. C. "A Comparison of Melanoidin and Humic Acid," *Carnegie Inst. Washington Yearbook* 72:682–690 (1973).

41. MacKinnon, M. D. "The Measurement of Organic Carbon in Seawater," in *Marine Organic Chemistry: Evolution, Composition, Interactions and Chemistry of Organic Matter in Seawater,* Elsevier Oceanography Series 31, E. K. Duursma and R. Dawson, Eds. (Amsterdam: Elsevier, 1981), pp. 415–444.

42. Vandenbroucke, M., R. Pelet, and Y. Debyser. "Geochemistry of Humic Substances in Marine Sediments," in *Humic Substances in Soil, Sediment, and Water: Geochemistry, Isolation, and Characterization,* G. R. Aiken, D. M. McKnight, R. L. Wershaw, and P. MacCarthy, Eds. (New York: John Wiley and Sons, 1985), pp. 249–274.

43. Mayer, L. M. "Geochemistry of Humic Substances in Estuarine Environments," in *Humic Substances in Soil, Sediment, and Water: Geochemistry, Isolation, and*

Characterization, G. R. Aiken, D. M. McKnight, R. L. Wershaw, and P. Mac-Carthy, Eds. (New York: John Wiley and Sons, 1985), pp. 211–232.

44. Krone, R. B. *Flume Studies of the Transport of Sediment in Estuarial Shoaling Processes—Final Report* (Berkeley, CA: Univ. California Hydraulic Eng. Lab. and Sanitary Eng. Research Lab., 1962).

45. Meade, R. H. "Transport and Deposition of Sediments in Estuaries," Memoir 133, Geological Society of America, Inc. (1972), pp. 91–120.

46. Sholkovitz, E. R. "Flocculation of Dissolved Organic and Inorganic Matter During the Mixing of River Water and Seawater," *Geochim. Cosmochim. Acta* 40:831–845 (1976).

47. Fox, L. E. "The Removal of Dissolved Humic Acid During Estuarine Mixing," *Estuarine Coastal Shelf Sci.* 16:431–440 (1983).

48. Eisma, D. "Flocculation and De-flocculation of Suspended Matter in Estuaries," *Netherlands J. of Sea Res.* 20(2/3):183–199 (1986).

49. Martin, J. P., K. Haider, and E. Bondietti. "Properties of Model Humic Acids Synthesized by Phenoloxidase and Autooxidation of Phenols and Other Compounds Formed by Soil Fungi," in *Proceedings of the International Meeting on Humic Substances, Nieuwersluis, 1972* (Pudoc, Netherlands: Wageningen, 1975), pp. 171–186.

50. Haddad, R. I., and C. S. Martens. "Biogeochemical Cycling in an Organic-Rich Coastal Marine Basin: 9. Sources and Accumulation Rates of Vascular Plant-Derived Organic Material," *Geochim. Cosmochim. Acta* 51:2991–3001 (1987).

51. Gibson, D. T., Ed. *Microbial Degradation of Organic Compounds,* Microbiology Series, Volume 13 (New York: Marcel Dekker, 1984).

52. Hatcher, P. G., E. C. Spiker, N. M. Szeverenyi, and G. E. Maciel. "Selective Preservation and the Origin of Petroleum-Forming Aquatic Kerogen," *Nature* 305:498–501 (1983).

53. Waite, T. D., and F. M. M. Morel. "Photoreductive Dissolution of Colloidal Iron in Natural Waters," *Environ. Sci. Technol.* 18(11):860–868 (1984).

54. Cunningham, K. M., M. C. Goldberg, and E. R. Weiner. "Mechanisms for Aqueous Photolysis of Adsorbed Benzoate, Oxalate, and Succinate on Iron Oxyhydroxide (Goethite) Surfaces," *Environ. Sci. Technol.* 22(9):1090–1097 (1988).

55. Kieber, D. J., and K. Mopper. "Photochemical Formation of Glyoxylic and Pyruvic Acids in Seawater," *Mar. Chem.* 21:135–149 (1987).

56. Paulson, G. D. "Sulfate Ester Conjugates: Their Synthesis, Purification, Hydrolysis, and Chemical and Spectral Properties," in *Bound and Conjugated Pesticide Residues,* ACS Symposium Series 29, D. D. Kaufman, G. G. Still, G. D. Paulson, and S. K. Bandal, Eds. (Washington, DC: American Chemical Society, 1976), pp. 86–102.

57. Robinson, T. *The Organic Constituents of Higher Plants,* 4th ed. (North Amherst, MA: Cordus Press, 1980).

58. Leenheer, J. A. "Effects of Salinity on Solubility and Stability of Humic Substances in Water," paper presented at the Penrose Conference of the Geological Society of America, Oxnard, CA, March 17, 1987.

59. Goldberg, S., and R. A. Glaubig. "Effect of Saturating Cation, pH, and Aluminum and Iron Oxide on the Flocculation of Kaolinite and Montmorillonite," *Clays and Clay Minerals* 35(3):220–227 (1987).

60. Leenheer, J. A., and U. M. Santos. "Consideracoes sobre os Processos de Sedi-

mentacao na Aqua Preta Acida do Rio Negro (Amazonia Central)," *Acta Amazonica* 10(2):343–355 (1980).

61. Carter, C. W., and I. H. Suffet. "Binding of DDT to Dissolved Humic Materials," *Environ. Sci. Technol.* 16(11):735–740 (1982).
62. White, A., P. Handler, and E. L. Smith. *Principles of Biochemistry* (New York: McGraw-Hill, 1964), pp. 129–130.
63. Means, J. C., and R. Wijayaratne. "Role of Natural Colloids in the Transport of Hydrophobic Pollutants," *Science* 215:968–970 (1982).
64. Yan, L., and R. F. Stallard. Personal Communication.
65. Vanderlaan, M., B. E. Watkins, and L. Stanker. "Environmental Monitoring by Immunoassay," *Environ. Sci. Technol.* 22(3):247–254 (1988).
66. Averett, R. C., J. A. Leenheer, D. M. McKnight, and K. A. Thorn, Eds. *Humic Substances in the Suwannee River, Georgia: Interactions, Properties, and Proposed Structures,* U.S. Geological Survey Open-File Report 87–557 (1989), pp. 231–310.
67. Preston, C. M., M. Schnitzer, and J. A. Ripmeester. "A Spectroscopic and Chemical Investigation on the De-ashing of Humin," *Soil Sci. Soc. Amer. J.* 53(5):1442–1447 (1989).
68. Philip, R. P., and T. I. Gilbert. "The Detection and Quantitation of Biological Markers by Computerized GC/MS," in *Biological Markers in the Sedimentary Record,* R. B. Johns, Ed. (Amsterdam: Elsevier, 1986), pp. 227–248.
69. Bracewell, J. M., K. Haider, S. R. Larter, and H.-R. Schulten. "Thermal Degradation Relevant to Structural Studies of Humic Substances," in *Humic Substances II: In Search of Structure,* M. H. B. Hayes, P. MacCarthy, R. L. Malcolm, and R. S. Swift, Eds. (Chichester, Eng.: John Wiley and Sons, 1989), pp. 181–222.

CHAPTER 2

The Importance of Humic Substance–Mineral Particle Complexes in the Modeling of Contaminant Transport in Sediment-Water Systems

R. L. Wershaw

INTRODUCTION

Sediments are, in general, complex mixtures of different mineral grains that result from the weathering of rocks. These mineral grains may be deposited on flood plains as soils, or they may be transported by streams to sedimentary basins. In either case, they are exposed to humic substances during all stages of their movement. Davis[1] and Ghosh and Varadachari[2] have shown that humic substances form coatings on the surfaces of hydrous metal oxides and clay minerals. The humic coatings on the hydrous metal oxides probably cover most of the exposed surfaces of the grains; however, on the clay minerals the humic coating may not be as uniform. The data of Ghosh and Varadachari indicate that humic molecules are bound mainly at the edge sites of kaolinite, but occupy both edge and sheet sites of illite.[2]

The sedimentary particles present in a natural water body interact with both hydrophobic and hydrophilic contaminants that are introduced into the water body. Karickhoff[3] and Chiou et al.[4] have pointed out that the sorption of uncharged organic compounds to composite sediment particles consisting of mineral grains coated by organic matter is primarily by a mechanism in which the organic molecules partition into the organic coating of the mineral grains. This partitioning is similar to the distribution of hydrophobic organic compounds between two different liquid phases.

The effect of organic coatings on sediment grains on the binding of metal ions by sediments has only been investigated in a few instances. Windom et al., in studies of the distribution of trace metal in estuarine and marine sediments, have found evidence that there is a correlation between cadmium and mercury concentrations and total organic content of the sediments.[5] They measured trace-metal concentrations released after total digestion of their samples, and therefore their data do not allow one to differentiate between trace metals bound in crystalline lattice positions and trace-metal ions occupying exchangeable sites either on the mineral surfaces or the organic coatings. Luoma and Bryan have used various different extraction solutions to attempt to differentiate between different types of binding sites for metals in sediments.[6,7] They have been able to get some indication of which metals are more strongly bound by humic substances in sediments than others; however, they have not measured metal-humic binding constants for any of the metals.

In order to provide a framework or model for the study of the effects of the influence of humic substances on the reactions that take place in natural water systems, I have proposed that humic substances constitute a number of separate hydrophobic and hydrophilic phases in natural water systems.[8] The rest of this chapter will deal with a description of the model and its significance in developing a thermodynamic description of physical-chemical interactions in sediment-water systems.

DESCRIPTION OF MODEL

General Description

I have proposed previously that humic substances are mixtures that cannot be represented by conventional structural diagrams of functional groups held together by covalent bonds, but, rather, they exist as membranelike or micellelike aggregates.[9] Furthermore, it appears that the physical and chemical properties of these humic aggregates are more a function of the structure of the aggregates than of the properties of the individual components. In terms of this model, the division of humic substances into humic acid, fulvic acid, and humin is a totally artificial division that tends to obscure the close interaction between the organic constituents of natural water systems.

The humic aggregates are composed of the partially degraded molecular components of plants. This partial degradation consists, in part at least, of the oxidation and formation of carboxylic acid groups. These carboxylic acid groups are attached, in general, to hydrophobic or less polar functional groups. Therefore, the constituent molecules of the humic aggregates are amphiphiles (see Figure 2.1 for a generalized diagram of an amphiphile). At concentrations above some critical value, amphiphiles will spontaneously arrange themselves in ordered aggregates, such as micelles or membranes, in which the interiors of the aggregates are made up of the more hydrophobic parts of the constituent molecules, and the exterior surfaces are composed mainly of polar groups such as carboxylic acid functionalities (Figures 2.2 and 2.3). The term *membrane* is used here in the same sense as it is used in the biochemical literature—to denote a planar bilayer structure composed of amphiphilic molecules of the type shown in Figure 2.3 (see Fendler[10]). The molecular segments in the hydrophobic interiors of these membranes or

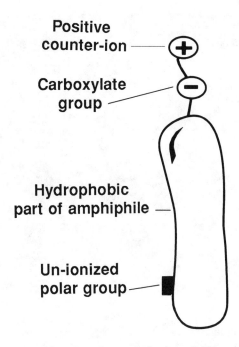

Positive counter-ion

Carboxylate group

Hydrophobic part of amphiphile

Un-ionized polar group

Figure 2.1.   Schematic representation of a generalized, nonlipid amphiphile humic material molecule. For simplicity only one carboxylate group and one un-ionized polar group are shown, but in some instances more of these groups may be present on the same molecule.

micelles have substantial freedom of motion, and therefore it is proposed that they constitute a separate liquidlike phase in sediment-water systems.

Humic Coatings on Sediments

A sediment-water system can be pictured as being made up of the following constituents (although the word *component* might be more familiar in this instance, we shall reserve it for use in the narrow, thermodynamic sense of the word to denote a single chemical species): bed sediment, suspended sediment, solvent (water), dissolved species, colloidal species, and surface films. In this classification each of the constituents may be composed of more than one phase and more than one component. Each of these constituents will interact with humic substances, as we shall discuss below.

Davis has concluded that the surfaces of hydrous aluminum oxides, hydrous iron oxides, and the edge sites of aluminosilicates are covered by humic substances in natural systems.[1] Davis pointed out that the adsorption of humic substances changes normally positively charged mineral surfaces to negatively charged surfaces. This is brought about by the complexation by mineral-surface hydroxyl groups of humic substances. Wershaw has proposed that these humic coverings are in the form of membranelike bilayers in which the

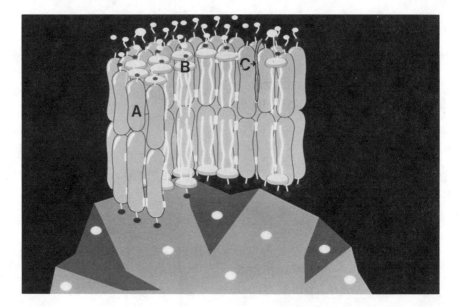

Figure 2.2. Schematic representation of humic membranelike structure (upper part of diagram) attached to a hydrous oxide surface (lower part of diagram); the hydrous oxide surface is pictured as being positively charged: *A*, nonlipid amphiphile; *B*, lipid amphiphiles (pictured as a diglyceride); *C*, intercalated pollutant molecules. Both humic acid and humin components are shown in this diagram.

outer surfaces are composed mainly of carboxylic acid groups with some phenolic groups, and the interiors are made up of the more hydrophobic parts of the constituent molecules.[9] In this model, the surfaces of the membranes can be considered as constituting one phase and the interior as another phase.

The high charge density at the surface of one of these membranes results in an increased concentration of counterions in the near vicinity of the surface. This effect is generally referred to as "ion condensation" (see Lindman et al.[11] and Manning[12]); it takes place at membrane surfaces and in the regions surrounding micelles and polyions. This condensation markedly decreases the activity coefficients of counterions in solutions in contact with membranes, micelles, or polyions.

Humic Colloids

Hunter and Lee have studied the surface activity of humic substances in river waters.[13] Their results indicate that when aquatic humic and fulvic acids are both present in a natural water they will interact to form humic acid–fulvic acid aggregates. In as much as both the humic and the fulvic acids they studied were surface active, it is reasonable to assume that the aggregates are micelle-like in structure. Hunter and Lee observed further evidence for the presence of

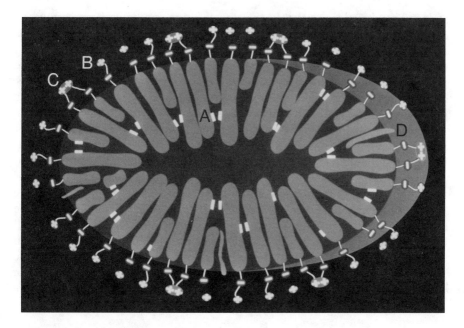

Figure 2.3. Schematic representation of humic micelle: *A*, amphiphile molecules; *B*, monovalent cations; *C*, divalent cations; *D*, pollutant molecules.

these micellelike aggregates from the enhanced solubility of relatively insoluble organic compounds such as PCB's and DDT. This enhanced solubility is most likely due to partitioning of the insoluble organic compounds into the interiors of the micellelike aggregates.

Interactions of Humic Micelles and Membranes

The behavior of both inorganic and organic constituents in natural waters will be altered by the presence of humic membranes and micelles; thus, in order to understand both the inorganic and organic reactions in natural waters, the effects of the humic aggregates must be taken into account. For example, in order to accurately represent the thermodynamics of metal-ion reactions in natural water systems, the effect of reduced ion activities brought about by humic membranes on the suspended sediments and by humic micelles in solution must be taken into account.

From the preceding discussion it is apparent that the existence of humic substances as membranes and micelles in natural systems will have a profound effect on the behavior of the other constituents in these systems. These effects can only be evaluated experimentally in systems in which the humic aggregates are present in the same forms as they are in nature. Similarly, model calculations must also take into account the presence of these aggregates.

Chiou et al. have studied the sorption of nonionic organic compounds by soils.[14] They have demonstrated that the sorption can be best represented by the same equations that are used to represent the partition of compounds between two liquid phases. The fact that these equations hold for the sorption of nonionic organics by soils is strong evidence for the presence of a liquidlike phase in soil humic substances as postulated in the membrane model. Further evidence for this liquidlike phase has been obtained from nuclear magnetic resonance (NMR) studies by Earl et al.[15]

Other types of interactions, such as metal binding and the solubilization of nonpolar organics in natural waters, generally are studied in systems containing isolated humic and fulvic acid extracts or filtered natural water samples. For these studies to yield results that can be extrapolated to actual field conditions, they must be carried out in the presence of all of the phases that are present in the natural systems. In the initial work, each of the phases may be studied separately, but the work should be extended to systems containing all of the phases as soon as possible. The metal ion binding data that have been collected on isolated aquatic fulvic acids may be of only very limited value because, as we have pointed out above, Hunter and Lee have shown that fulvic acids form aggregates with humic acids in natural waters,[13] and it is these aggregates that interact with metal ions in natural systems.

Phase Concept

The phase concept for humic substances allows one to more readily characterize those physical and chemical properties of humic substances that are important in determining how humic substances interact with the other organic and inorganic constituents of natural water systems. In a given natural water system, the humic substances will be present in several different phases:

1. the liquidlike interior regions of the humic membranes coating mineral grains
2. the charged surface regions of these membranes
3. the liquidlike interiors of the humic micelles in solution or in colloidal suspension in the water
4. the charged surface regions of the micelles

For simplicity, we shall assume that, although there are a very large number of discrete humic membranes or micelles in a given natural system, the humic substances in the system are uniform enough so that each of the four regions enumerated above may be considered as a single phase.

In addition to these four phases, the components of the humic phases will also be present as dissolved monomeric units in the aqueous solution phase. These monomeric units, and the other components of the solution phase, are free to enter into the aggregated humic phases. Examples of this are the movement of metal ions in and out of the charged surface phases of the humic membranes and micelles, partitioning of hydrophobic molecules in and out of the liquidlike interior phases, and the association of monomeric units to micelles and the dissociation of micelles to monomers. In all of these examples, dynamic equilibria are set up which can be characterized by appropriate equilibrium constants. Evaluation of these equilibrium constants will allow us to characterize and model the interactions between humic substances and the rest of the components of natural water systems.

THERMODYNAMIC RELATIONSHIPS

General Relationships

To develop the thermodynamic relationships in systems containing the various humic phases, the affinity variable $\mathbf{A}$, which is a vector quantity, will be used (for a more complete discussion of this variable, see Prigogine and Defay[16]). As discussed in Prigogine and Defay (p. xxiii),[16] "affinity may be regarded formally as the driving force of a chemical reaction." The affinity $\mathbf{A}$ is defined by the relationship:

$$dQ' = \mathbf{A}d\xi \geq 0 \tag{2.1}$$

where Q' is the uncompensated heat that arises from irreversible changes taking place within a system, and ξ is the extent of change or extent of reaction. Thus, for changes in a system at constant temperature T,

$$dS = \frac{dQ}{T} + \frac{dQ'}{T} \tag{2.2}$$

where S is the entropy of the system, and Q is the heat received by the system from outside the system. For reversible changes in the system, $dQ' = 0$; for irreversible changes, $dQ' > 0$.

The systems that we shall consider are those that are, in the terminology of Prigogine and Defay,[16] in partial equilibrium; that is, equilibrium exists with respect to temperature and pressure, but not with respect to redistribution of matter. Therefore, in such systems, chemical reactions can take place between the various components of the system, and matter can move from one phase to another in the system. In such systems, $\mathbf{A}$ is a function of the physical variables (e.g., temperature, T; pressure, p; or volume, V) of the system and the extent of reaction, ξ; therefore, $\mathbf{A}$ may be written as $\mathbf{A}(x,y,\xi)$, where x and y designate any two physical variables. The advantage of using $\mathbf{A}$ as a thermodynamic function of state is that it is a function of state that is associated with chemical reactions, as opposed to functions of state that apply to the whole system (such as internal energy) or those that apply to a single component (such as chemical potential). In this scheme, the velocity or rate of a reaction, $\mathbf{v}$, is given by the equation:

$$\frac{d\xi}{dt} = \mathbf{v}\,(T,p,\xi). \tag{2.3}$$

where t is time.

The use of the function **A** allows one to adapt the standard equilibrium chemical potential terms to irreversible (partial equilibrium) conditions, as shown by the following set of differential equations:

$$dU = TdS - pdV - \mathbf{A}d\xi \tag{2.4}$$

$$dH = TdS + Vdp - \mathbf{A}d\xi \tag{2.5}$$

$$dF = -SdT - pdV - \mathbf{A}d\xi \tag{2.6}$$

$$dG = -SdT + Vdp - \mathbf{A}d\xi \tag{2.7}$$

where U is the internal energy, H is the enthalpy, F is the Helmholz free energy, and G is the Gibbs free energy. The chemical potentials of the components in a given reaction are related to the affinity of the reaction by the equation:

$$A_p = \sum_i \nu_{ip}\mu_i. \tag{2.8}$$

where $\mathbf{A}_p$ is the affinity for reaction ρ, ν_{ip} is the stoichiometric coefficient of component i in reaction ρ, and μ_i is the chemical potential of component i. At equilibrium, $\mathbf{A}_{\rho = 0}$.

Wershaw has shown how the membrane (micelle) model of humic substances provides a framework for understanding the various types of interactions that take place between humic substances and other components in natural water systems.[17] We shall now consider two examples of the various types of reactions that humic membrane and micelle structures can undergo in natural water systems and briefly outline the thermodynamic relationships that describe these reactions.

Partitioning Between Phases

Hydrophobic organic compounds will partition into the various hydrophobic humic phases. At equilibrium this partitioning may be represented by the Nernst distribution law:

$$K_i(T,p) = \frac{x_i'' \, \tau_i''}{x_i' \, \tau_i'} \tag{2.9}$$

where $K_i(T,p)$ is the distribution coefficient of component i, x_i'' is the equilibrium mole fraction of component i in the first phase, τ_i'' is the activity coefficient of i in the first phase, x_i' is the equilibrium mole fraction of i in the second phase, and τ_i' is the activity coefficient of i in the second phase. If more than two phases are present in the system, then a distribution equation must be written for every combination of phases. Simultaneous solution of the equations weighted for the relative amounts of each phase will yield the distribution of each of the components between the various phases. In a sediment-water system, there will be at least two separate hydrophobic humic phases: (1) the interiors of humic membranes coating the sediment grains and (2) the interiors of the humic micelles in suspension in the water phase. Enfield et al. have shown in soil column studies that both of the humic phases must be considered in order to properly model the distribution of hydrophobic organic components in the systems they studied.[18]

Ionic Interactions

The negatively charged surfaces of the humic membranes and micelles will interact with positively charged ionic species. Two general types of interactions will take place, which are designated, in the terminology of Manning,[19] as *site binding* and *territorial binding*. Site binding may be either a reversible reaction or an irreversible reaction. If the binding is reversible, then the model developed by Marinsky may be used for both soluble and insoluble humic charged phases.[20,21] The Marinsky model was developed for weakly acidic polymeric gels. Marinsky showed that there are two types of gels: those that are permeable to low molecular weight salts and those that are impermeable to low molecular weight salts. The distribution of charge in the surfaces of gels in the Marinsky model is very similar to that on a membrane or micelle, and therefore his development should be applicable to the model discussed here. Furthermore, Marinsky concluded from his observations that polymeric gel molecules, even when completely dissolved, must be treated as a separate phase from the bulk solution.

Marinsky pointed out that acid-base titration curves of most weakly acidic or basic gels are very sensitive to the concentration of low molecular weight background electrolyte present in the system. He attributed this dependence to a Donnan potential that arises from the permeability of the gel phase to the background electrolyte. Therefore, he included a Donnan potential term in his calculations of the ionization constants of weakly ionic polymeric gels. The acid-base titration curves of some ionic gels, however, are not strongly dependent on background electrolyte concentration. Marinsky proposed that these are gels that are impermeable to the background electrolyte, and therefore no Donnan potential exists. Marinsky and Ephraim[21] and Ephraim et al.[22] have found that the titration curves of isolates of aquatic humic substances can be explained by the Marinsky model and that both types of behaviors are encountered.

The Marinsky model is a thermodynamic equilibrium model; if, however, the binding reactions between the charged humic surfaces and a cationic species, i, in solution deviate from reversibility, then the chemical potential of the cation in solution, μ_i'', will not be equal to that of the cation in the humic phase, μ_i'. The difference between the two chemical potentials will be given by Equation 2.8:

$$\mathbf{A} = \mu_i'' - \mu_i' \qquad (2.10)$$

The affinity for this or any other reaction may be evaluated calorimetrically as outlined by Prigogine and Defay[16] (pp. 50–51). Examples of nonreversible reactions between humic surfaces and cationic species include oxidation-reduction (redox) of metal ions. These redox reactions may or may not be photochemically mediated.

Organic ionic species also interact with charged humic surfaces.[17] A number of different types of ionic interactions have been reported between humic substances, including the binding of organic cations such as protonated amino acids and triazine herbicides. Enzymes also form complexes with humic substances. The presence of metal ions appears to be necessary for the binding to take place.[17] It is likely that both reversible and irreversible reactions take place.

The model discussed in this chapter allows one to identify the various humic phases that are present in a sediment-water system. The interactions of each of these phases with each of the components in the system may be experimentally evaluated separately in order to measure the thermodynamic parameters of the reaction of each phase with a given component. These thermodynamic parameters then may be used in one of the multicompartment transport models, such as the multimedia transport model developed by Cohen and his group to model chemical interactions in a natural water system.[23]

CONCLUSIONS

The concepts developed in this chapter should allow one to begin to develop predictive models of the distribution of the organic and inorganic components of natural water systems between the various humic phases. These relationships can be represented graphically by phase diagrams in the same way as the distribution of inorganic ionic components between solution and mineral phases can be represented by phase diagrams. Preparation of the appropriate phase diagrams and evaluation of the constants of the chemical equilibria of the distribution of the components between the various humic phases then will allow one to develop predictive mathematical models for chemical interactions in natural water systems. In the final step of the modeling, the thermodynamic relationships may be combined with an appropriate transport model to account for the movement of the components through the systems.

REFERENCES

1. Davis, J. A. "Adsorption of Natural Dissolved Organic at the Oxide/Water Interface," *Geochim. Cosmochim. Acta* 46(11):2381–2393 (1982).
2. Ghosh, K., and C. Varadachari. Unpublished results presented at U.S. Geol. Survey, September 1989.
3. Karickhoff, S. W. "Organic Pollutant Sorption in Aquatic Systems," *J. Hydraul. Eng.* 110(6):707–735 (1984).

4. Chiou, C. T., L. J. Peters, and V. H. Freed. "A Physical Concept of Soil-Water Equilibria for Nonionic Organic Compounds," *Science* 206(16):831–832 (1979).

5. Windom, H. L., S. J. Schropp, F. D. Calder, J. D. Ryan, R. G. Smith, Jr., L. C. Burney, F. G. Lewis, and C. H. Rawlinson. "Natural Trace Metal Concentrations in Estuarine and Coastal Marine Sediments of the Southeastern United States," *Environ. Sci. Technol.* 23(3):314–320 (1989).

6. Luoma, S. N., and G. W. Bryan. "A Statistical Assessment of the Form of Trace Metals in Oxidized Estuarine Sediments Employing Chemical Extractants," *Sci. Total Environ.* 17:165–195 (1981).

7. Luoma, S. N. "A Comparison of Two Methods for Determining Copper Partitioning in Oxidized Sediments," *Mar. Chem.* 20:45–59 (1986).

8. Wershaw, R. L. "Humic Substances as Separate Phases in Soil and Water Systems – A New Hypothesis for Thermodynamic Modeling in Natural Water Systems," paper presented at the 1989 International Chemical Congress of Pacific Basin Societies, Honolulu, Hawaii, December 17–22, 1989.

9. Wershaw, R. L. "A New Model for Humic Materials and Their Interactions with Hydrophobic Organic Chemicals in Soil-Water or Sediment-Water Systems," *J. Contaminant Hydrol.* 1:29–45 (1986).

10. Fendler, J. H. "Membrane Mimetic Chemistry," *Chem. Eng. News* 62(1):25–38 (1984).

11. Lindman, B., G. Lindbloom, H. Wennerstrom, N.-O. Persson, H. Gustavsson, and A. Khan. "Molecular and Ionic Behaviour of Water-Amphiphile Interfaces," in *Magnetic Resonance in Colloid and Interface Science,* J. P. Fraissard and H. A. Resing, Eds. (Dordrecht: D. Reidel Publishing Co., 1980), pp. 307–320.

12. Manning, G. S. "Counterion Binding in Polyelectrolyte Theory," *Accounts Chem. Res.* 12:443–449 (1979).

13. Hunter, K. A., and K. C. Lee. "Polarographic Study of the Interaction between Humic Acids and Other Surface-Active Organics in River Waters," *Water Res.* 20:1489–1491 (1986).

14. Chiou, C. T., P. E. Porter, and D. W. Schmeddling. "Partition Equilibria of Nonionic Organic Compounds between Soil Organic Matter and Water," *Environ. Sci. Technol.* 17:227–231 (1983).

15. Earl, W. L., R. L. Wershaw, and K. A. Thorn. "The Use of Variable Temperature and Magic-Angle Sample Spinning in Studies of Fulvic Acids," *J. Magnetic Resonance* 74:264–274 (1987).

16. Prigogine, I., and R. Defay. *Chemical Thermodynamics* (London: Longmans Green and Co., 1954).

17. Wershaw, R. L. "Application of a Membrane Model to the Sorptive Interactions of Humic Substances," *Environ. Health Perspect.* 83:191–203 (1989).

18. Enfield, C. G., G. Bengtsson, and R. Linqvist. "Influence of Macromolecules on Chemical Transport," *Environ. Sci. Technol.* 23:1278–1286 (1989).

19. Manning, G. S. "Counterion Binding in Polyelectrolyte Theory," *Accounts Chem. Res.* 12:443–449 (1979).

20. Marinsky, J. A. "An Interpretation of the Sensitivity of Weakly Acidic (Basic) Polyelectrolyte (Cross-Linked or Linear) Equilibria to Excess Neutral Salt," *J. Phys. Chem.* 89:5294–5302 (1985).

21. Marinsky, J. A., and J. Ephraim. "A Unified Physicochemical Description of the Protonation and Metal Ion Complexation Equilibria of Natural Organic Acids (Humic and Fulvic Acids). 1. Analysis of the Influence of Polyelectrolyte Proper-

ties on Protonation Equilibria in Ionic Media: Fundamental Concepts," *Environ. Sci. Technol.* 20:347–354 (1986).

22. Ephraim, J., S. Alegret, A. Mathuthu, M. Bicking, R. L. Malcolm, and J. A. Marinsky. "A Unified Physicochemical Description of the Protonation and Metal Ion Complexation Equilibria of Natural Organic Acids (Humic and Fulvic Acids). 2. Influence of Polyelectrolyte Properties and Functional Group Heterogeneity on the Protonation Equilibria of Fulvic Acid," *Environ. Sci. Technol.* 20:354–366 (1986).

23. Cohen, Y. "Organic Pollutant Transport," *Environ. Sci. Technol.* 20:538–544 (1986).

CHAPTER 3

Composition of Humin in Stream Sediment and Peat

James A. Rice and Patrick MacCarthy

INTRODUCTION

Humin is defined as the fraction of humic substances which is insoluble in an aqueous solution at any pH value. It is typically the largest fraction of the organic matter in recent sediments. Humin organic carbon can constitute as much as 80% of the total organic carbon in a sediment.[1-3] It is an enigmatic material even by comparison to the other humic substances. Its chemical nature has proved difficult to describe, and despite the development of numerous models to summarize its characteristics,[4-12] there is no general consensus on what comprises the components of humin. Humin has been described as a high molecular weight polymer,[4] humic acid bound to clay,[5] humic acid bound to inorganic soil colloids,[8] a lignoprotein,[10] a melanin,[12] and decomposing plant and fungal residues.[13] Each of these views may represent a humin isolated from a particular source, but none of them can be generally applied to all humin samples.

The differences in the models for humin arise, at least in part, from the method by which humin has usually been isolated. Humin is traditionally obtained as the residue which remains after a sample is stirred with an alkaline aqueous solution and centrifuged.[4,14] This residue contains not only humin but also extraneous mineral matter and undecomposed plant tissue. Recently we described a novel method for the isolation of humin.[15] In the methyl isobutyl ketone (MIBK) isolation method, the humic substances in a sample are partitioned back and forth between an aqueous phase and MIBK as a function of pH. In this procedure, humin is actually isolated from a substrate and is not simply the residue remaining after extracting the other organic matter fractions. The humin obtained by the MIBK isolation method is free from nonhumic substances (both organic and inorganic) that are mixed with a humin sample isolated in the traditional manner. A method that permits the isolation of humin free from other insoluble, but nonhumic, materials facilitates the study of humin. But even with this capability, it must be recognized that humin usually consists of a relatively small amount of organic material dispersed in a much larger amount of inorganic material. To concentrate these organic components, humin samples are usually digested with a mixture of HF/HCl to dissolve mineral matter.[14] This procedure almost certainly alters the nature of the organic components of humin from their form in the isolated humin. These problems have recently been discussed elsewhere.[15] An extension of the MIBK method not only eliminates the need for HF/HCl digestion but also performs a preliminary separation of humin's organic components into recognized compound classes.[15] To achieve this fractionation, humin is first exhaustively extracted with an organic solvent (such as chloroform or a mixture of chloroform and methanol) in a Soxhlet apparatus. The material extracted from the humin in this manner is referred to as "bitumen". The humin following Soxhlet extraction, in the salt form and suspended in MIBK, is then shaken with deionized water. This disaggregates humin into three simpler components:

1. a bound lipid fraction which cannot be Soxhlet extracted, but is released by the MIBK method
2. a brown material that is soluble in alkali and insoluble in acid, which is referred to as bound humic acid
3. an insoluble fraction, which is usually inorganic in nature

None of these three fractions, or the bitumen, by itself conforms to the operational definition of humin.

The purposes of this chapter are to:

1. describe the characteristics of the bitumen, bound lipids, and the bound humic acid
2. present a model for the nature of humin consistent with these characteristics
3. discuss some implications of this model for the fate of contaminants in a sedimentary environment

MATERIALS AND METHODS

Samples

The materials from which humin was isolated were a stream sediment and a peat collected on Guanella Pass in Clear Creek County, Colorado. The stream sediment was collected as random grabs from the top 5 cm of undisturbed sediment in South Clear Creek.[16] In this area, South Clear Creek is an unpolluted, perennial stream which drains alpine bogs on Guanella Pass. The peat sample was collected as random grabs from the upper horizons of a boggy soil classified as a *Cryohemist*.[17] The organic material which comprises the peat consists primarily of decomposing sedges (*Carex* sp.), willows (*Salix* sp.), rushes, and grasses.

The stream sediment contained 7.0% organic carbon by weight (moisture free), of which 58.4% was humin organic carbon. The peat contained 40.1% organic carbon by weight (moisture free), of which 24.3% was humin organic carbon.

Isolation of Humin, Humic Acid, and Fulvic Acid

Humin, humic acid, and fulvic acid were isolated from the stream sediment and peat using the MIBK method.[15,16] The humic acid from each sample was taken through two additional cycles (for a total of three cycles) of the MIBK procedure and converted to the hydrogen form. Fulvic acid was desalted and converted to the hydrogen form by the method of Thurman and Malcolm.[18]

Disaggregation of Humin

One gram of each humin (in the hydrogen form) obtained by the MIBK isolation method was then extracted for 72 hours in a Soxhlet apparatus with a benzene-methanol azeotrope (~ 3:1, v:v). The material extracted was collected by evaporation of the solvent and weighed. This fraction, which consists of a mixture of lipids, is referred to as the bitumen. Each solvent-extracted humin was then disaggregated into bound humic acid, bound lipid, and insoluble residue fractions using the MIBK method[15] and converted into the hydrogen form as described by Rice and MacCarthy.[16] Each fraction was then weighed. Triplicate samples of each humin were treated in this manner.

Characterization of Bound Humic Acid, Humic Acid, and Fulvic Acid

Elemental analysis of the bound humic acid, humic acid, and fulvic acid consisted of the direct determination of C, H, N, and S contents; O was determined by difference. Atomic H:C and O:C ratios were calculated from the results. Elemental contents are presented on an ash-free, moisture-free basis. Rice and MacCarthy discuss limitations on the applicability of elemental data in the characterization of humic materials.[19] Acidic functional group determinations consisted of the barium hydroxide titration for total acidity[20] and a modified acetate method for the determination of the carboxyl content.[1] Following convention, the weakly acidic hydroxyl content was taken as the difference between the total acidity and carboxyl content.[4] Liquid-state ^{13}C NMR spectra were obtained for the bound humic acid, humic acid, and fulvic acid from the stream sediment and peat using an inverse-gated decoupled pulse sequence. In this experiment a long pulse delay (8 sec) ensures that all of the carbon atoms in the sample are permitted adequate time to relax.[21] This experiment yields quantitative determinations of the aliphatic (0–50 ppm), carbohydrate (50–105 ppm), aromatic (105–160 ppm), and carboxyl (160–187 ppm) carbon in a humus sample. Preston and Blackwell describe the application of this technique to the study of humic materials.[21]

Table 3.1. Yields of the Four Fractions Obtained by Disaggregating Humins from Stream Sediment and Peat

	Bitumen	Bound Lipids	Bound Humic Acid	Insoluble Residue	Total
Stream sediment	26.3% ± 2.1%	1.7% ± 0.7%	11.0% ± 1.9%	67.8% ± 2.7%	107%
Peat	56.2% ± 9.7%	1.8% ± 0.4%	16.3% ± 1.2%	18.2% ± 0.5%	93%

Note: All values are expressed as weight percents on a moisture-free basis. Errors are absolute standard deviations.

Characterization of the Bitumen and Bound Lipids

Each bitumen and bound lipid fraction was separated into hexane-, benzene-, and methanol-soluble eluates using column chromatography on activated silica. The methanol-soluble eluate from each fraction was methylated with 14% BF_3-methanol. The derivatized samples were then analyzed by gas chromatography–mass spectrometry (GC-MS). The gas chromatograph was programmed from 70 to 300°C at 4°C/min, with an initial hold of 1.5 min and a final hold of 10 min. All samples were injected in the split mode with a split ratio of 5:1. The carrier gas was He at a linear flow rate of 34 cm/sec through a 30 m DB-5 bonded stationary-phase capillary column. The mass spectrometer was operated in the electron impact mode at an acceleration voltage of 4 kV, electron energy of 70 eV, a source temperature of 200°C, and low resolution conditions.

RESULTS AND DISCUSSION

Disaggregation of Humin

Table 3.1 describes the composition of the stream sediment and peat humins in terms of humin's four disaggregation products. The totals may reflect the inadequate removal of salt from the humin prior to drying (in the case of the stream sediment) and difficulty in collecting all of the fibrous material, which would have been classified as insoluble residue (in the case of the peat).

The bitumen component represents a larger fraction of each humin than any of the other organic components. The bound lipids represent a small fraction of the total organic matter in each sample. Once the bitumen has been extracted, however, the bound lipids represent 13.4% and 10.0% of the remaining organic matter in the stream sediment and peat humins, respectively.

Elemental Analysis

Figure 3.1 compares the atomic H:C and O:C ratios for the humic acid, fulvic acid, and bound humic acid from the stream sediment and peat. When identified by source, the three fractions show the same interrelationships. The bound humic acids plot toward the upper-middle portion of the diagram, humic acids lie in the lower-left portion of the diagram, and fulvic acids tend

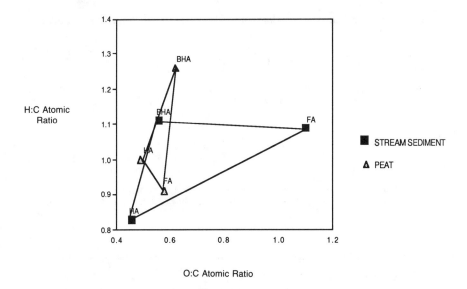

Figure 3.1. Elemental contents of stream sediment and peat humic materials.

to fall to the right of both. Bound humic acid exhibits a higher H:C ratio than the humic acid from the corresponding source. This suggests that the bound humic acid component of humin is more aliphatic than the humic acid from the same source. Bound humic acid also displays a greater O:C ratio than the humic acid from the same source. These observations indicate that bound humic acid is not simply unextracted humic acid, but rather it is a distinct organic matter fraction.

Acidic Functional Group Analysis

Table 3.2 gives the acidic functional group contents of the bound humic acid, humic acid, and fulvic acid from the stream sediment[16] and peat. Bound humic acid exhibits total acidity values and carboxyl and weakly acidic hydroxyl contents that are higher than those of the corresponding humic acid. The acidity values for the bound humic acid from peat are more similar to those of fulvic acid than humic acid. These observations again indicate that bound humic acid is not merely unextracted humic acid but, instead, represents a distinct fraction of organic matter.

^{13}C NMR Spectra

The solution-state, inverse-gated decoupled ^{13}C NMR spectra of the bound humic acid, fulvic acid, and humic acid from the stream sediment[16] and peat are given in Figure 3.2. All of the spectra are dominated by aromatic carbon signals (105–160 ppm). The humic acid and fulvic acid spectra also display a prominent signal at 30 ppm. This resonance has been attributed to the methyl-

Table 3.2. Acidic Functional Group Contents of Humic Materials from Stream Sediment and Peat

	Total Acidity	Carboxyl Content	Weakly Acidic Hydroxyl Content
Stream sediment			
Bound humic acid	4.84 ± 0.69	2.61 ± 0.30	2.23
Humic acid	3.61 ± 1.39	1.47 ± 0.11	2.14
Fulvic acid	9.13	4.85	4.28
Peat			
Bound humic acid	7.55 ± 4.60	4.86 ± 0.64	2.69
Humic acid	5.00 ± 0.62	2.54 ± 0.16	2.46
Fulvic acid	8.75	5.19	3.56

Note: Contents are reported in milliequivalents per gram. Errors are absolute standard deviations from at least three measurements; fulvic acid values represent one measurement.

ene units of long alkyl chains.[22,23] The bound humic acid spectra exhibit more pronounced carboxyl carbon resonances (160–187 ppm), relative to the aromatic carbon region, than either humic or fulvic acid, indicating that bound humic acid should have a greater carboxyl content than humic acid. This observation is consistent with the acidic functional group data. The 30-ppm resonance in the bound humic acid spectra does not appear with the same intensity as it does in the spectra of the humic and fulvic acids. In fact, the whole aliphatic region seems diminished with respect to the other regions in the spectrum. The carbohydrate region (50–105 ppm), on the other hand, appears to have become more significant relative to the other regions. This region displays a prominent resonance at 70 ppm, attributed to the ring carbons of carbohydrates.[24] These results suggest that the greater H:C ratios of the bound humic acid are the result of a diminished aromatic carbon character and an increased carbohydrate carbon character relative to the humic acid from the same sample.

Characteristics of the Bitumen and Bound Lipids

Figure 3.3 gives the distribution of the normal, saturated monobasic fatty acid methyl esters (FAMEs) formed by methylation of the methanol-soluble fraction of the bitumen and bound lipids from the stream sediment[16] and peat. The distribution of the stream sediment and the peat bitumen FAMEs are in both cases bimodal. Both distributions are dominated by the methyl ester of palmitic acid (C_{16}) and secondarily by the methyl ester of lignoceric acid (C_{24}).

The distribution of the FAMEs in the bound lipid fractions of the peat and stream sediment humins are significantly different from those of the corresponding bitumen fractions. The bound lipid distribution is dominated by C_{16} and C_{18} FAMEs. Though the observed bound lipid FAMEs range from C_{14} to C_{30}, and their distribution also appears to be bimodal, the higher–carbon number FAMEs are considerably less abundant in the bound lipids than in

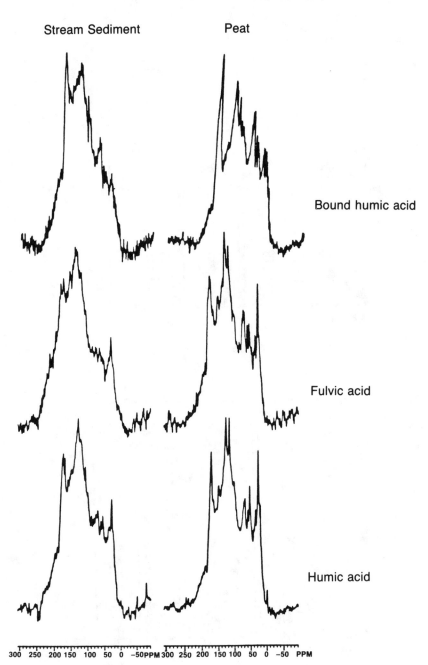

Figure 3.2. Solution-state, inverse-gated decoupled ^{13}C NMR spectra of bound humic acid, fulvic acid, and humic acid from a stream sediment and peat.

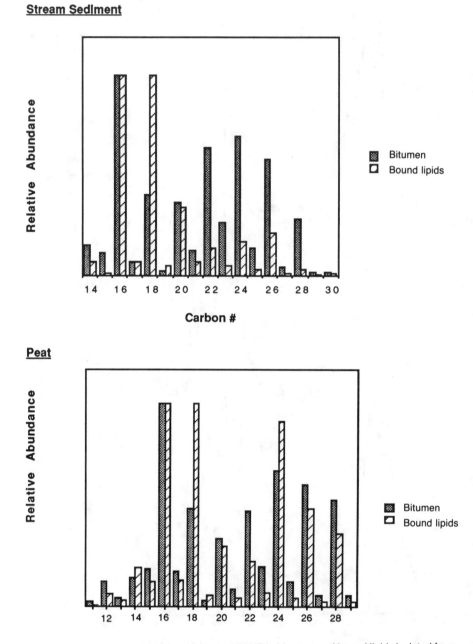

Figure 3.3. Distribution of normal, saturated FAME in bitumen and bound lipids isolated from stream sediment and peat humins.

bitumen. These results indicate that the bound lipids are not merely unextracted bitumen, but are in fact distinct organic matter fractions.

Summary of Results

This study describes the characterization of a previously unrecognized organic geochemical material: the humic component of humin free from non-humic material and associated insoluble residues. This separation is achieved by an extension of the MIBK method. The quantitative isolation of organic components of humin is accomplished by a procedure which is no more drastic than the traditional alkali extraction procedure. Three organic fractions have been identified in this study: bound humic acid, bitumen, and bound lipids.

A humic component of humin has been labeled "bound humic acid" because it conforms to the operational definition of humic acid and is released only by the disaggregation of humin with the MIBK method. However, it is not simply unextracted humic acid. Bound humic acid displays elemental contents, acidic functional group contents, and a ^{13}C NMR spectrum different from humic acid or fulvic acid isolated from the same sample by the MIBK method.

The bitumen (lipids obtained by Soxhlet extraction with organic solvents) and the bound lipids (lipids released only by the disaggregation of humin with the MIBK method) are also distinct fractions within humin. The FAME distributions of the bitumen and bound lipids are distinctly different. This indicates that the bound lipids are a distinct component of humin and are not simply unextracted bitumen.

A MODEL FOR HUMIN

It is proposed that humin be viewed as an aggregation of bitumen, bound humic acid, bound lipids, and some insoluble material. None of these fractions, by itself, conforms to the operational definition of humin. The exact manner in which these components interact to form humin has not yet been identified but is currently being studied. Preliminary experiments have shown that remixing of the separated bitumen, bound lipids, bound humic acid, and the insoluble residue in the presence of MIBK under acidic aqueous conditions, followed by evaporation of the liquid phases, leads to the formation of an acid- and alkali-insoluble material similar to humin. Humin that is obtained by the traditional alkali extraction method can also be separated into these four components when the isolated humin is subjected to Soxhlet extraction with organic solvents and followed by fractionation by the MIBK procedure. There is no certainty that humin isolated by the MIBK method is identical to humin isolated by the traditional method or to humin as it occurs in the environment. It is possible that the humin actually disaggregates partially or completely during the MIBK extraction procedure, and that it aggregates into a reconstituted humin during evaporation of the MIBK. However, because all humic materials are operationally defined, this caution must be constantly borne in mind during studies involving humic materials. There is no guarantee that any isolated humic material actually represents the substance as it existed in nature.

This model has similarities to previously proposed models of humin but, nevertheless, is quite different from all of them. It presents a simple picture of humin and provides a basis for understanding its properties in terms of the four components that can be isolated by the MIBK method. This model is described in detail in a recent publication.[25]

IMPLICATIONS FOR THE FATE OF ENVIRONMENTAL CONTAMINANTS

The bitumen and bound lipid fractions identified in this model would contribute hydrophobic character to the humin and, by doing so, provide the ability to retain nonpolar compounds. The bound humic acid fraction would add some humic characteristics, such as acidity and metal-complexing properties, to the humin. Humin has been shown to exhibit a marked affinity for compounds such as the triazine herbicides,[26-28] picloram,[29] chlorfenvinphos,[30] and carbofuran.[31] The clay minerals in the humin sample from the stream sediment would contribute cation exchange capacity to those humin samples in addition to that provided by the bound humic acid. Accordingly, this model provides a basis for accounting for many of the known properties of different humin samples and for predicting their behavior in other situations. The MIBK method allows the components of humin to be disaggregated in a mild manner, apparently without any significant chemical decomposition, so that all four major constituents can be studied separately and the nature of their interaction with environmental contaminants studied.

REFERENCES

1. Rice, J. A. "Studies on Humus: I. Statistical Studies on the Elemental Composition of Humus. II. The Humin Fraction of Humus," PhD Dissertation, Colorado School of Mines, Golden, CO (1987).
2. Vandenbrouke, M., R. Pelet, and Y. Debyser. "Geochemistry of Humic Substances in Marine Environments," in *Humic Substances in Soil, Sediment, and Water: Geochemistry, Isolation, and Characterization,* G. R. Aiken, D. M. McKnight, R. L. Wershaw, and P. MacCarthy, Eds. (New York: Wiley-Interscience, 1985) pp. 249–273.
3. Stuermer, D. H., K. E. Peters, and I. R. Kaplan. "Source Indicators of Humic Substances and Protokerogen," *Geochim. Cosmochim. Acta* 42:989–997 (1978).
4. Stevenson, F. J. *Humus Chemistry: Genesis, Composition, Reactions* (New York: Wiley-Interscience, 1982).
5. Shah, R. K., M. R. Choski, and B. C. Joshi. "Development Studies on Soil Organic Matter: Humin," *Chem. Era* 6:1–3 (1975).

6. Shah, R. K., M. R. Choski, and B. C. Joshi. "Development Studies on Soil Organic Matter: Humus," *Chem. Era* 4:31–34 (1975).
7. Banerjee, S. K. "Humin Fraction of Soil Humus," *J. Indian Chem. Soc.* 56:1094–1097 (1979).
8. Theng, B. K. G. *Formation and Properties of Clay-Polymer Complexes* (Amsterdam: Elsevier, 1979).
9. Cloos, P., C. Badot, and A. Herbillon. "Interlayer Formation of Humin in Smectites," *Nature* 289:391–393 (1981).
10. Somani, L. L., and S. N. Saxena. "Studies on the Distribution of Sulfur in Humus Fractions of Some Soils of Rajasthan," *Agrochimica* 26:95–103 (1982).
11. Kononova, M. M. *Soil Organic Matter,* T. A. Nowakowski and A. C. D. Newman, Transl. (Oxford: Pergamon Press, 1966).
12. Russell, J. D., D. Vaughan, D. Jones, and A. R. Fraser. "An IR Spectroscopic Study of Soil Humin and Its Relationship to the Soil Humic Substances and Fungal Pigments," *Geoderma* 29:1–12 (1983).
13. Anderson, D. W., E. A. Paul, and R. J. St. Arnau. "Extraction and Characterization of Humus with Reference to Clay-Associated Humus," *Can. J. Soil Sci.* 54:317–323 (1974).
14. Hatcher, P. G., I. A. Breger, G. E. Maciel, and N. Szeverenyi. "Geochemistry of Humin," in *Humic Substances in Soil, Sediment, and Water: Geochemistry, Isolation, and Characterization,* G. R. Aiken, D. M. McKnight, R. L. Wershaw, and P. MacCarthy, Eds. (New York: Wiley-Interscience, 1985) pp. 275–302.
15. Rice, J. A., and P. MacCarthy. "Isolation of Humin by Liquid-Liquid Partitioning," *Sci. Total Environ.* 81/82:61–69 (1989).
16. Rice, J. A., and P. MacCarthy. "Characterization of a Stream Sediment Humin," in *Aquatic Humic Substances: Influence on Fate and Treatment of Pollutants,* Advances in Chemistry Series No. 219, I. H. Suffet and P. MacCarthy, Eds. (Washington, DC: American Chemical Society, 1989), pp. 41–54.
17. Moore, R. U.S. Department of Agriculture, personal communication (1986).
18. Thurman, E. M., and R. L. Malcolm. "Preparative Isolation of Aquatic Humic Substances," *Environ. Sci. Technol.* 15:463–466 (1981).
19. Rice, J. A., and P. MacCarthy. "Statistical Studies on the Elemental Composition of Humic Substances," *Org. Geochem.,* in press.
20. Schnitzer, M., and U. C. Gupta. "Determination of Acidity in Soil Organic Matter," *Soil Sci. Soc. Am. Proc.* 29:274–277 (1965).
21. Preston, C. M., and B. A. Blackwell. "Carbon-13 Nuclear Magnetic Resonance for a Humic and Fulvic Acid: Signal to Noise Optimization, Quantitation and Spin-Echo Techniques," *Soil Sci.* 139:88–96 (1985).
22. Preston, C. M., and M. Schnitzer. "Effects of Chemical Modifications and Extractants on the Carbon-13 NMR Spectra of Humic Materials," *Soil Sci. Soc. Am. J.* 48:305–311 (1984).
23. Hatcher, P., R. Rowan, and M. A. Mattingly. "^{1}H and ^{13}C NMR of Marine Humic Acids," *Org. Geochem.* 2:77–85 (1980).
24. Wilson, M. A. *NMR Techniques and Applications in Geochemistry and Soil Chemistry* (Oxford: Pergamon Press, 1987).
25. Rice, J. A., and P. MacCarthy. "A Model for Humin," *Environ. Sci. Technol.* 24:1875–1877.
26. Kloskowski, R., and F. Führ. "Aged and Bound Herbicide Residues in Soil and Their Bioavailability. Part 1: Uptake of Aged and Non-Extractable (Bound) [3-^{14}C]

Metamitron Residues by Sugar Beets," *J. Environ. Sci. Health* B22:509–535 (1987).

27. Kloskowski, R., and F. Führ. "Aged and Bound Herbicide Residues in Soil and Their Bioavailability. Part 2: Uptake of Aged and Nonextractable (Bound) [Carbonyl-^{14}C] Methabenzthiazuron Residues by Maize," *J. Environ. Sci. Health* B22:623–643 (1987).

28. Kloskowski, R., F. Führ, and W. Mittelstaedt. "The Uptake of Nonextractable Soil-Bound Pesticide Residues by Roots," in *Pestic. Sci. Biotechnol., Proc. Int. Congr. Pestic. Chem., 6th, 1986,* R. Greenhalgh and T. R. Roberts, Eds. (Oxford: Blackwell, 1986), pp. 405–410.

29. Nearpass, D. C. "Adsorption of Picloram by Humic Acids and Humin," *Soil Sci.* 121:272–277 (1976).

30. Dec, J., E. Czaplicki, J. Giebel, and B. Zielinska-Psuja. "Bound Residues of Carbon-14-chlorfenvinphos in Soil and Plants," in *Quantif. Nat. Bioavailability Bound ^{14}C-Pestic. Residues Soil, Plants Food, Proc. Final Res. Co-ord. Meet. 1985* (Vienna: IAEA, 1986), pp. 83–92.

31. Hussain, A., F. Azam, and K. A. Malik. "Bound Residues of Carbon-14-carbofuran in Soil," in *Quantif. Nat. Bioavailability Bound ^{14}C-Pestic. Residues Soil, Plants Food, Proc. Final Res. Co-ord. Meet. 1985* (Vienna: IAEA, 1986), pp. 23–29.

CHAPTER 4

Particulate and Colloidal Organic Material in Pueblo Reservoir, Colorado: Influence of Autochthonous Source on Chemical Composition

James F. Ranville, Richard A. Harnish, and Diane M. McKnight

INTRODUCTION

Particulate and colloidal organic material in natural waters is composed of living and senescent organisms, cellular exudates, and partially to extensively degraded detrital material, all of which may be associated with mineral phases. Determination of the nature of this material has recently been an active research area because of the importance of particulates in the transport of trace metal and organic contaminants. Further, aquatic ecologists have studied particulate organic carbon (POC) because carbon cycling strongly influences ecosystem structure and function. The aspects of POC that are most important in an ecological approach are source, size, and nutritional value. Organic material can be characterized as *allochthonous,* transported to a lake or stream from the watershed, or *autochthonous,* derived from organic material produced in the lake or stream by photosynthesis or other autotrophic processes. The premise of this chapter is that source, size, and nutritional value also have important implications for the chemical composition of POC and its affinity for contaminants.

Allochthonous particulate organic material is chiefly comprised of plant fragments and woody debris,[1] and may be physically or chemically degraded before and during transport to the aquatic system. Allochthonous material may be rich in plant structural and cell wall material, i.e., lignins and cellulose.

Autochthonous particulate organic carbon can be derived from either aquatic plants (macrophytes) or algae (either phytoplankton or periphyton) depending on the defined ecosystem boundaries. For algal-dominated systems, lipids, pigments, proteins, and polysaccharides will be more abundant compo-

Table 4.1. Measurable Chemical Characteristics of Particulate and Colloidal Organic Material of Significance in Sorbing Contaminants

Contaminant Class	Dominant Process	Characteristics
Trace metals	Sorption	1. Surface charge 2. Carboxylic acid group content 3. Nitrogen content

Hydrophobic organic compounds	Partitioning	1. Lipid content
		2. Carbon distribution
		    a. ^{13}C NMR
		    b. IR

nents of the particulate organic carbon than recalcitrant plant structural components.[2]

Some measurable chemical characteristics of POC that may be important for contaminant transport are listed in Table 4.1. For trace metals, sorption on particle surfaces is probably the dominant interaction. Most particles in natural waters are negatively charged, reflecting their content of biogenic materials and aluminosilicates. Negatively charged particles can potentially sorb cationic metal ions. Specific metal binding sites can also be involved in sorption. Carboxylic acid groups are the major acidic functional groups in dissolved and particulate aquatic humic substances; these groups can form charge-transfer complexes with several trace metals (Cu, Fe) and strong complexes with other metals (Al). Another class of metal binding sites are amino acids, which also form strong trace metal complexes. For this reason, nitrogen content may be a significant characteristic and is expected to vary between plant and algal sources.

For hydrophobic organic contaminants, recent research has shown that the interaction is best represented as a partitioning into the organic phases of the particles.[3] The extent of this partitioning will depend upon the nonpolar nature of particulate organic material. Lipids, some other aliphatic moieties, and aromatic moieties can contribute to the nonpolar nature of particles. Lipid content can be measured by extraction and by spectroscopy, and aromaticity can be measured by ^{13}C NMR spectroscopy. A greater abundance of lipids may be expected in algal-derived material because lipids can accumulate as energy storage products when algae are not growing rapidly. A greater aromaticity may be expected for material derived from plant lignins. In addition to partitioning, reactions involving organic contaminants can occur at particle surfaces that influence their transformation and fate in aquatic ecosystems, as recently reviewed by Voudrias and Reinhard[4] and Zepp and Wolfe.[5]

We conducted a study of organic material collected from Pueblo Reservoir, Pueblo, Colorado, to examine the relationship between sources of organic material and the chemical characteristics of particulate and colloidal material. Although Pueblo Reservoir is an oligotrophic system, we expected that autochthonous algal sources would account for most of the POC, as we have observed in an alpine Rocky Mountain lake.[6] We characterized particulate and colloidal material to confirm an algal source and to demonstrate that the characteristics listed in Table 4.1 were influenced by this source.

CARBON FLUX IN OLIGOTROPHIC LAKE/RESERVOIR ECOSYSTEMS

The biological utilization of allochthonous and autochthonous organic material varies from one ecosystem to another. A schematic representing carbon cycling in an oligotrophic lake or reservoir ecosystem is presented in Figure 4.1a, and, for contrast, carbon cycling in a headwater stream ecosystem is presented in Figure 4.1b. Figure 4.1a may represent Pueblo Reservoir, and Figure 4.1b may represent the headwater streams that flow into the reservoir.

For the lake/reservoir ecosystem, the allochthonous organic material comes from the watershed, predominantly as dissolved organic carbon (DOC) in inflow streams and shallow groundwater runoff. Watershed POC inputs may be very limited.[7] This allochthonous DOC may be the most abundant organic pool in the lake or reservoir, but the turnover of this biologically recalcitrant material by heterotrophic uptake may be relatively slow, such that it is not significant in overall carbon cycling. In the euphotic zone, where light intensities are sufficient for photosynthesis to exceed respiration, production of autochthonous POC results from growth of planktonic algae. In Sky Pond, a lake in Rocky Mountain National Park, POC was strongly correlated with chlorophyll a, a measure of algal biomass.[6,8] Similar temporal changes in POC concentration and chlorophyll a in Sky Pond are shown in Figure 4.2. Some additional DOC is released from algal growth, but generally this DOC is assimilated by bacteria more rapidly than the allochthonous DOC and is not the major source for the total DOC pool.[9]

Algal POC is removed from the euphotic zone by zooplankton grazing and by settling of dead or senescent algae.[10] The zooplankton are commonly filter feeders, such as *Daphnia*, and may selectively graze particles in a certain size range. Rotifers can also be important grazers of fine particles and of larger diatoms. For example, the anatomy of some rotifers make them capable of consuming large diatoms by breaking the silica frustule and ingesting the cellular contents.[11] Intact algal and bacterial cells and debris from inefficient grazing settle slowly through the water column and often accumulate at the thermocline (a transition zone between warmer, well-mixed surface waters and colder bottom water). During settling, bacterial degradation of POC occurs. In general, autochthonous POC settles more slowly and undergoes greater degradation in the water column than allochthonous POC from the watershed.[7] Degradation continues once the POC reaches the sediments. Commonly, lake sediments become anoxic at some depth, and sediment interstitial water is enriched in DOC and colloidal organic material, which may reenter the water column by diffusion or groundwater flushing.[12]

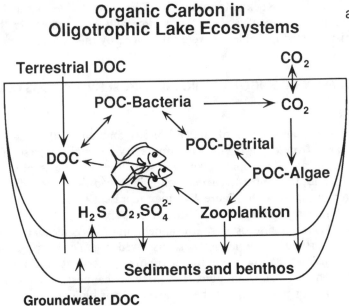

Organic Carbon in Oligotrophic Lake Ecosystems

a

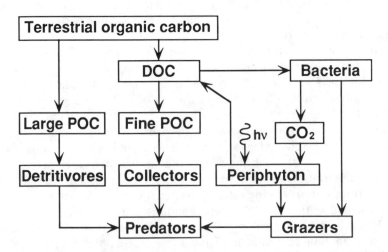

Organic Carbon in Headwater Stream Ecosystems

b

Figure 4.1. Schematic diagrams representing carbon cycling *(a)* in an oligotrophic lake or reservoir and *(b)* in a headwater stream.

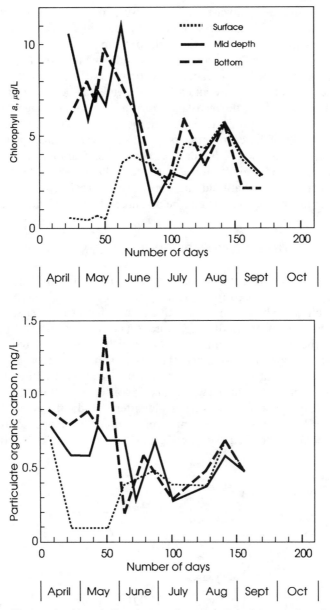

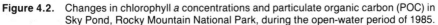

Figure 4.2. Changes in chlorophyll *a* concentrations and particulate organic carbon (POC) in Sky Pond, Rocky Mountain National Park, during the open-water period of 1985.

Carbon cycling in headwater stream ecosystems (Figure 4.1b) is different from carbon cycling in lake/reservoir ecosystems in several ways. First, the input of allochthonous POC in the form of fallen leaf litter is an important food source.[1] Some benthic invertebrates are adapted to shredding such coarse POC and are classified as shredders or detritivores.[13] Finer POC is produced by the shredding of the leaves, and a general trend may be that the average size of POC decreases in the downstream direction.[14] Also, shading of the stream by the canopy limits productivity of the periphyton, the algae growing on streambed rocks. Therefore grazers, benthic invertebrates that graze on periphyton, may not be as abundant as shredders. The organic surface layer composed of periphyton, bacteria, and extracellular organic material (e.g., muccopolysaccharides that hold microorganisms to rocks) can be sloughed off by scouring or grazing actions and be a source of POC in the stream water.

SITE DESCRIPTION AND SAMPLE COLLECTION

Pueblo Reservoir (Figure 4.3) is the water supply for the city of Pueblo, Colorado, and is also used for recreation and fishing. The region is semiarid. The reservoir is fed by the Arkansas River, which has headwater streams in the Rocky Mountains near Leadville, Colorado (110 miles northwest of Pueblo). Several of the streams entering the Arkansas River have elevated trace metal concentrations from mine drainage.[15] The poor survival of trout in the Arkansas River has been attributed to elevated concentrations of Pb, Cu, and Zn. The inorganic chemistry of the water sample collected for this study, presented in Table 4.2, is similar to results from a long-term study of Pueblo Reservoir.[16] The high sulfate concentration may reflect mine drainage sources in the Arkansas River watershed, weathering of Cretaceous shales, or the general

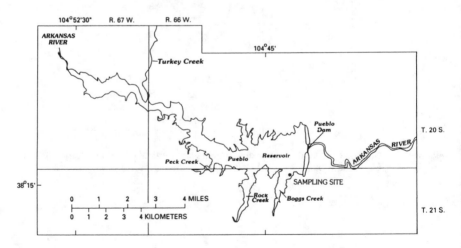

Figure 4.3. Map showing Pueblo Reservoir and the sampling site.

Table 4.2. Chemistry of Pueblo Reservoir Surface Water Sample Collected on February 23, 1990 and of the Organic Colloid Concentrate*

Chemical Constituent	Filtered Water	Organic Colloid	Ratio
Ca	62	92	1.48
Mg	19	24	1.23
Na	27	32	1.19
K	3	NA	—
Si	7	18	2.50
Sr	0.68	0.86	1.26
HCO_3	149		
SO_4	152		
DOC (mg C/L)	2.4		
Trace Metals (mg/L)			
Al	<2	NA	—
Fe	<3	1,600	550
Mn	<1	99	99
Zn	8	161	21

*Sample was filtered through 100 k polysultone membranes by cross-flow ultrafiltration.

aridity of the watershed. The phytoplankton was dominated by cryptophytes (*Chroomonas* sp. and *Cryptomonas erosa*) and diatoms (*Asterionella formosa* and *Fragilaria crotonensis*) (Table 4.3). The pattern of low phytoplankton densities and dominance by cryptophytes and diatoms is typical for winter/spring phytoplankton assemblages in Colorado reservoirs.[17] The planktonic diatoms were more abundant in the upper water column, and although they comprised a lesser portion of the total cells, because of their larger size they probably accounted for the greater portion of the algal biomass.

We sampled Pueblo Reservoir on February 23, 1990. A surface water sample was collected from a dock on the shore of one of the southern arms of the reservoir, as indicated in Figure 4.3. The sample was collected using plastic buckets, transferred into six 10-gal (39-L) stainless steel milk cans, and transported to the laboratory in Arvada, Colorado, on the same day.

METHODS

Sample Fractionation

The 220-L sample was size-fractionated using settling, crossflow ultrafiltration, and centrifugation (Figure 4.4). The sample was allowed to settle for 42 hours in the stainless steel milk cans, which provided a settling distance of 0.5 m. This resulted in a size cutoff of 2–5 μm, based on Stokes settling and assuming a particle density range of 1.5–2.5 g/cm³. Each milk can was then decanted, using a positive displacement pump fitted with Teflon* tubing, leaving the last 3–4 L of sample containing the settled silt. The silt fractions were combined for a total of 23.8 L and allowed to settle again, then decanted

*Use of brand names is for identification purposes only and does not constitute endorsement by the U.S. Geological Survey.

Table 4.3. Species List and Density (Cells/mL) of Phytoplankton Taxa Collected from Pueblo Reservoir, March 1990

TAXA	Integrated Depth	
	0–1.5m	2–3m
BACILLARIOPHYTA (Diatoms)		
Order Centrales		
Melosira italica	116	—
Stephanodiscus dubius	—	17
Stephanodiscus sp.	87	—
Order Pennales		
Asterionella formosa	203	34
Fragilaria crotonensis	540	101
CHLOROPHYTA (Green algae)		
Chlorococcum sp.	17	17
Oocystis elliptica	68	—
CHRYSOPHYTA (Golden-brown algae)		
Diachros sp.	17	17
Ochromonas minuta	101	68
Unknown flagellate	—	34
CYANOPHYTA (Blue-green algae)		
Chroococcus sp.	—	34
Dactylococcopsis fascicularis	17	—
PYRROPHYTA (Dinoflagellates)		
Glenodinium sp.	34	—
CRYPTOPHYTA (Cryptomonads)		
Chroomonas sp.	371	371
Cryptomonas erosa	169	287
Cryptomonas ovata	17	17
TOTAL CELLS/mL	1,757	997
NUMBER OF SPECIES	13	11

to a volume of approximately 1200 mL, put in a Teflon bottle, and allowed to settle a final time. The silt fraction was decanted a final time to a volume of approximately 42 mL and lyophilized.

The colloid sample was isolated using crossflow ultrafiltration. This method has been applied by a number of workers to concentrate phytoplankton[18] and suspended sediments from rivers[19,20] and marine waters.[21] A Millipore Pellicon system with six polysulfone membranes (size cutoff of 100,000 daltons) were used to provide a total filter surface area of 3 ft². During the filtration, which took 14 hours to process the 194.1 L, the crossflow rate ranged from 2.5 to 3.5 L/min, and the filtrate rate ranged from 200 to 350 mL/min. The final colloid concentrate (retentate) volume was 455 mL. The filter was disassembled and colloids were removed from the filters by placing a filter in a Teflon bag along with some retentate and gently rubbing the colloids off. This process was repeated for all six filters used. Previous work has shown this method provides high recovery of sediment mass. The colloid concentrate was centrifuged at 1600 rpm for 30 min. This resulted in three separate layers, which resulted

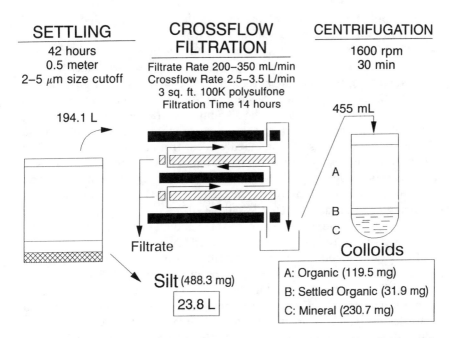

SETTLING
42 hours
0.5 meter
2–5 μm size cutoff

194.1 L

CROSSFLOW FILTRATION
Filtrate Rate 200–350 mL/min
Crossflow Rate 2.5–3.5 L/min
3 sq. ft. 100K polysulfone
Filtration Time 14 hours

CENTRIFUGATION
1600 rpm
30 min

455 mL

A
B
C

Filtrate

Silt (488.3 mg)

23.8 L

Colloids

A: Organic (119.5 mg)
B: Settled Organic (31.9 mg)
C: Mineral (230.7 mg)

Figure 4.4. Diagram showing the operation of crossflow ultrafiltration for collection of the colloid sample.

from size and density differences in the colloid sample. The water layer was a pale white suspension, which was carefully pipetted off. The sediment layer at the bottom of the centrifuge tube consisted of an upper thin white layer, which appeared similar to the material in the water layer, and a densely packed grey claylike layer at the bottom. The water (395 mL) and sediment layers were lyophilized separately. The white sediment layer after lyophilization was physically distinct and was separated from the claylike material. These three fractions were termed *organic colloid* (water layer), *settled organic colloid* (thin white sediment layer), and *mineral colloid* (grey sediment layer).

Physical and Chemical Characterization

Particle size was determined by photon correlation spectroscopy (PCS), using a Brookhaven BI-2030 AT spectrometer on an aliquot of the whole water sample taken prior to settling. This technique is based on measuring the time dependence of intensity fluctuations of scattered light (wavelength = 633 nm) resulting from Brownian motion of the particles.[22] Surface charge—more properly electrophoretic mobility—was determined by electrophoretic light scattering, using a Malvern Zetasizer IIc with an applied field strength of 20 V/cm. The velocity of particle drift under a given applied electrical field is proportional to the charge on the particle. The velocity is determined by measuring the shift in frequency of illumination of particles under motion.[22,23]

The electrophoretic mobility of particles in the whole water and each of the concentrates was determined. Scanning electron microscopy–energy dispersive X-ray (SEM-EDX) analysis was performed on filtered particulates from whole water and each of the fractions. Approximately 7 mL was taken from the whole water sample by dipping a filter holder, containing a 0.1-μm Nuclepore filter into the sample and then vacuum-filtering with minimum vacuum. This approach was taken to minimize the disruption of fragile aggregates present in the sample.[24] Samples of each of the colloid fractions were also filtered using 0.1-μm Nuclepore filters. A small area was cut from each filter, mounted to a carbon SEM stub using carbon paint, and gold-coated for SEM analysis.

Elemental composition (C, N) of the silt and colloid fractions was determined using various combustive techniques by Huffmann Laboratories, Wheatridge, Colorado.[25] A 60-mL sample of colloid concentrate was taken prior to lyophilization, and from this a small aliquot (5–7 mL) was taken from the upper layer, representing the concentrated organic colloid fraction. Analysis of this sample and a sample of the filtrate was performed by inductively coupled plasma–atomic emission spectroscopy (ICP-AES). Samples of silt, organic colloid, and mineral colloid were digested with a 50:50 mixture of 70% perchloric acid and concentrated nitric acid, followed by 30% hydrogen peroxide to destroy residual organic matter. Residual silicates were dissolved by adding 48% hydrofluoric acid. The sample bottles were tightly capped after the addition of hydrofluoric acid to minimize loss of volatile silicon fluoride complexes. Boric acid was added to the digestions to complex residual fluoride ion, prior to ICP-AES analysis. ICP-AES analysis was performed using the method of standard additions. Infrared analysis was performed on the lyophilized silt, organic colloid, and mineral colloid. All the samples showed significant bands at 1450 cm^{-1} indicative of carbonates. The samples were treated with HCl to remove this interference, which allowed investigation of organic bands present in the same region. The samples were run in KBr pellets on a Perkin-Elmer 580 infrared spectrophotometer.

Organic carbon functionality was determined using solid state ^{13}C nuclear magnetic resonance spectroscopy with cross-polarization/magic angle spinning (CPMAS-NMR) on a Nicolet NT-150 wide-bore NMR spectrometer at the Colorado State University Regional NMR Center. The data was collected for 8 hours for the mineral colloid and 12 hours for the organic colloid and silt fractions. The line broadening was 50 Hz for the mineral colloid and 20 Hz for the organic colloid and the silt. Larger broadening improved signal-to-noise ratio at the expense of an increased line width.

Lipids were extracted from the samples with a chloroform: methanol (2:1) mixture in a small Soxhlet apparatus. The lipid sample solution was transferred to preweighed aluminum foil, evaporated to dryness, and reweighed. Lipid residue was removed from the foil by extracting with chloroform, and the volume reduced under dry N$_2$ in a calibrated glass vial. The lipid extracts had a green color resulting from the extraction of chlorophyll. The sample was placed on a small silica gel column and the nonpolar fraction was eluted with

Table 4.4. Physical Properties of Particulate and Colloidal Material in Surface Water Sample from Suspended Sediments from Pueblo Reservoir, February 23, 1990

| Sample | Electrophoretic Mobility[a] (μm/sec/volt/cm) | Particle Size (PCS) | |
		Effective Diameter (μm)	Polydispersity
Whole Water	-0.88 ± 0.08	2.6	1.2
Silt	-0.90 ± 0.04		
Organic Colloid	-0.93 ± 0.01		
Mineral Colloid	-0.88 ± 0.07		

[a]Mean values.

hexane, benzene, and methylene chloride. The nonpolar fraction was analyzed by gas chromatography–mass spectrometry (GC-MS) on a 30-m, 0.25-mm i.d., 0.25 μm Rt$_x$-5 column capillary GC using a linear temperature program. Data were collected with one-second MS scans from mass 50 to 450, under electron impact mode.

RESULTS

Physical and Inorganic Chemical Characteristics

Size fractionation of the sample into silt and colloids by settling has two advantages—the first being its simplicity. A more important reason is the hydrologic significance for this approach. This method fractionates the sample into a component that has a high probability of settling through the water column to the bed sediments (silt), and a component that may be transported through the reservoir (colloids). From this we can infer how these particulates may effect the transport and fate of associated contaminants. The mass recovery of silt and colloidal material from the Pueblo Reservoir sample is shown in Figure 4.4. The silt fraction contained most of the sediment mass (56.1%), and the settled organic colloid fraction contained the least (3.7%). The results of particle size (PCS), electrophoretic mobility, and SEM analysis are given in Table 4.4. The effective hydrodynamic mean diameter was 2.6 μm with a polydispersity of 1.2, indicating a wide size distribution. The settling conditions should have caused particles greater than 2–5 μm to be included in the silt fraction, so that the size range for the colloids was between 2–5 μm and 100,000 daltons (0.005 μm). The total recovery of suspended material was 870.4 mg, corresponding to a concentration of 4 mg/L in the whole water sample. The distribution of mass in the silt (488.3 mg) and colloid (382.1 mg) fractions also indicate a mass distribution centered around 2–5 μm, comparable to the PCS result. Roughly half the mass of the suspended sediments will be expected to remain in suspension for sufficient periods of time to allow transport through the reservoir system.

SEM analysis of the whole water sample indicated that the particulates were aggregated to a large extent prior to sample processing (Figure 4.5). Aggregates up to 70 μm, composed of much smaller particles, are evident. Close examination of an aggregate (Figure 4.5d) suggests that the inorganic particles (clays) may be held together by gelatinous organic material. EDX analysis showed that these particles were primarily composed of Si and Al, with much smaller amounts of Ca, K, and Fe, indicating the particles were predominantly clays and other detrital aluminosilicates. SEM analysis of the mineral colloid fraction (Figure 4.6) indicated that additional aggregation occurred during the crossflow ultrafiltration procedure, and that the surfaces of the diatoms in the water sample (Table 4.3) may have been substrates for aggregation. Figure 4.6a shows aggregation around a colony of the diatom *Fragillaria crotonensis*. Figure 4.6b shows a colony of *Fragillaria crotonensis* alongside an aggregate of remarkably similar dimensions which may have been formed on the diatom colony and been dislodged in the process of preparing the filter for SEM analysis.

The negative electrophoretic mobilities are in keeping with organic-rich aluminosilicate particles[26] and organisms[27] common to natural waters. The magnitude of the negative charge is at the low range of mobilities reported for most freshwaters and is typical for hard waters where Ca^{2+} and Mg^{2+} reduce the excess negative charge on the organic matter.[28] Electrophoretic mobilities of the whole water particulates and the size fractions are nearly identical despite differences in other bulk properties, most likely because charged organic functional groups are dominated by carboxylic acid groups which may more or less uniformly coat particle surfaces. Coagulation between different particle types (clays, oxides, and organic matter) may also result in a generally uniform particle electrophoretic mobility.

Comparison of the inorganic chemistry of the 100,000-dalton filtrate to the organic colloid concentrate is shown in Table 4.2. Most of the major elements show little enrichment in the colloid concentrate, indicating they are primarily dissolved components. Iron, manganese, and zinc, however, show significant enrichment in the organic colloid fraction, indicating they are in some way associated with this fraction. Iron shows a concentration factor of over 500 in the organic colloid fraction. Comparison of this result to the volume reduction factor of 485 indicates that most of the iron in the sample was concentrated in this fraction. Manganese and zinc concentration factors were greater than 99 and 20.1, respectively, also indicating association with this fraction. Results from the total digestions are given in Table 4.5. The ratio of Fe, Mn, and Zn to Si is much higher in the organic colloid than in either the mineral colloid or silt. This result suggests that trace metals in the organic colloid are not simply a result of the presence of detrital minerals in the organic colloid, but implies additional trace metal association with this phase.

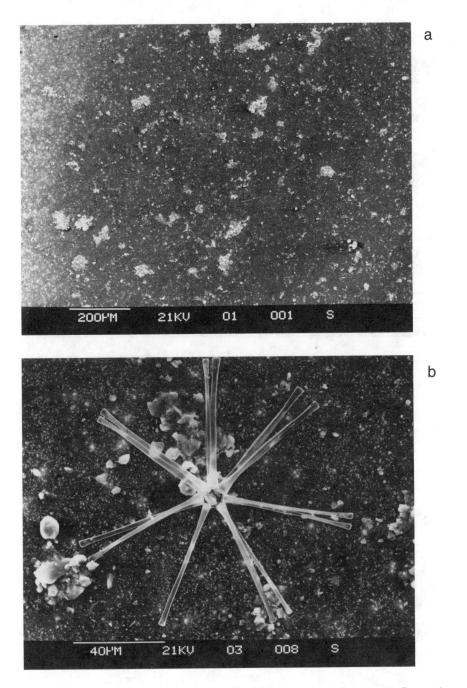

Figure 4.5.   SEM micrographs of particles in the whole water sample from Pueblo Reservoir.

Figure 4.6. SEM micrographs of particles in the mineral colloid fraction from Pueblo Reservoir: *a,* colony of *Fragillaria crotonensis* with aggregated material; *b,* colony of *Fragillaria crotonensis* alongside an aggregate of similar dimensions.

Table 4.5. Chemical Composition of the Fractions from Pueblo Reservoir

	Silt	Organic Colloid	Settled Organic Colloid	Mineral Colloid
Organic Constituents				
Organic carbon (wt %)	3.58	20.12	9.27	6.98
Nitrogen (wt %)	0.45	3.08	1.68	1.14
Atomic C:N	9.3	7.6	6.5	7.1
Lipids (wt %)	0.56	6.66	<0.01	1.01
Inorganic Constituents				
Si (wt %)	27.6	2.5	NA	35.3
Fe (wt %)	1.5	0.8	NA	3.3
Mn (ppm)	450	1800	NA	1250
Zn (ppm)	200	350	NA	370
Fe:Si	5.4×10^{-2}	32.0×10^{-2}	—	9.3×10^{-2}
Mn:Si	0.16×10^{-2}	7.2×10^{-2}	—	0.35×10^{-2}
Zn:Si	0.07×10^{-2}	1.4×10^{-2}	—	0.10×10^{-2}

General Organic Chemical Characteristics

The results of the analysis for organic carbon and nitrogen are presented in Table 4.5. These results confirm that the organic colloid fraction is indeed enriched in organic C compared to the other fractions. Even though the organic colloid fraction comprises only 13.7% of the total sediment mass, it comprises 40.6% of the total particulate organic carbon (Figure 4.7). If we assume that the organic material in the organic colloid fraction has a carbon content of 50% (which is typical for aquatic humic substances), then the organic carbon content of 20.12% corresponds to the organic colloid fraction being 40.2% organic material. The settled organic colloid fraction had an organic carbon content intermediate between the organic colloid and mineral colloid fractions, and the silt fraction had the lowest organic carbon content. These data were used to calculate that the concentration of particulate and colloidal organic carbon was 0.28 mg C/L. The DOC concentration was 2.4 mg C/L, an order of magnitude greater, which is consistent with the general picture of carbon cycling presented in Figure 4.1a.

The N content was distributed among the fractions in a similar manner to the organic carbon content, with the organic colloid fraction accounting for about 41% of the total N (Figure 4.7). Comparison of carbon-to-nitrogen ratios shows that the settled organic colloid fraction has the most nitrogen-enriched organic material, the silt fraction has the least, and the organic and mineral colloid fractions have similar ratios (Table 4.5). All the fractions contain organic material that is more enriched in nitrogen than aquatic fulvic acids (where the C:N ratios range from 20 to 80). Lower values are found for freshwater fulvic acids derived from autochthonous algal sources. Therefore, the nitrogen contents of the particulate and colloidal organic material in Pueblo Reservoir are consistent with autochthonous algal and microbial biomass as the major source.

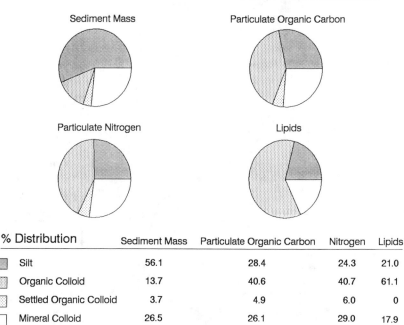

% Distribution	Sediment Mass	Particulate Organic Carbon	Nitrogen	Lipids
▨ Silt	56.1	28.4	24.3	21.0
▥ Organic Colloid	13.7	40.6	40.7	61.1
▦ Settled Organic Colloid	3.7	4.9	6.0	0
☐ Mineral Colloid	26.5	26.1	29.0	17.9

Figure 4.7. Relative mass distribution (in weight percent) of sediment mass, particulate organic carbon, nitrogen, and lipids in the silt, organic colloid, settled organic colloid, and mineral colloid fractions from Pueblo Reservoir.

Algal and microbial cells contain several N-containing classes of organic compounds, including amino acids and chlorophylls. The most abundant amino acids in cell wall material are glycine, alanine, glutamic acid, and aspartic acid. Chlorophyll *a* was detected in the silt and mineral colloid fractions at concentrations of 0.19 and 0.10 μg/L in water and 0.86 and 0.098 μg/mg in the solid, respectively. These chlorophyll concentrations are too low to account for a significant portion of the N content of these fractions.

The results of the infrared analysis are shown in Figure 4.8, and the major peaks present are summarized in Table 4.6. The silt and mineral colloid show strongest bands resulting from mineral matter (see Table 4.6) due to their lower proportion of organic matter than the organic colloid. The spectra of all three samples are similar with respect to the bands resulting from organic matter. The organic colloid may have a slightly stronger amide II band (1540 cm^{-1}), indicative of proteins and in keeping with the higher percent N in this fraction. However, comparison of fractions is difficult due to the large 0-H band (1640 cm^{-1}) resulting from adsorbed water.

The solid-state ^{13}C CPMAS-NMR spectra for the organic colloid, silt, and mineral colloid fractions are shown in Figure 4.9. The spectrum for the organic colloid fraction shows excellent resolution of the major carbon moieties. The other two spectra are more poorly resolved and are probably influenced by the lower carbon content and paramagnetic metals (Fe and Mn) in the

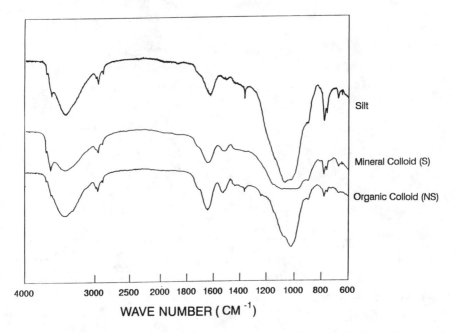

Figure 4.8.   IR spectra of particulate and colloidal fractions from Pueblo Reservoir.

Table 4.6. Infrared Adsorption Bands Present in Silt and Colloid Fractions (cm⁻¹)

	Covalent Bond	Infra-absorption Bands
Alumino –Silicates	Si-O-S: Al-O-H	1050 (900–1300) 3690, 3620
Quartz	Si-O-Si	800, 780
H_2O	O-H	3400, 1650
Organic	C-H C=O C-N	2960, 2920, 2850 1725 1540

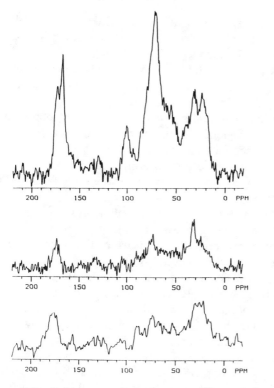

Organic colloid

Silt

Mineral colloid

Figure 4.9. Solid-state natural abundance ^{13}C NMR spectra of particulate and colloidal fractions from Pueblo Reservoir, Pueblo, Colorado. These spectra are comprised of major broad peaks characteristic of aquatic humic substances, which are described in the text.

Table 4.7. Relative Peak Areas for Carbon Moieties in Natural Abundance ^{13}C-CPMAS-NMR Spectra of Particulate and Colloidal Fractions from Pueblo Reservoir, Pueblo, Colorado. (The carbon moieties represented by the major bands are explained in the text. Replicate spectra show a variation in the integration of about 0.5%)

| | Relative Peak Areas | | | | | |
Sample	AL-I (0–62)	AL-II (62–90)	AL-III (90–110)	AR (110–165)	C-I (165–185)	C-II (185–220)
Organic Colloid	38	34	7	8	13	–
Silt	50	25	5	9	11	–
Mineral Colloid	50	20	5	8	15	2

inorganic portion of the fractions. There are six broad peaks that are typically present in spectra of natural organic material: The first aliphatic peak (AL-I), 0 to 62 ppm, represents carbons bonded to other carbons (methyl, methylene, and methine) as well as some carbons bonded to N or S. The second aliphatic peak (AL-II), 62 to 90 ppm, represents carbons bonded to oxygen (carbohydrates, alcohols, and ethers). The third aliphatic peak (AL-III), 90 to 110 ppm, represents anomeric carbons. The aromatic peak (AR), 110 to 165 ppm, represents primarily sp^2-hybridized carbons (aromatic carbons and other double-bonded carbons). The first carbonyl peak (C-I), 165 to 185 ppm, represents primarily carboxylic carbons, and the second carbonyl peak (C-II), 185 to 220 ppm, primarily represents ketones. The peak areas for the three fractions are summarized in Table 4.7.

In the CPMAS-NMR spectrum for the organic colloid fraction, the C-II peak is not present, and the AR peak is present only to a very limited extent, accounting for only 8% of the total area of the spectrum. The minor presence of aromatic moieties suggests that this material is not derived from degradation of plant materials in the watershed. The ratio of the AL-III peak and the AL-II peak is consistent with carbohydrates comprising a large portion of the organic carbon in the organic colloid fraction.[29] Means and Wijayaratne found that colloidal organic material in an estuarine system was dominated by carbohydrates (35–60%).[30] The AL-I peak represents 38% of the total area, and lipids are represented in this region. The C-I peak includes carboxylic acids as well as carbonates in the inorganic portion of the organic colloid fraction. The presence of carboxylic acid groups indicates that some of the carbohydrates may be present as sugar acids.[29]

The spectra for the silt and mineral colloid fractions are poorly resolved. However, a discrete peak in the upper portion of the AL-I region of the silt fraction is evident. This peak may correspond, in part, to lipids. The overall carbon distributions for the silt and mineral colloid fractions are very similar, and these fractions are more enriched in noncarbohydrate aliphatic material than the organic colloid fraction. Although this should be interpreted cautiously because of the poor resolution of these spectra, these results indicate that for the organic colloid fraction, the greater heteroatom content and polarity of the organic material may limit its sorption on mineral surfaces, as proposed by Means and Wijayaratne.[30]

Characterization of Lipids

The mass of lipids extracted from each of the fractions, in weight percent, is given in Table 4.5. The organic colloid fraction was highest in lipids, followed by the mineral colloid, the silt, and finally the settled organic colloid. The bulk of the total lipids (61.1%) is contained in the organic colloid fraction, which was 6.66% by weight lipids (Figure 4.7). In contrast, Means and Wijayaratne found that lipids were less than 1% of the estuarine organic colloids.[30] However, based on the relative peak area of the Al-I region of the ^{13}C NMR spectra, the silt and mineral colloid are the more lipid rich. There are two possible explanations for this discrepancy. For one, it is possible that sorption of lipids to mineral surfaces and interlayers in the clay-rich fractions (silt and mineral colloids) may prevent extraction. Another possibility is that the Al-I region includes aliphatic carbons in molecules other than lipids which are not soluble in chloroform-methanol.

Analysis of the n-alkane distribution as biomarkers in the lipid extract of the silt and organic colloid fractions gives us some further insight into the source and nature of POC. Full scan gas chromatography–mass spectrometry of the samples indicated the presence of phthalate esters resulting from materials used in the filtering and sampling equipment. By utilizing the mass 85 ion for alkanes, interference from the phthalate esters was avoided. The results, shown in Figure 4.10, indicate a unimodal distribution of n-alkanes beginning at approximately C-15 with slight odd-over-even predominance from C-25 to C-35. The n-alkane distribution (C-15 to C-32) is consistent with organic matter that has undergone little or no degradation. Comparison of the alkane molecular weight distribution indicates that the organic colloid has a slightly lower molecular weight distribution (maximum at C-21) than the silt (maximum at C-23). A strong odd-over-even preference (10 or greater) has been observed for organic matter originating from higher plants with a distribution mode at C-27 to C-31.[31] This is in contrast to a distribution mode of C-15 to C-19 for an algal-derived material. Our results point toward a mixed source with the dominance of algae. The greater abundance of C-27, C-29, and C-31 in the silt indicates a larger contribution of lipids from higher plants in this fraction.

Implications for Contaminant Transport

A detailed characterization of particulate and colloidal organic material in every aquatic ecosystem where contaminant transport is a concern is not always practical. However, from the general knowledge of a lake or stream, it may be possible to identify the main sources of dissolved, particulate, and colloidal organic material. If, in turn, there is a consistent relationship between the source and the chemical characteristics, this relationship could be useful in studies of transport of trace metal or organic contaminants. The characterization of the particulate and colloidal material from Pueblo Reservoir supports this approach: Several characteristics were consistent with a predominantly autochthonous source as we had hypothesized. More information on particulate and colloidal material in other aquatic ecosystems, such as a headwater stream ecosystem (Figure 4.1b), are needed to further develop this approach. Another verification of this approach is the similarity of the Pueblo Reservoir samples in several ways to samples from other environments where autochthonous algal sources predominate. For example, the ^{13}C CPMAS-NMR spectrum for the organic colloid fraction (Figure 4.9) is very similar to

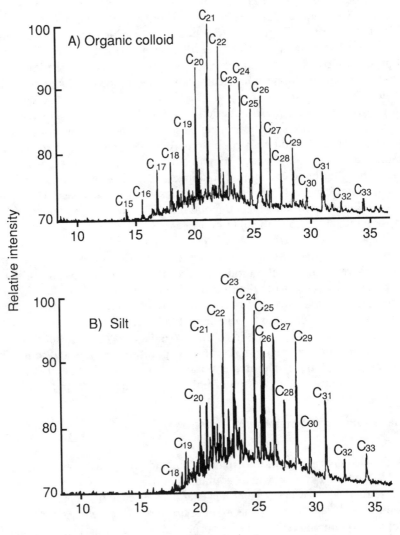

Figure 4.10. Distribution of n-alkanes in lipid extracts of the organic colloid and silt fractions from Pueblo Reservoir.

the ^{13}C NMR spectrum for organic material isolated by ultrafiltration from the interstitial waters of an algal-rich lake in Bermuda.[32]

A number of factors may make POC a significant transport phase for organic contaminants. In the sorption of hydrophobic organic compounds, the single most important characteristic of the sorbent phase is its organic matter content.[33] The particulate samples isolated from Pueblo Reservoir, and in particular the organic colloid fraction, are quite rich in organic material. The sorption of hydrophobic compounds can be viewed as a partitioning into the nonpolar region of the organic phase of the particle, and is described by each compound's octanol-water partition coefficient. The significant presence of lipids in the organic and mineral colloids is in keeping with the algal source of POC in Pueblo Reservoir. Among algae, diatoms (the greatest biomass of algae in Pueblo Reservoir) are lipid rich and can be up to 70% by weight lipid.[31] This lipid phase may represent a suitable nonpolar region for the sorption of hydrophobic contaminants. Estuarine colloids have been shown to be important in hydrophobic partitioning of the herbicides atrazine and linuron.[34] They found that this estuarine colloidal organic material was 10–35 times better as an absorbent than soil or sediment organic material. The sorption of polycyclic aromatic hydrocarbons (PAHs) by these colloids was found to be about 10 times greater on colloidal organic material than on soil or sediment organic material.[35] The presence of colloidal material was required to model the partitioning of polychlorobiphenyl congeners (PCBs) in Lake Superior.[36] This study indicated that a three-phase model, including nonfilterable colloids, best explained the data, and it also suggested that colloid-associated contaminants may be the dominant species in surface waters.

Specific interactions between contaminants and functional groups on natural organic matter may be important in addition to hydrophobic sorption. These mechanisms include hydrogen bonding, charge transfer complexes through aromatic pi-electrons, cation exchange, and conjugate formation through biochemical processes.[37] The primary source of particulate and colloidal organic material in aquatic ecosystems will influence which of these mechanisms will predominate. For Pueblo Reservoir, we have shown a primarily autochthonous source which yields organic matter low in aromaticity, and rich in carbohydrates, nitrogen content, and aliphatic lipids. The high content of carbohydrates in the organic colloid fraction indicates that this phase may hydrogen-bond to contaminants. Hydrogen-bonding between amino hydrogens of substituted herbicides and model absorbents containing carbonyl groups has been demonstrated.[38] Repulsion between -OH groups on the clay surface and hydroxyl groups on the carbohydrates may explain why the organic colloid fraction did not aggregate during concentration.[30] The low aromaticity indicates that the formation of charge transfer complexes by pi-bonding may be unimportant. This contrasts with ecosystems where charge transfer complexes may be important because organic inputs are primarily allochthonous material with greater aromaticity resulting from lignin degradation. The sorption of benzidine and toluidine onto estuarine colloids was enhanced over that expected for hydrophobic partitioning.[39] The pH dependence of sorption indicated that under conditions favoring the cationic form of the amine, sorption (presumably by cation exchange) was enhanced. Conversely, the relatively high nitrogen content of the colloidal organic material indicates a strong potential for cation exchange between organic material and anionic contaminants.

In the Pueblo Reservoir sample, several trace metals (Fe, Mn, and Zn) were concentrated in the organic colloid fraction (Table 4.5). Zinc is toxic to aquatic organisms at dissolved concentrations on the order of 100 ppb and is present upstream at much higher concentrations in the acid mine drainage streams of the Arkansas River watershed. The biological availability of colloidal and particulate metals is poorly understood.[40] The association of these metals with the organic colloids is consistent with previous results showing the importance of organic complexation in trace-metal transport.[41] Eisenreich et al. showed in a study of the upper Mississippi River that the highest concentrations of Cu, Cd, and Pb occurred in the 1–10,000K molecular weight fractions and correlated with organic carbon content.[42] In analysis of colloidal particles in the Patuxent River, Sigleo and Helz found that As, Sb, Se, and Zn were enriched in the colloid fraction during the summer when the carbon content also increased.[43] In a field study of settling particles, Sigg investigated the importance of various phases for metal transport.[44] Based on the Redfield ratio, Sigg used phosphorous as a measure of autochthonous algal material and found a correlation between concentrations in settling material of P, Cu, and Zn.

The size distribution of Fe and Mn in three freshwater lakes was determined by Laxen and Chandler.[45] The median diameters for iron ranged from about 0.3 to 4.5 μm and were similar for Mn, with the exception that more Mn was dissolved (< 0.005 μm). Salbu et al. investigated surface and groundwater colloid size distribution using filtration, centrifugation, in situ dialysis, and hollow-fiber ultrafiltration.[46] They investigated the size range 0.005–0.45 μm and found for surface water that Cd, Cr, and Zn were primarily in the 0.005–0.1 μm size range. They found Cu primarily in the 0.1–0.45 μm range. Iron was in the > 0.45 μm, and Mn was bimodally distributed in the > 0.45 and < 0.005 μm fractions. Neither of these two studies investigated the role of organic carbon in metal distribution.

In laboratory studies, it has been shown that sorption of natural organic material on Al, Fe, and Mn oxides enhances the sorption of copper and cadmium.[28,47] The proposed mechanism for adsorption on metal oxides and trace metal complexation involve the formation of inner sphere complexes between the metal and primarily carboxyl groups on the organic matter. It should be expected then that the surface and nature of organic matter will affect trace-metal transport.

ACKNOWLEDGMENTS

We wish to thank A. Jurkiewicz for the 13C CPMAS spectra, Colleen Rostad for GC-MS spectra, Paul Briggs for ICP data on the sediments, and also Briant Kimball for providing the cross-flow filtration equipment. We also thank J. Leenheer for providing the impetus for this study.

REFERENCES

1. Cummins, K. W., J. R. Sedell, F. J. Swanson, G. W. Minshall, S. G. Fisher, C. E. Cushing, R. C. Petersen, and R. L. Vannote. "Organic Matter Budgets for Stream Ecosystems: Problems in their Evaluation." pp 299–353. In: Barnes, J. R. and Minshall, G. W. (Eds.), Stream Ecology: testing general ecological theory. Plenum. N.Y.
2. Stewart, W. D. P. *Algal Physiology and Biochemistry* (Berkeley: University of California Press, 1974).
3. Chiou, C. T. "Partition Coefficients of Organic Compounds in Lipid-Water Systems and Correlations with Fish Bioconcentration Factors," *Environ. Sci. Technol.* 19:57–62 (1985).
4. Voudrias, E. A., and M. Reinhard. "Abiotic Organic Reactions at Mineral Surfaces," in *Geochemical Processes at Mineral Surfaces,* ACS Series 323, J. A. Davis and K. F. Hayes, Eds. (Washington, DC: American Chemical Society, 1986), pp. 462–486.
5. Zepp, R. G., and N. L. Wolfe. "Abiotic Transformation of Organic Chemicals at the Particle-Water Interface," in *Aquatic Surface Chemistry,* W. Stumm, Ed. (New York: Wiley-Interscience, 1987), pp. 423–455.
6. McKnight, D. M., C. Miller, and R. L. Smith. "Phytoplankton Populations in Lakes in Loch Vale, Rocky Mountain National Park, Colorado: Sensitivity to Acidic Conditions and Nitrate Enrichment," U.S. Geological Survey Water-Resources Investigations Report 88–4115 (1988).
7. Wetzel, R. *Limnology* (Philadelphia: W. B. Saunders Company, 1983).
8. Baron, J., personal communication.
9. Chrost, R. J., U. Munster, H. Rai, A. Dieter, P. K. Witzel, and J. Overbeck. "Photosynthetic Production and Exoenzymatic Degradation of Organic Matter in the Euphotic Zone of a Eutrophic Lake," *J. Plank. Res.* 11:223–242 (1989).
10. Crumpton, W. G., and R. G. Wetzel. "Effects of Differential Growth and Mortality in the Seasonal Succession of Phytoplankton Populations in Lawrence Lake, Michigan," *Ecology* 63:729–739 (1982).
11. Pennak, R. W. *Fresh-Water Invertebrates of the United States,* 2nd ed. (New York: John Wiley, 1978).
12. Aiken, G. R., D. M. McKnight, and L. Miller. "Evidence for the Diffusion of Aquatic Fulvic Acid from the Sediments of Lake Fryxell, Antarctica," Chapter 5, this volume.
13. Merritt, R. W., and K. W. Cummins, Eds. *An Introduction to the Aquatic Insects of North America* (Dubuque, IA: Kendall Hunt Publishing Company, 1978).
14. Vannote, R. L., G. W. Minshall, K. W. Cummins, J. R. Sedell, and C. E. Cushing. "The River Continuum Concept," *Can. J. Fish. Aquat. Sci.* 37:130–137 (1980).
15. Kimball, B. A., K. E. Bencala, and D. M. McKnight. "Research on Metals in Acid Mine Drainage in the Leadville, Colorado, Area," in *U.S. Geological Survey Toxic Substances Hydrology Program—Proceedings of the Technical Meeting, Phoenix, Arizona, September 26–30, 1989,* G. E. Mallard and S. E. Ragone, Eds., U.S. Geological Survey Water-Resources Investigations Report 88–4420 (1989), pp. 65–70.
16. Edelman, P. personal communication.
17. Conklin, D. J., Chadwick, and Associates, written communication.
18. Barthel, K. G., G. Schneider, R. Gradinger, and J. Lenz. "Concentration of Live

Pico-and Nanoplankton by Means of Tangential Flow Filtration," *J. Plank. Res.* 11:1213–1221 (1989).

19. Hernandez, L. K., and R. F. Stallard. "Sediment Sampling through Ultrafiltration," *J. Sed. Petrol.* 58:768–769 (1988).

20. Leenheer, J. A., R. H. Meade, H. E. Taylor, and W. E. Periera. "Sampling, Fractionation, and Dewatering of Suspended Sediment from the Mississippi River for Geochemical and Trace-Contaminant Analysis," in *U.S. Geological Survey Toxic Substances Hydrology Program: Proceedings of the Technical Meeting, Phoenix, Arizona, September 26–30, 1988,* G. E. Mallard and S. E. Ragone, Eds., U.S. Geological Survey Water-Resources Investigations Report 88–4220 (1989).

21. Whitehouse, B. G., G. Petrick, and M. Ehranrdt. "Crossflow Filtration of Colloids from Baltic Sea Water," *Water Res.* 20:1599–1601 (1986).

22. Rees, T. F. "A Review of Light-Scattering Techniques for the Study of Colloids in Natural Waters," *J. Contam. Hydrol.* 1:425–439 (1987).

23. Ranville, J. F., K. S. Smith, D. L. Macalady, and T. F. Rees. "Electrophoretic Light Scattering: A Rapid Technique for Measuring Electrophoretic Mobilities of Suspended Particles," paper presented at the Symposium on Environmental Analytical Chemistry, Jekyll Island, GA, May 22–24, 1989.

24. Gibbs, R. J., and L. N. Konwar. "Effect of Pipetting on Mineral Flocs," *Environ. Sci. Technol.* 16:119–121 (1982).

25. Huffman, E. W. D., Jr., and H. A. Stuber. "Analytical Methodology for Elemental Analysis of Humic Substances," in *Humic Substances in Soil, Sediment and Water,* G. R. Aiken, D. M. McKnight, R. L. Wershaw, and P. MacCarthy, Eds. (New York: John Wiley, 1985), pp. 433–458.

26. Hunter, R. J., and P. S. Liss. "The Surface Charge of Suspended Particles in Estuarine and Coastal Waters," *Nature* 282:823–825 (1979).

27. Gerritsen, J., and S. W. Bradley. "Electrophoretic Mobility of Natural Particles and Cultured Organisms in Freshwaters," *Limnol. Oceanogr.* 32:1049–1058 (1987).

28. Tipping, E., and D. Cooke. "The Effects of Adsorbed Humic Substances on the Surface Charge of Goethite (α-FeOOH) in Freshwaters," *Geochim. Cosmochim. Acta* 46:75–80 (1982).

29. Leenheer, J. A., personal communication.

30. Means, J. C., and R. D. Wijayaratne. "Chemical Characterization of Estuarine Colloidal Organic Matter: Implications for Adsorptive Processes," *Bull. of Marine Sci.* 35:449–461 (1984).

31. Tissot, B. P., and D. H. Welte. *Petroleum Formation and Occurrence* (Berlin: Springer-Verlag, 1984).

32. Orem, W. H., and P. G. Hatcher. "Solid-State ^{13}C-NMR Studies of Dissolved Organic Matter in Pore Waters from Different Depositional Environments," *Org. Geochem.* 11:73–82 (1987).

33. Karickoff, S. W. "Organic Pollutant Sorption in Aquatic Systems," *J. Hydrol. Eng.* 110:707–735 (1984).

34. Means, J. C., and R. D. Wijayaratne. "Role of Natural Colloids in the Transport of Hydrophobic Pollutants," *Science* 215:968–970 (1982).

35. Wijayaratne, R. D., and J. C. Means. "Sorption of Polycyclic Aromatic Hydrocarbons by Natural Estuarine Colloids," *Mar. Environ. Res.* 11:77–89 (1984).

36. Baker, J. E., P. D. Capel, and S. J. Eisenreich. "Influence of Colloids on

Sediment-Water Partition Coefficients of Polychlorobiphenyl Congeners in Natural Waters," *Environ. Sci. Technol.* 20:1136–1143 (1986).

37. Leenheer, J. A. "Organic Substance Structures That Facilitate Contaminant Transport and Transformations in Aquatic Sediments," Chapter 1, this volume.

38. Ward, T. M., and K. Holly. "The Sorption s-Triazines by Model Nucleophiles as Related to Their Partitioning between Water and Cyclohexane," *J. Colloid. Interface Science* 22:221–230 (1966).

39. Means, J. C., and R. D. Wijayaratne. "Sorption of Benzidine, Toluidine, and Azobenzene on Colloidal Organic Matter," in *Aquatic Humic Substances,* Advances in Chemistry Series 219, J. H. Suffet and P. MacCarthy, Eds. (Washington, DC: American Chemical Society, 1989), pp. 209–222.

40. Turner, D. "Relationships between Biological Availability in Chemical Measurements," in *Metal Ions in Biological Systems,* H. Siegel, Ed. (New York: Marcel Dekker, 1984).

41. Buffle, J. *Complexation Reactions in Aquatic Systems* (Chichester, Eng.: Ellis Horwood, 1988), pp. 304–330.

42. Eisenrich, S. J., M. R. Hoffman, D. Rastetter, E. Yost, and W. J. Maier. "Metal Transport Phases in the Upper Mississippi River," in *Particulates in Water, Characterization, Fate, Effects and Removal,* Advances in Chemistry Series 189, M. C. Kavanaugh and J. O. Leckie, Eds. (Washington, DC: American Chemical Society, 1980), pp. 135–176.

43. Sigleo, A. C., and G. R. Helz. "Composition of Estuarine Colloidal Material: Major and Trace Elements," *Geochim. Cosmochim. Acta* 45:2501–2509 (1981).

44. Sigg, L. "Surface Chemical Aspects of the Distribution and Fate of Metal Ions in Lakes," in *Aquatic Surface Chemistry,* W. Stumm, Ed. (New York: John Wiley and Sons, 1987), pp. 319–349.

45. Laxen, D. P. H., and I. M. Chandler. "Size Distribution of Iron and Manganese Species in Freshwaters," *Geochim. Cosmochim. Acta* 47:731–741 (1983).

46. Salbu, B., H. E. Bjornstad, N. S. Lindstrom, E. Lydersen, E. M. Brevik, J. P. Rambaek, and P. E. Paus. "Size Fractionation Techniques in the Determination of Elements Associated with Particulate or Colloidal Material in Natural Fresh Waters," *Talanta* 32:907–913 (1985).

47. Davis, J. A. "Complexation of Trace Metals by Adsorbed Natural Organic Matter," *Geochim. Cosmochim. Acta* 48:679–691 (1984).

CHAPTER 5

Evidence for the Diffusion of Aquatic Fulvic Acid from the Sediments of Lake Fryxell, Antarctica

George Aiken, Diane McKnight, Robert Wershaw, and Laurence Miller

INTRODUCTION

Sources of dissolved organic carbon (DOC) in lakes can be categorized as (1) allochthonous, entering the lake from the terrestrial watershed, and (2) autochthonous, being derived from algae or aquatic macrophytes growing in the lake. The chemical characteristics of the DOC are not only influenced by the source material but are also influenced by all the biogeochemical processes involved in carbon cycling within the lake. These processes include allochthonous inputs of organic carbon to the lake from the watershed, autochthonous carbon fixation by algae and aquatic plants, transformation and degradation of both autochthonous and allochthonous organic material by heterotrophic microbial activity, transport of particulate organic material to the sediments, remobilization of DOC from the sediments, and photodegradation by incident UV light. Each of these processes is likely to affect the DOC chemistry in a different way. Even though some of these processes are restricted to certain zones in the lake, physical mixing in most lakes results in spatially uniform DOC chemistry. DOC resulting from degradation processes occurring in the sediments is a potentially significant contribution to the total DOC pool that has been largely overlooked, in part, because of the complexity of these systems.

Permanently ice-covered lakes in the Antarctic deserts offer unique opportunities to study the internal production and degradation of organic material in a lake ecosystem because the sources of organic carbon in the lakes are relatively simple, restricted to autochthonous microbial growth, and because the lakes are amictic. These lakes are located in one of the most arid and barren desert environments on earth, where, in addition to the absence of plants, the microflora of the soils are quite sparse,[1] and the organic content of the soils is less than 0.1%.[2] Furthermore, organic compounds derived from higher plants are absent from the dissolved organic material in the lakes.[3] The lakes are fed by glacial meltwater streams, which flow only during the brief austral summer. Most of the streams have low concentrations of particulates and dissolved material.[4] In the lakes phytoplankton are typically abundant just above the transition from O_2-enriched to anoxic zones, and benthic algal mats are abundant in littoral zones.[5] In terms of the sources and chemistry of dissolved organic materials, the Antarctic desert lakes have unique features that are not replicated by lakes on any other continent.

The year-round ice cover of these lakes greatly restricts internal circulation within the lake. In addition, several of the lakes are characterized by significant salinity gradients within the water column, which further stabilizes the water column.[5] An underlying hypothesis for this study was that the extreme stability of the water column would make it possible to interpret the vertical profiles of chemically and biologically reactive constituents in terms of processes occurring at specific depths and in terms of interactions between processes occurring in the major redox zones. We conducted this study in Lake Fryxell, located in the Taylor Valley in the McMurdo Dry Valleys. We present here the results from water column stability calculations, measurement of DOC depth profiles, and chemical characterization of dissolved fulvic acid, a major fraction of the DOC. Based on these results, the potential role of sediments in the generation of DOC within the lake is discussed.

METHODS

Limnological Measurements and Sample Collection

The data reported here were collected during December 1987 from a sampling site on the Lake Fryxell ice sheet located over the deepest area near the center of the lake. Water samples for analysis of dissolved ions and DOC were obtained by pumping from a measured depth through Tygon tubing. Samples for analysis of dissolved ions were filtered through 0.4-μm Nuclepore filters using an Antlia hand pump and were collected in 250-mL plastic bottles rinsed with distilled water. Samples for analysis of cations and metals were acidified with 0.5 mL of Ultrex nitric acid. DOC samples were filtered through 0.4-μm Selas silver membrane filters using a pressurized Gelman stainless-steel filtration unit and were collected in 250-mL precombusted glass bottles. Conductivity and dissolved oxygen measurements were obtained from probes deployed in the water column, a YSI probe for conductivity and an Orbisphere probe for dissolved oxygen. Measurements of in vivo fluorescence were obtained onsite shortly after sample collection (1–2 hr) using a Turner Designs fluorometer.

Large-volume water samples (up to 1800 L) are required for isolation of preparative quantities of dissolved fulvic acid. These large volumes were obtained by pumping the sample from a given depth through Teflon-lined stainless-steel flexible tubing and two stainless-steel Balston filtration units with cylindrical 1-μm and 0.3-μm glass-fiber depth filters in sequence. The filtered water was collected in 38-L stainless-steel milk cans.

Analysis of Dissolved Solutes

On the return of the samples to the United States several months after the field season, cations, including sodium and calcium, were analyzed using a Jarrel Ash 975 inductively coupled plasma spectrometer. Samples from the more saline bottom waters were diluted as necessary. DOC concentrations were measured within several days of sample collection using an Oceanographic International carbon analyzer operated at the Eklund Biological laboratory at McMurdo Station.

Isolation of Fulvic Acid Samples

Fulvic acid was isolated using 4-L columns of XAD-8 and XAD-4 resins connected in series by Teflon tubing. Filtered water was acidified to pH 2 in glass carboys, and 240 L of sample were then passed through the XAD-8/XAD-4 column pair. Fulvic acid was retained on the XAD-8 column, and a fraction of the hydrophilic acids was retained on the XAD-4 resin. Each column was back-eluted separately with 8 L of 0.1 N NaOH. NaOH eluates were immediately acidified with concentrated HCl to pH 2 to minimize alteration of the sample at high pH. Eluates were reconcentrated on the appropriate resin, H-saturated using AG-MP 50 cation exchange resin obtained from Biorad, and lyophilized. The chemistry of the hydrophilic acid fraction will not be discussed in this chapter.

Resin

The Amberlite XAD resins were obtained from Rohm and Haas. The resins were cleaned by first washing the beads (20 to 50 mesh) in 0.1 N NaOH and then rinsing with distilled water. The resin was then placed in a Soxhlet extractor and sequentially extracted for 48 hr each with methanol and acetonitrile. This sequence was repeated twice. Clean resin was stored in methanol. Glass columns were packed with a H_2O-resin slurry and rinsed with distilled water to remove methanol. The resin was further cleaned with 3 sets of alternating 0.1 N NaOH and 0.1 N HCl rinses immediately before using.

Sample Analyses

Samples were characterized by elemental, molecular weight, and ^{13}C NMR analyses. A review of the methods used for the determination of each element has been published by Huffman and Stuber.[6] Molecular weights were determined by vapor pressure osmometry with water as solvent. Details of the method and correction of the data for dissociation have been published by

Aiken and Malcolm.[7] The NMR spectra were measured on solutions of
approximately 100 mg/mL of the sodium salt of each of the samples dissolved
in H_2O-D_2O (3:1) mixtures in 10 mm tubes on a Varian XL 300 spectrometer at
75.429 MHz. For the nonquantitative spectra the transmitter was set for a 45°
tip angle, and no pulse delay was used. The acquisition time was 0.2 sec, and
the sweep width was 30,000 Hz; continuous broadband decoupling of protons
by the WALTZ method was employed. All of the spectra were recorded with a
line broadening of 25 Hz. These conditions were chosen to obtain maximum
signal-to-noise ratio for optimum structural group characterization. The quantitative spectra were measured using inverse gated decoupling in which the
proton decoupler was on only during the acquisition of the free induction
decay curve (FID); an 8-sec delay time was used.

RESULTS AND DISCUSSION

Limnological Characteristics and Dissolved Constituents

Depth profiles for a number of dissolved constituents in Lake Fryxell are
presented in Figures 5.1 to 5.4. The lake is essentially freshwater at the top and
increases in salinity with depth; the specific conductance of the bottom waters
is 14 millisiemens, with the most abundant species being Na^+ and Cl^-. The
density gradient, caused by chemical stratification, and the year-round ice
cover result in stability of the water column from year to year. The depth
profiles that we measured in 1987 for Ca^{+2} and Na^+ (Figure 5.1), for instance,

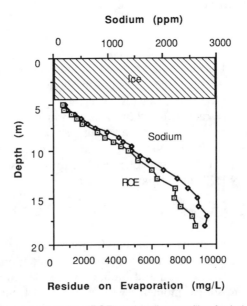

Figure 5.1.   Residue on evaporation (ROE) and sodium profiles for Lake Fryxell.

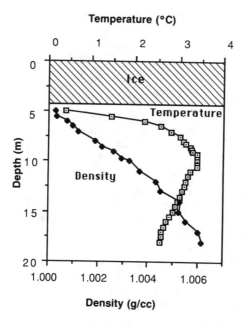

Figure 5.2. Temperature and density profiles for Lake Fryxell.

are very similar to those measured during the 1979–80 summer by Lawrence and Hendy,[8] and the temperature profile (Figure 5.2) has not changed significantly since the lake was first studied in 1963.[8,9]

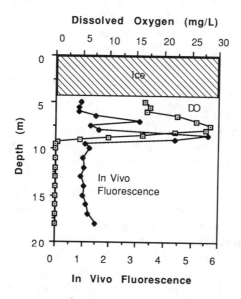

Figure 5.3. In vivo fluorescence and dissolved oxygen depth profiles for Lake Fryxell.

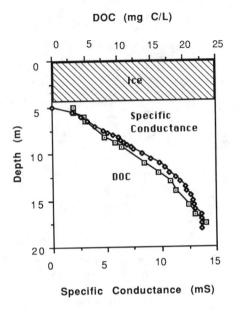

Figure 5.4. Dissolved organic carbon and specific conductance depth profiles for Lake Fryxell.

The microflora of Lake Fryxell have been studied in detail and are generally composed of blue-green algae, green algae, bacteria, yeasts, and fungi.[10-15] Despite the low light intensities caused by the 4.5-m thick ice cover, abundant algal populations develop in the oxic zone of the water column above the 9.5-m depth, as demonstrated by in vivo fluorescence data (Figure 5.3), which are an indirect estimate of phytoplankton abundance. In contrast to the sparse distribution of plants and microflora in the surrounding desert soils,[1,2] the microbial productivity of these lakes is comparable to that of temperate oligo-trophic lakes.

The depth profile for DOC in Lake Fryxell, however, is quite different from the depth profile for in vivo fluorescence (Figure 5.4). The DOC concentration increases with depth throughout the oxic and anoxic zones to a maximum of 25 mg C/L at the bottom of the lake (18 m). This profile is generally similar to the depth profiles for specific conductance and major inorganic species such as sodium (Figure 5.1) that have been attributed, in part, to upward diffusion of ions from the saline bottom water.[8] The chemically induced density gradient in Lake Fryxell is common to many Dry Valley Lakes.[8] Wilson first suggested that the density profile in Lake Vanda, located in the Wright Valley, Antarc-tica, resulted from the mixing of evaporitic brines with inflowing freshwater from the melting of glaciers.[16] Molecular diffusion was assumed to be the mechanism for the mixing, whereby the residual brine diffused into the overly-ing freshwater.

Water Column Stability and DOC Profile

In order to characterize the mixing regime of Lake Fryxell, we determined the stability of the water column as a function of its density. Included in the density calculation are the contributions of the major dissolved solutes and gases to the mass and volume of water at the in situ temperature. The density of pure water at the in situ temperature was obtained from tables in the Handbook of Chemistry and Physics.[17] The major element composition of Lake Fryxell[18] at 12 m was used to determine the molar concentration of the major dissolved solutes using the chemical equilibrium program HYDRAQL.[19] The ratio of the residue on evaporation (ROE) to the molar concentration was determined at 12 m and was assumed to be constant throughout the water column. Thereafter, the solute-corrected density was obtained simply as a function of ROE by converting ROE to an average molar concentration, and including the partial molar volume (we used $v_{sea\ salt} = 13.47$ cm^3/eq) and the added mass (ROE) in the density.[20] A final correction to the density in Lake Fryxell was made by summing the mass of dissolved N_2, Ar, O_2, and CH_4 at each depth and correcting for the cumulative partial molar volume at the in situ pressure.[21,22] The results of these density calculations can be contrasted using 5-m and 18-m water. The pure water densities were 0.99987 and 0.99995 g/cm^3, respectively, at 0.40 and 2.55°C. The solute corrected densities were 1.00030 and 1.00614 g/cm^3, and the final dissolved gas corrected densities were 1.00026 and 1.00612 g/cm^3.

A linear regression of density with depth in Lake Fryxell (Figure 5.2) resulted in a slope of 4.83×10^{-6} g/cm^{-4} and coefficient of determination of 0.977 (r^2). The density gradient was used to calculate the stability of the water column as the rate of change of density with depth, $N^2 = (g/\rho)(\Delta\rho/\Delta Z)$, where N is known as the Brunt-Vaisala frequency, g is gravitational acceleration, ρ is density, and Z is height above the bottom of the lake. N^2 can in turn be used to calculate K_z, the vertical eddy diffusion coefficient, which is well described in regions of high stratification in temperate lakes.[23,24] By comparing calculated density gradients with K_z measured using tritiated water as a tracer, Quay derived the empirical relationship $K_z = 1 \times 10^{-6} (N^2)^{-0.76}$ for Canadian Lake 227, where both temperature and dissolved solids contributed to the density profile.[24] Applying this relationship to Lake Fryxell resulted in an estimated K_z of 5.86×10^{-5} cm^2/sec. K_z values for stratified lakes range from 10^{-2} to 10^2 cm^2/sec.[25] While the value obtained for Lake Fryxell is three orders of magnitude lower than the lowest of these values, it is within an order of magnitude of the molecular diffusion coefficients for dissolved salts in water, and only a factor of 4 to 5 greater than the molecular diffusion rates of gases in Lake Fryxell. These results strongly suggest that molecular diffusion, and not eddy diffusion, is the driving force for mixing in Lake Fryxell.

For a system driven by molecular diffusion from a plane, the solution to Fick's second law of diffusion relating the concentration to distance from a source is given by

$$C = M/2(\pi Dt)^{0.5} \exp(-Z^2/4Dt)$$

where C is the concentration of the diffusing substance, Z is the distance from the plane (in this case the distance above the bottom of the lake), D is the diffusion coefficient for the diffusing substance, t is time, and M is the total mass of diffusing substance.[26] Application of this model to Lake Fryxell assumes that the bottom of the lake is flat and that the sides of the lake are steep. Samples were collected from the deepest part of the center of the lake, and it is reasonable to assume that these assumptions are justified given the bathymetry of the lake.[8] By plotting log C vs Z^2 it is unnecessary to specify values for D, t, and M. The only constraint is that the effective D for DOC be constant throughout the water column. Application of this diffusion model to the DOC profile in Lake Fryxell gives an excellent fit (Figure 5.5) with a coefficient of determination of 0.986. This result strongly suggests that diffusion from the sediments is the dominant source of DOC in Lake Fryxell, and that the application of this simple model of diffusion to the DOC profile is appropriate in this system.

Chemical Characteristics of DOC

Comparison of the results of DOC fractionation analyses for samples collected from three depths within the lake (Table 5.1) indicates that the composition of DOC with respect to different compound classes at each depth is essentially constant. Aquatic fulvic acid is a major fraction of the organic material in the lake, accounting for 40% of the DOC. These samples of fulvic acid exhibit a number of distinctive characteristics as a result of their being derived solely from algae and bacteria. In Table 5.2, the elemental compositions of fulvic acids from Lake Fryxell are presented along with those from

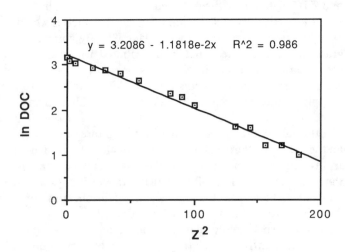

Figure 5.5. Application of Fick's second law of diffusion to the DOC profile in Lake Fryxell.

Table 5.1. Composition of Dissolved Organic Carbon from Three Depths in Lake Fryxell

Depth (meters)	DOC (mg C/L)	Percent Fulvic Acid	Percent Hydrophilic Acid
5.5	3.0	40	12
7.5	5.2	41	14
18	25	39	10

other aquatic environments—including Lake Hoare, another lake located in the Taylor Valley—that were isolated using comparable methods of resin adsorption chromatography. In general, the data for fulvic acids from the two Dry Valley lakes are more similar to each other than to any of the other data sets. While the carbon and hydrogen contents of the Lake Fryxell and Lake Hoare samples are comparable with those for fulvic acids isolated from temperate lakes and rivers, the nitrogen and sulfur contents are higher relative to these samples. The sulfur content is about 1.3% in all the samples except for the Lake Fryxell sample from the sulfide-rich bottom water, which may be artificially high due to reactions with elemental sulfur during the alkaline elution of the resin column. The molecular weights of the Lake Fryxell samples (463–468 daltons) are less than those reported for aquatic fulvic acids isolated from surface waters in North America (500–950 daltons).[7]

The large nitrogen contents of the fulvic acids from Antarctic lakes result in C:N ratios that are significantly lower than for other aquatic fulvic acid samples, including those isolated from marine environments by similar chromatographic methods. These differences, in part, are the result of differences in the nitrogen contents of the precursor biomolecules. Of particular importance are lignin-derived compounds—resulting from the degradation of lignin-

Table 5.2. Elemental and Molecular Weight Data for Fulvic Acids Isolated from a Variety of Aquatic Environments

Sample	C (%)	H (%)	O (%)	N (%)	S (%)	Ash (%)	Molecular Weight (daltons)
			Ash Free				
Lake Fryxell:							
5.5 meters	54.9	5.5	34.9	3.3	1.2	2.3	463
7.5 meters	55.0	5.5	34.9	3.1	1.3	1.0	na
18 meters	52.6	5.4	31.8	2.4	8.0	0.1	468
Lake Hoare:							
5.6 meters	55.2	5.7	35.5	2.8	0.8	1.8	na
12.5 meters	54.9	5.5	35.7	2.9	0.9	1.9	na
Other aquatic environments:							
Merrill Lake, Washington	52.9	5.2	40.7	0.7	0.5	na	na
Suwannee River, Georgia	54.2	3.9	38.0	0.7	0.4	0.2	840
Missouri River, Iowa	55.4	5.3	35.0	1.3	0.8	0.1	540

containing plants—that have been recognized as components of aquatic fulvic acids isolated from temperate lakes and streams.[27] Lignin does not contain N, and its presence in the precursor pool for fulvic acid would lead to lower overall N content.

Other distinctive characteristics of these samples are illustrated by quantitative ^{13}C NMR spectroscopy, which provides important structural information for organic molecules. In Figure 5.6 the liquid-state spectrum for the fulvic acid sample from the chlorophyll maximum zone (7.5 m) of Lake Fryxell is presented. This sample has the following characteristics:

1. Aliphatic carbons (0–62 ppm) are more abundant than aromatic carbons, with the region representing methylene carbons ($\approx$ 20 ppm) being prominent.
2. The aromatic carbon region (110–160 ppm) is well defined with no side peaks.
3. The carboxyl peak (160–190 ppm) is also narrow with no side peaks.

The integration data for the NMR spectra for Lake Fryxell samples isolated from five depths are compared in a bar graph format in Figure 5.7, along with the data for Merrill Lake, a pristine mountain lake located in a forest dominated by Douglas fir in the state of Washington. The Lake Fryxell samples are very similar to each other and differ significantly from other aquatic fulvic acids, having greater amounts of aliphatic carbon (0–62 ppm) and lesser amounts of aromaticity (110–160 ppm). This comparison indicates that the different sources of organic material in Lake Fryxell and Merrill Lake result in very different molecular compositions for these two fulvic acids.

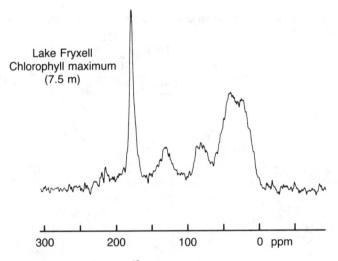

Lake Fryxell
Chlorophyll maximum
(7.5 m)

300   200   100   0 ppm

Figure 5.6. Quantitative liquid-state ^{13}C NMR spectra for aquatic fulvic acid isolated from Lake Fryxell, Antarctica (7.5 m depth).

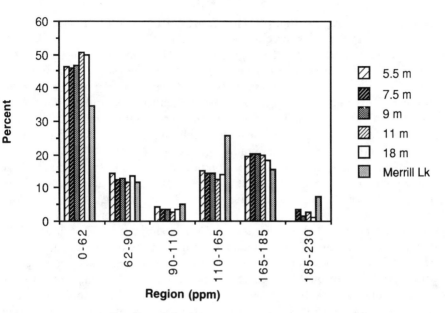

Figure 5.7. Comparison of ^{13}C NMR integration data for samples isolated from five depths in Lake Fryxell and a sample from Merrill Lake, Washington.

Sources of DOC

Comparison of the samples isolated from three depths within Lake Fryxell provides some indication of the processes that may control the generation of DOC in the water column. As noted above, the fulvic acid samples isolated from these depths vary little in elemental composition, and ^{13}C NMR analyses of these samples indicate that there are few structural differences between these samples as well. It is also significant that the DOC profile does not match the in vivo fluorescence profile, and that no compositional differences are noted for the fulvic acid sample collected from 7.5 m, which is a zone of high algal activity. This observation indicates that excretion of organic compounds from viable algae does not exert a strong influence on the distribution or nature of the fulvic acid.

On the other hand, the similarity in the chemical composition of the fulvic acid samples, and the similarity between the DOC and specific conductance profiles, suggest that a major source of DOC in Lake Fryxell is the degradation of particulate organic carbon, derived from algae and bacteria, in the sediments or bottom waters of the lake, with subsequent diffusion of the more refractory components into the water column. A large gradient in DOC exists between the sediment interstitial water (120 mg C/L) and the bottom water of the lake (25 mg C/L). This gradient would drive the diffusive transport of dissolved organic matter, including fulvic acid, into the water column. These results strongly suggest that diffusive transport of organic compounds from sediments is the major source of DOC in Lake Fryxell.

The potential influence of sediments on the DOC of overlying waters has not been studied in detail for lakes in general, due in part to mixing of the water column. Because of the stability of the water column and because the sources of organic matter in the lake are limited to algae and bacteria, Lake Fryxell is an ideal natural laboratory to study processes related to the chemistry of microbially derived organic matter in the absence of factors that complicate the interpretation of data obtained from other aquatic systems. The diffusion of microbially derived organic compounds from the sediments into the water column may be an important process controlling the amount and chemistry of the DOC in lakes in general and, hence, would be an important component of the carbon cycle as it pertains to large water bodies.

The results presented in this chapter also have implications for understanding the processes controlling contaminant transport between sediments and the water column. Capel and Eisenreich, for instance, have studied the potential transport of hydrophobic organic compounds from the sediment into the water column as dissolved species and in association with organic colloids.[28] These authors have identified diffusion from the sediment as an important process in controlling the fate of hydrophobic organic chemicals in lakes. The cotransport of contaminants, in association with humic substances being generated in the sediments and diffusing into the water column, may provide a possible mechanism for moving contaminants from sediments to the water column.

DISCLAIMER

Use of trade names in this chapter is for identification purposes only and does not constitute endorsement by the U.S. Geological Survey.

ACKNOWLEDGMENTS

We gratefully acknowledge the assistance in the field and laboratory provided by C. Bowles, R. Smith, E. Andrews, J. Duff, M. Brooks, and R. Harnish. We also thank the members of the VXE-6 helicopter squadron for assistance in transporting large-volume water samples. This research was supported by NSF-Division of Polar Programs, grant number NSF8613607, through an interagency agreement with the U.S. Geological Survey.

REFERENCES

1. Cameron, R. E., J. King, and C. N. David. "Microbiology, Ecology and Microclimatology of Soil Sites in Dry Valleys of Southern Victoria Land, Antarctica," in

Antarctic Ecology, Vol. 2, M. W. Holdgate, Ed. (London: Academic Press, 1970), p. 702.

2. Horowitz, N. H., R. E. Cameron, and J. S. Hubbard. "Microbiology of the Dry Valleys of Antarctica," *Science* 176:242–245 (1972).

3. Matsumoto, G., T. Torri, and T. Hanya. "Vertical Distribution of Organic Constituents in an Antarctic Lake: Lake Vanda," *Hydrobiologia* 111:119–126 (1984).

4. Howard-Williams, C., and W. F. Vincent. "Ecosystem Properties of Antarctic Streams," *New Zealand Antarctic Record* 6(Special Supplement):11–20 (1985).

5. Heywood, R. B. "Antarctic Inland Waters," in *Antarctic Ecology,* Vol. 1, R. M. Laws, Ed. (London: Academic Press, 1984), p. 279.

6. Huffman, E. W. D., and H. A. Stuber. "Analytical Methodology for Elemental Analysis of Humic Substances," in *Humic Substances in Soil, Sediment and Water,* G. R. Aiken, D. M. McKnight, R. L. Wershaw, and P. MacCarthy, Eds. (New York: John Wiley and Sons, 1985), p. 433.

7. Aiken, G. R., and R. L. Malcolm. "Molecular Weight of Aquatic Fulvic Acids by Vapor Pressure Osmometry," *Geochim. Cosmochim. Acta* 51:2177–2184 (1987).

8. Lawrence, M. J. F., and C. H. Hendy. "Water Column and Sediment Characteristics of Lake Fryxell, Taylor Valley, Antarctica," *New Zealand Journal of Geology and Geophysics* 28:543–552 (1985).

9. Hoare, R. A., K. B. Popplewell, D. A. House, R. A. Henderson, W. M. Prebble, and T. A. Wilson. "Solar Heating of Lake Fryxell, a Permanently Ice Covered Antarctic Lake," *J. Geophys. Res.* 70:1555–1558 (1965).

10. Benoit, R., R. Hatcher, and W. Green. "Bacteriological Profiles and Some Chemical Characteristics of Two Permanently Frozen Antarctic Lakes," in *The Structure and Function of Freshwater Microbial Communities,* J. Cairns, Ed. (Blacksburg, VA: Virginia Polytechnic and State University, 1971), p. 281.

11. Vincent, W. F., M. T. Downes, and C. L. Vincent. "Nitrous Oxide Cycling in Lake Vanda, Antarctica," *Nature* 292:618–620 (1981).

12. Koob, D. D., and G. L. Leister. "Primary Productivity and Associated Physical, Chemical, and Biological Characteristics of Lake Bonney, a Perennially Ice-Covered Lake in Antarctica," *Antarctic Research Series* 20:51–68 (1972).

13. Parker, B. C., J. T. Whitehurst, and R. C. Hoehn. "Observations of In Situ Concentrations and Production of Organic Matter in an Antarctic Meromictic Lake," *Virginia J. Sci.* 25:136–140 (1974).

14. Parker, B. C., R. C. Hoehn, R. A. Paterson, J. A. Craft, L. S. Lane, R. W. Stavros, H. G. Sugg, Jr., J. T. Whitehurst, R. D. Fortner, and B. L. Weand. "Changes in Dissolved Organic Matter, Photosynthetic Production, and Microbial Community Composition in Lake Bonney, Southern Victoria Land, Antarctica," in *Adaptations within Antarctic Ecosystems,* G. A. Llano, Ed. (Washington, DC: Smithsonian Institute, 1977), p. 873.

15. Vincent, W. F. "Production Strategies in Antarctic Inland Waters: Phytoplankton Ecophysiology in a Permanently Ice-Covered Lake," *Ecology* 62:1215–1224 (1981).

16. Wilson, A. T. "Evidence from Chemical Diffusion of a Climatic Change in the McMurdo Dry Valleys 1200 Years Ago," *Nature* 201:176–177 (1964).

17. Weast, R. C., Ed. *CRC Handbook of Chemistry and Physics,* 50th ed. (Cleveland, OH: Chemical Rubber Company, 1970), p. F-4.

18. Green, W. J., M. P. Angle,, and K. E. Chave. "The Geochemistry of Antarctic

Streams and Their Role in the Evolution of Four Lakes in the McMurdo Dry Valleys," *Geochim. Cosmochim. Acta* 52:1265–1274 (1988).

19. Papelis, C., K. F. Hayes, and J. O. Leckie. "HYDRAQL: A Program for the Computation of Chemical Equilibrium Composition of Batch Systems Including Surface-Complexation Modeling of Ion Adsorption at the Oxide/Solution Interface," Stanford University, Environmental Engineering and Science, Department of Civil Engineering, Technical Report No. 306 (1988).

20. Duedall, I. W., and P. K. Weyl. "The Partial Equivalent Volumes of Salts in Seawater," *Limnol. Oceanogr.* 12: 52–59 (1967).

21. Tippel, E. W., and K. E. Gubbins. "Partial Molal Volumes of Gases Dissolved in Electrolyte Solutions," *J. Phys. Chem.* 76:3044–3049 (1972).

22. Enns, T., P. F. Scholander, and E. D. Bradstreet. "Effect of Pressure on Gases Dissolved in Water," *J. Phys. Chem.* 69:389–391 (1965).

23. Jassby, A., and T. Powell. "Vertical Patterns of Eddy Diffusion during Stratification in Castle Lake, California," *Limnol. Oceanogr.* 20:530–543 (1975).

24. Quay, P. D. "An Experimental Study of Turbulent Diffusion in Lakes," PhD Dissertation, Columbia University, New York, NY (1977).

25. Lerman, A. *Geochemical Processes Water and Sediment Environments* (New York: John Wiley and Sons, 1979).

26. Crank, J. *The Mathematics of Diffusion,* 2nd ed. (Oxford, Eng.: Oxford University Press, 1975).

27. Ertel, J. R., J. I. Hedges, and E. M. Perdue. "Lignin Signature of Aquatic Humic Substances," *Science* 223:485–487 (1984).

28. Capel, P., and S. Eisenreich. "Sorption of Organochlorines by Lake Sediment Porewater Colloids," in *Aquatic Humic Substances: Influence on Fate and Transport of Pollutants,* Advances in Chemistry Series 219, I. H. Suffet and P. MacCarthy, Eds. (Washington, DC: American Chemical Society Publications, 1989), p. 185.

CHAPTER 6

Change in Properties of Humic Substances by H$_2$SO$_4$ Acidification

Egil T. Gjessing, Harry Efraimsen, Magne Grande, Torsten Källqvist, and Gunnhild Riise

INTRODUCTION

Natural organic acids affect the speciation of inorganic solutes in water and the bioavailability of micropollutants in general. During the last 25 years, common conceptions of the chemical nature and properties of these organics in water, humic substances (HS), has changed. In the late 1950s Shapiro reported that the molecular weight of the colored organic matter in natural water was 456, that halogen and sulfur were not integral parts of HS, and that nitrogen was an impurity.[1] During the following two decades, gel filtration and the ultramembrane filtration techniques suggested that the molecular size was several orders of magnitude higher. In addition, elemental analysis suggested that S and N are parts of the organic matter.[2] As the chemical tools for isolation and characterization of organic compounds in water have been improved, reports on the composition of humic substances suggest C, H, O, N, and S to be the major elements, that the molecular weight is in the range of $1-3 \times 10^3$ daltons, and that most of the acidic properties are due to carboxylic groups.[3]

It is safe to assume from the vast number of reports on chemical and biological properties of humic substances that the physicochemical nature of HS in the aquatic phase — including size, element composition, acidity, and their potentials to mobilize water-insoluble micropollutants — will change substantially with the general composition of the ambient water. Factors such as age, the time the HS have been in the water phase, the concentration, ionic strength, and general composition of the water, including pH, will affect the nature of humic substances. This means that the biological importance of humic substances will change. Our views of the role that humic substances play, together with pollutants in the aquatic environment, must then be revised.

The role of natural organic matter during acidification of surface water is

Table 6.1. Sulfate in Acidified Samples of Humic Substances (18.3 mg C/L) Before and After Removal of Organic Matter

Treatment	Before After UV/H_2O_2 Oxidation SO_4 mg/L		DIFF SO_4 mg "Org."	
	Before	After	/L	/mg C
HNO_3	5.1	6.0	0.9	0.05
H_2SO_4	893	920	27	1.5
HCl	3.2	4.1	0.9	0.05

also under discussion. With regard to the potential change in chemical and biological properties in acidified water, data reported from Norway indicate that HS from acidified areas are more toxic to fish than HS from less acidified areas.[4,5] The experiments behind these observations are based on water samples where mineral acids are removed by equilibrium (membrane) dialysis, prior to the toxicity test. The purpose of this chapter is to confirm these findings and to offer an explanation for the change in properties.

EXPERIMENTAL MATERIALS AND METHOD

1. Three aliquots of filtered humus water (Hellerud, July 1986, identical to that used to isolate Nordic reference humus[3]) were acidified to pH 2.0 with HNO_3, H_2SO_4, and HCl, respectively. After a reaction time of one week, each of the three samples were divided into two halves. One half was analyzed directly for SO_4 concentration; the other half was mineralized, using UV/H_2O_2, prior to the analysis for SO_4. The results are shown in Table 6.1.

2. A 60-mg sample of concentrated humus water (NIVA-conc., low-temperature, low-pressure evaporate of Hellerud 1986) was dissolved in a small volume of diluted H_2SO_4 (pH 2.5). After 14 days at room temperature the samples were freeze-dried. The freeze-dried samples were then dissolved in 1 L distilled water. One part of this sample was UV/H_2O_2-oxidized and analyzed for sulfate. The other part was analyzed directly. As a control 60 mg NIVA-conc. was used, following the same procedure, except that distilled water was used instead of H_2SO_4. In addition, several samples from an acidified area (Risdalsheia) were analyzed for the SO_4 content (colorimetrically) before and after UV/H_2O_2 oxidation. The results are listed in Table 6.2.

3. Concentrated humus water (NIVA-conc., 180 mg material in 25 mL of distilled water) was fractionated on Sephadex G-15 (diameter 1.8 cm, length 50 cm). The first fractions of the first peak (see Figure 6.1) were collected and refractioned on the same Sephadex column, and the water representing the purified part of the first peak (5 and 6 in Figure 6.1) was analyzed for SO_4. The SO_4 analyses were performed both before and after UV/H_2O_2 mineralization. The results are shown in Figure 6.1 *(left)* and in Table 6.3. The same procedure was repeated, except the concentrate (180 mg in 25 mL) was acidified with H_2SO_4 to pH 2.7 and kept at that low pH for a week prior to neutralization (pH 4.0) and fractionation; the results are illustrated in Figure 6.1 *(right)* and given in Table 6.3.

Table 6.2. Difference in SO₄ Content of "Natural" and Artificial H₂SO₄-Acidified Humus Samples Before and After UV/H₂O₂ Oxidation

		After UV–H₂O₂	Before Treatment	Diff. mg SO₄		TOC "Org."
		mg SO₄ /L		/L	/mg C	mg C/L
1 H₂SO₄-treated	NIVA-conc.	5.5	4.2	1.3	0.07	18.3
2 Not treated	NIVA-conc.	3.4	3.2	0.2	0.01	18.3
3 Hellerud 1988,	not acidif.	6.8	6.2	0.6	0.04	12.9
4 Risdalsheia	acidif.	2.4	2.0	0.4	0.02	17.7
5 Risdalsheia	acidif.	8.0	6.8	1.2	0.09	13.8
6 Risdalsheia	acidif.	6.6	5.2	1.4	0.10	13.7
7 Risdalsheia	acidif.	8.9	8.0	0.9	0.10	9.2
8 Risdalsheia	acidif.	6.1	5.2	0.9	0.11	7.8
9 Risdalsheia	acidif.	7.0	6.6	0.4	0.06	6.4
10 Risdalsheia	acidif.	6.2	5.6	0.6	0.10	5.9

4. Five L of water with humus (TOC 21.7 mg/L) were acidified with H₂SO₄ to pH 2.00 (1 mL H₂SO₄ concentration per 5.5-L sample) and stored for 5 days at room temperature; pH increased to 2.31 while stored. The sample was then dialyzed (using 4-m dialyzing tube, SpectraPor membrane tubing 3787-D22, Thomas Scientific, Philadelphia, PA: diameter, 1.6 cm; tube volume, ~ 900 mL; tube area, 0.2 m²) until the change in conductivity in the sample was less

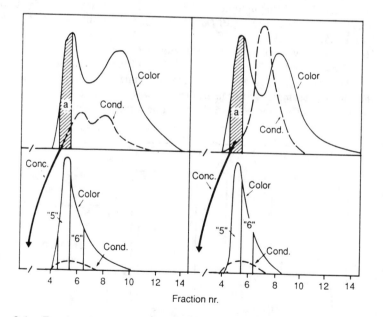

Figure 6.1. Fractionation of concentrated humus on Sephadex G-15. The high-molecular-weight fraction (shaded fraction *a*) is purified by re-elution on the same column two times. Distilled water is used as eluant, and color of the eluting water is measured. The fractions "5" and "6" are analyzed for TOC and SO₄ (see Table 6.3). *Left*, nonacidified HS concentrate; *right*, H₂SO₄-acidified HS concentrate.

Table 6.3. SO$_4$ in High-Molecular-Weight HS Fraction (HMWF) Before and After H$_2$SO$_4$ Acidification

Sample		TOC mg C/L	COND m S/m	pH	SO$_4$ mg/L	After UV/H$_2$O$_2$		"Org." SO$_4$ mg/L
						TOC mg C/L	SO$_4$ mg/L	
HMWF ord. con.	"5"	34.4	1.87	4.61	2.2	1.4	2.5	0.3
	"6"	10.0	1.54	4.89	0.5	1.5	0.7	0.2
HMWF acid. conc.	"5"	35.3	1.61	4.91	1.2	1.0	1.8	0.6
	"6"	8.5	1.14	5.18	1.2	0.8	1.8	0.6

Notes: SO$_4$ is determined colorimetrically before and after UV/H$_2$O$_2$ mineralization. Conc. humus water from Hellerud, fractionated and refractionated (cleaned) on Sephadex G-15. "5" & "6" are fractions, see Figure 6.1.

than 0.01 mS/m/hr. The experimental setup is illustrated in Figure 6.2. As a control, 5 L of the same water without H$_2$SO$_4$ addition was used. The dialysis patterns are illustrated in Figure 6.3. The acidified and nonacidified samples were also extracted with octanol, and the octanol-water partition coefficient determined.[6] The toxicity of these two dialyzed water samples were tested, using yearlings of Atlantic salmon *(Salmo salar* L., 1–2 g): Each of the two dialyzed samples was divided into three 1-L aliquots and placed in a 2-L beaker. Into each of these six beakers were put two salmon. The time of survival during a period of 4 days was observed. The results are illustrated in Figure 6.4. After being used in the fish test, the same waters (after filtration) were used in a toxicity test with *Daphnia magna.* The results are illustrated in Figure 6.5.

5. Humus water, with color 110 mg Pt/L and TOC 13 mg C/L, was acidified to pH 2.2 and stored for 2 months at 4°C. The acid sample was then neutralized to pH 6.8 with NaOH (sample A). At the same time, Na$_2$SO$_4$ corresponding to the content in sample A after neutralization was added to a second sample of the original water (sample B). A third sample was the original water, with pH

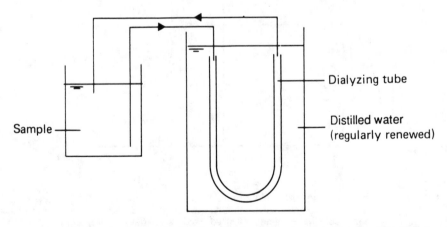

Figure 6.2. Equipment for "equilibrium" dialyzation of humic water. Dialyzing tube (Spectra-Por membrane tubing 3787-D22): diameter, 1.6 cm; length, 4 m; volume, 900 mL; surface, 0.2 m^2.

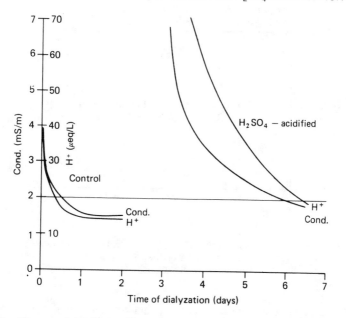

Figure 6.3. Change of conductivity and pH (μeq H⁺/L) with time of dialyzation (days) in acidified and nonacidified (control) samples of humus water.

adjusted to pH 6.8 (sample C). These three pH 6.8–adjusted water samples were filtered through 0.45 μm membrane filters. The samples were enriched with K_2HPO_4 and $NaNO_3$ to give a concentration of 20 Mg P/L and 200 μg N/L. The nutrient-enriched samples were inoculated with 10^6 cells/L of the alga *Selenastrum capricornutum* in 100-mL flat-bottom flasks. The cultures were

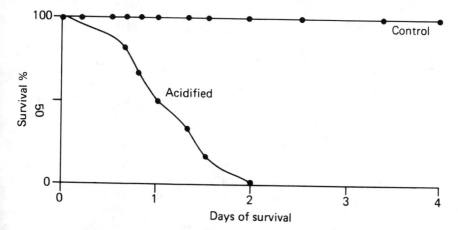

Figure 6.4. Toxicity test with yearlings of Atlantic salmon *(Salmo salar* L.) with six individuals for each of the two qualities. The tests are performed on normal (control) and H₂SO₄-acidified HS water after removal of mineral acid by membrane dialyzation (see Figures 6.2 and 6.3 and Table 6.4).

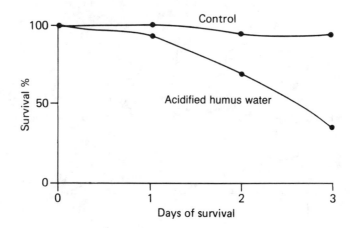

Figure 6.5.   Toxicity test with *Daphnia magna*. The survival of 10 individuals after 1, 2, and 3 days is given. The test is done on the same water sample as those in Table 6.4 (and Figure 6.4)—however, after pH adjustment to about pH 7.

incubated on a shaking table at 20°C under continuous illumination from fluorescent tubes (60 μE/cm^2 sec). The growth of algae was followed by counting cell number and measuring size distribution and total volume of algae during 8 days. All growth experiments were carried out in triplicate. The growth response of the green algae with time is illustrated in Figure 6.6. The same three pH-adjusted membrane-filtered samples were tested on *Daphnia magna* following the same procedure as the experiment on dialyzed water (experiment 4).

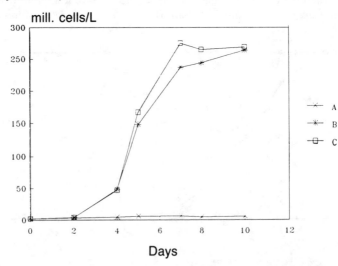

Figure 6.6.   Growth of alga *Selenastrum capricornutum* as a function of time. *(A)* in humus water H$_2$SO$_4$-acidified and neutralized with NaOH, *(B)* in humus water with Na$_2$SO$_4$ added and neutralized with NaOH, *(C)* in humus water neutralized with NaOH.

RESULTS AND DISCUSSION

Sulfur Associated with HS

The degree to which sulfur is associated with HS and the type of linkage is indeed unclear. In relation to the discharge of SO$_2$ to the atmosphere and acidification of surface water, an important question is whether one may expect that there has been an increase in the content of HS sulfur during the last decades. Perhaps the content of S in the late 1950s as reported by Shapiro[1] was below detectable limits, with a significant increase during the last 30 years. The analytical problems related to distinguishing between the S fraction that is associated with HS and the S fraction that appears as the anion SO$_4^{2-}$ are difficult to overcome due to the same charge (negative) of the two components. With regard to sulfur in sediments, Kukkonen and Tolonen[7] have demonstrated that organic S, which constitutes a major part of the total S, has increased during the last three decades. In lakes with humus, C-bond S apparently has increased more than the portion bound in sulfate esters.[7]

The results in Tables 6.1 and 6.2 show that the content of SO$_4$, determined colorimetrically, increases after removal of the organic matter by UV/H$_2$O$_2$ treatment. The results indicate that in natural water samples the HS sulfur is higher in water from acidified areas (Risdalsheia): 0.08 as compared to 0.03 mg SO$_4$/mg in less acidified waters. The SO$_4$ determination is based on reaction with thorin, giving a yellowish color. In water with HS, the SO$_4$ concentration may therefore be overestimated. This means that this indicated increase of organic sulfur is underestimated.

The results of experiment 3, where the high-molecular-weight fraction using Sephadex has been purified by re-eluation on the same column (Figure 6.1 and Table 6.3), suggest that some sulfur is associated with the high-molecular-weight part at the HS, and that this content of "organic S" apparently increases with H$_2$SO$_4$ acidification. Even though the humus color will have an interfering effect on the results, since sulfate is determined by measuring in color decrease, such an interference could, at the most, explain 20% of the observed differences. The results in Table 6.3 also suggest that S combines with HS in water and that this S, at least partly, originates from H$_2$SO$_4$.

Biological Response to ''Sulfated'' HS

In relation to acidification of surface water and biological effects, it has been observed that HS have a buffering effect since organisms survive at lower pH in this type of water. On the other hand, it has been suggested that the undissociated part of the organic acids is more toxic.[8] One should expect, therefore, that as pH decreases, the undissociated form increases, thereby making the HS more biologically active. Experiment 4, where the toxicity to fish of dialyzed water before and after acidification with H$_2$SO$_4$ is determined, shows a significantly higher toxicity of the acidified sample (Figure 6.4). It

Table 6.4. Dialyzed Samples (Hellerud July 1986), Before (Control) and After Acidification with H_2SO_4

	pH	TOC mg/L	Ca mg/L	Mg mg/L	Na mg/L	SO_4 mg/L	Cl mg/L
Control	4.8	17.6	0.99	0.21	0.19	1.4	<0.2
Acidification	4.7	8.8	0.78	0.17	<0.05	4.5	<0.2

	NO_3 $\mu g/L$	NLAL $\mu g/L$	LAL $\mu g/L$	Fe $\mu g/L$	K_{how}
Control	27	189	80	670	4.6×10^{-2}
Acidification	13	155	89	420	8.4×10^{-2}

Notes: NLAL = nonlabile Al; LAL = labile Al; K_{how} = humic substances octanol-water partition coefficient (see Petersen[8]).

appears from the chemical composition given in Table 6.4 that this sample has a slightly lower pH compared to the control, and that the TOC and Ca concentration is lower. From our experiences with the fish species used (salmon) and water of the quality as described in Table 6.4, it is unlikely that these differences in composition alone should account for the observed differences in toxicity. It is interesting to note that the octanol-water partition coefficient (as shown in the lower right part of Table 6.4) increased by a factor of two by the acid treatment. From our experiences with humus water and the effect of pH on the octanol-water partition coefficient,[9] it is not likely that the small pH decrease will explain this increase. The significance of this in relation to the illustrated difference in biological activity is, however, not known. The results, given in Figure 6.5, show that the survival of *Daphnia magna* was reduced in the acidified water compared to the controls (which were not influenced by acidity of the samples, since this test involved an adjustment of pH to 7.2–7.4). The findings suggest, therefore, that H_2SO_4 acidification results in HS that are more biologically active.

Experiment 5 (in which the rate of growth of algae in acidified water with HS, neutralized with NaOH to pH 6.8, is compared with that of original water at same pH) shows a clear difference in the effect (Figure 6.6 and Table 6.5): H_2SO_4 treatment of humus water will, after some time of reaction (in this case, 2 months at 4°C), apparently form a "product" that has a negative effect on the growth of *Selenastrum capricornutum*.

Table 6.5. Composition of Humus Water (Hellerud 1988) After H_2SO_4 Acidification to pH 2.2 and Neutralization to pH 6.8 with NaOH (A), Addition of Na_2SO_4 and NaOH to pH 6.8 (B), and Only Neutralization to pH 6.8 with NaOH (C)

Sample	pH	COND. mS/m	TOC	Al mg	LAL Al/L	SO_4 mg/L	UVabs	COLOR mg Pt/L
(A) + H_2SO_4 + NaOH	6.8	105	12.4	304	192	450	0.545	103
(B) + Na_2SO_4 + NaOH	6.8	190	11.8	227	136	870	0.554	103
(C) + NaOH	6.8	3.9	12.0	213	69	7	0.549	103

Note: LAL = Labile Al = Reactive Al – Organically bound Al.

The findings suggest changes in chemical characteristics and biological activity and, therefore, a reaction between HS and sulfate. Since HS are reported to contain OH groups, both phenolic and alcoholic OH, we suggest that H$_2$SO$_4$ may undergo esterification with HS, resulting in an "internal" rearrangement of the HS. These changes can result in an increase of undissociated HS. The present results support this; for instance, the pH of the purified high-molecular-weight fraction (HMWF) of acidified and unacidified humus concentrate, given in Table 6.3, suggests that H$_2$SO$_4$ treatment reduces H$^+$ ion concentration by one half. It is our experience also, from comparing charge density (CD) data of humus water, that the CD of water from acidified areas generally is lower than that of water from less acidified regions.[5] With regard to biological action, it is known that the activity of HS increases with the increase of the undissociated fraction.[8]

The biological consequence of the indicated increase in lipophilicity by H$_2$SO$_4$ treatment (Table 6.4, increase in octanol-water partition coefficient) is not known; however, it might be assumed that this is a result of the reduced electronegativity of the HS, which in turn increase the potentials for the HS to penetrate cell membranes. We conclude, therefore, that the H$_2$SO$_4$-modified HS are more bioavailable.

SUMMARY AND CONCLUSION

Based on the hypothesis that H$_2$SO$_4$ reacts with the humic substances to form a "sulfated" HS with other properties, a series of experiments were performed. The experiments involved determination of SO$_4$ in artificial and natural H$_2$SO$_4$-acidified humus water before and after mineralization of the organic matter with UV-H$_2$O$_2$. SO$_4$ was found in the high-molecular-weight fraction of humus water, using Sephadex G-15. The S:C ratio in this fraction apparently increases in acidified waters. Finally H$_2$SO$_4$-(and HCl-) acidified humus water were tested for toxicity to yearlings of Atlantic salmon and *Daphnia magna,* after dialyzation or neutralization. These experiments indicate that H$_2$SO$_4$ reacts with HS to form organically associated S, "an esterified HS" that apparently is more toxic to the aquatic organisms.

ACKNOWLEDGMENTS

This work was initiated by a grant from the Surface Water Acidification Program (SWAP) Project: "Natural Organic Acids: Their Role in Freshwater Acidification and Aluminium Speciation."

REFERENCES

1. Shapiro, J. "Chemical and Biological Studies of the Yellow Organic Acids in Lake Water," *Limnol. Oceanogr.* 11:161–172 (1957).

2. Gjessing, E. T. *Physical and Chemical Characteristics of Aquatic Humus* (Ann Arbor, MI: Ann Arbor Science, 1976).
3. Aiken, G. R., D. M. McKnight, R. L. Wershaw, and P. MacCarthy. *Humic Substances in Soil, Sediment and Water* (New York: John Wiley and Sons, 1985).
4. Gjessing, E. T. "Mechanisms and Effects of Reactions of Organic Acids with Anions," in *Organic Acids in Aquatic Ecosystems,* E. M. Perdue and E. T. Gjessing, Eds. (New York: John Wiley and Sons, 1990).
5. Gjessing, E. T., M. Grande, and E. Røgeberg. "Natural Organic Acids: Their Role in Freshwater Acidification and Aluminium Speciation," Acid Rain Research, Report 15/1988, NIVA, Oslo, Norway (1988).
6. Petersen, R. C., and A. Kullberg. "The Octanol/Water Partition Coefficient of Humic Material and Its Dependance on Hydrogenian Activity," *Vatten* 41:236–239 (1985).
7. Kukkonen, P., and K. Tolonen. "Analysis of Organic and Inorganic Sulfur Constituents and 34S-Isotopes in Dated Sediments of Fresh Lakes in Southern Finland," *Water Air Soil Poll.* 35:157–170 (1987).
8. Petersen, R. C. "Effects of Changes in Ecosystems (e.g., Acid Status) on Formation and Biotransformation of Organic Acids," in *Organic Acids in Aquatic Ecosystems,* E. M. Perdue and E. T. Gjessing, Eds. (New York: John Wiley and Sons, 1990), p. 151.
9. Gjessing, E. T., G. Riise, R. C. Petersen, and E. Andruchow. "Bioavailability of Aluminium in the Presence of Humic Substances at Low and Moderate pH," *Sci. Total Environ.* 81/82:683–690 (1989).

CHAPTER 7

Polynuclear Aromatic Hydrocarbon Binding to Natural Organic Matter: A Comparison of Natural Organic Matter Fractions

Gary L. Amy, Martha H. Conklin, Houmao Liu, and Christopher Cawein

INTRODUCTION

The movement of hydrophobic organic compounds in groundwater systems is controlled by their partitioning between the aqueous (mobile) and solid (immobile) phases. If there is another nonsorbing compound in the aqueous phase that interacts with the contaminant (e.g., by hydrophobic interactions), the hydrophobic organic compound may preferentially remain in solution[1] with its transport facilitated.[2] The other interacting compound can function as a miscible cosolvent within the aqueous phase; the behavior of aquatic natural organic matter (NOM) falls within this context. The movement of the solute will be attenuated, however, if the interacting compound is sorbed on the aquifer media; the behavior of soil NOM falls within this realm. One objective of this research was to identify important characteristics of aquatic NOM that affect partitioning and transport of polynuclear aromatic hydrocarbons (PAHs). Another objective of the research was to delineate the roles of dissolved vs mineral-bound NOM in the partitioning and transport of hydrophobic organic contaminants in groundwater.

In this work, the partitioning between one PAH, phenanthrene, and each of two sources of groundwater NOM has been studied in the presence and absence of a model mineral surface. The behavior of a soil-derived fulvic acid sorbed to the mineral surface has also been investigated.

The underlying assumptions behind this work are twofold: first, PAH partitioning into dissolved NOM can be considered as a solvent partitioning process in which water and NOM micelles function as two solvents,[3] and second, PAH partitioning between water and mineral-bound NOM can be represented as linear partitioning similar to that demonstrated for partitioning between water and soil organic matter,[4-6] as characterized by fraction organic carbon, f_{oc}. (The term "micelle" is used somewhat broadly here, ranging from molecular aggregates[3] to the intramolecular regions of a single humic/fulvic acid molecule). Rigorously, "dissolved" NOM molecules may actually be stable colloidal suspensions in aquatic solutions that act as a discrete physicochemical phase similar to mineral-bound NOM.

MATERIALS

Laboratory-scale batch experiments have been emphasized. Most experiments have been conducted in triplicate using glassware without Teflon. Control experiments have been performed to quantify any loss of PAH to glass surfaces.[7] Where needed, high-purity, organic-free water (Milli-Q) was used.

Model PAH

This work has focused on phenanthrene, having a log K_{ow} of 4.4 and a water solubility of 75 $\mu g/L$. We have used fluorescence for detecting free phenanthrene in solution (Hitachi F-3010 Spectrofluorimeter). Standard curves for phenanthrene in both organic-free water and methanol were developed for quantification purposes.

Aqueous dissolution of phenanthrene was accomplished in four steps:

1. weighing a known amount
2. dissolving it in a known volume of acetone contained in a graduated cylinder
3. evaporating the acetone
4. adding a known volume of water to the PAH residue

Control experiments indicated little loss of phenanthrene due to adsorption on glassware.

NOM Sources and Characterization

Two sources of aquatic NOM have been evaluated: (1) groundwater derived from a basin underlying Orange County (OC), California, and (2) groundwater obtained from the Biscayne Aquifer (BA), Florida. These sources have different origins and characteristics. The OC NOM is collected from a deep well (1200 ft) perforating a layer of high organic content, presumably due to redwood deposits. The BA NOM is collected several kilometers downgradient of a recharge area, so although the dissolved organic carbon (DOC) is high, it contains a significantly lower humic fraction. We have obtained two different samples from each source; the second series is identified with an asterisk (*). Only the evaluation of mineral-bound NOM has involved the second series; all other results are based on the first series. We also have evaluated a commercially available soil fulvic acid (Contech, Inc., Ottawa, Canada) as a model of mineral-bound NOM.

Both aquatic NOM sources were filtered through prewashed 0.45-μm membrane filters, separating dissolved organic matter (DOM) from particulate NOM. As expected for these groundwater sources, there was virtually no

Table 7.1. Important Characteristics of NOM Sources

Source	DOC (mg/L)	UV Abs (cm^{-1})	pH	Cond. (μmhos/cm)	Avg. MW	% Humic	Spec. Abs (UV:DOC)
OC	5.2	0.238	7.95	420	1700	80	0.046
BA	5.7	0.120	8.05	480	1600	60	0.021
OC*	3.7	0.185	8.26	280	n/a	n/a	0.050
BA*	7.8	0.135	7.75	410	n/a	n/a	0.017
Soil-FA	24.9	0.335	8.00	—	n/a	n/a	0.013

measurable particulate NOM; hereafter, the terms NOM and DOM are used synonymously. The DOC content was measured on a Shimadzu Model TOC-500 carbon analyzer. The UV absorbance at 254 nm was measured on a Shimadzu UV-160A Spectrophotometer. Ultrafiltration (UF) was used to determine the MW distributions and average MW of the different sources used.[8] Since UF is affected by pH and ionic strength, it provides an operational definition of apparent molecular weight (AMW). Control experiments were performed to demonstrate no sorptive losses associated with the UF membranes. UF was also used to produce discrete molecular weight fractions, resulting in permeates and retentates which were reconstituted back to original pH and ionic strength conditions. Using the technique of Thurman and Malcolm,[9] XAD-8 resin adsorption was used to produce humic and nonhumic fractions of NOM; the humic fraction was reconstituted back to the original pH and ionic strength. The soil fulvic acid, determined to have an organic carbon content of 51%, was prepared by dissolution in a 0.01 M solution of NaHCO$_3$, adjustment to a pH of 8.0, and filtration through a 0.45-μm filter.

The unfractionated NOM is referred to as bulk NOM. Important characteristics of the bulk sources are shown in Table 7.1. The aquatic NOM associated with the OC source exhibited a substantially higher percentage of humic material than the BA source, and a slightly larger average MW.

Model Mineral

Alumina, α-Al$_2$O$_3$, has been used as a model mineral phase (Union Carbide, Linde SF6). The average size (volume average diameter) was reported to be 1.85 μm, and the surface area was 6.5 m^2/g. To determine surface charge and the pH$_{zpc}$ for the α-Al$_2$O$_3$, the electrophoretic mobility (EPM) was measured with a Rank Brothers Mark II microelectrophoresis unit. The presence of HCO$_3^-$ has been found to influence the electrophoretic mobility (and pH$_{zpc}$) of α-Al$_2$O$_3$: the pH$_{zpc}$ ranged from 7.2 (with HCO$_3^-$ present) to 8.5 (without HCO$_3^-$ present).

METHODS COMPARISON

Several sets of experiments have been performed to determine the best method for determining free PAH concentrations and NOM-bound PAH con-

Table 7.2. Binding Constants, K_{doc}, Derived from Different Methods

Source	K_{doc} ($\times 10^4$)			
	FQ	Sep-Pak	Dialysis	Avg.
OC-Bulk	9.4 ± 0.4	7.1 ± 2.0	7.7	8.1
BA-Bulk	5.7 ± 0.3	8.4 ± 0.8	11.0	8.4

centrations in the aqueous phase. Three techniques were evaluated: fluorescence quenching, reverse-phase (Sep-Pak) separations, and dialysis bags.

Batch experiments were conducted by employing a contact time of 3 days to establish equilibrium. The NOM-bound PAH (PAH-NOM) is defined as the initial PAH_o minus the PAH at equilibrium. Expressing the concentration of NOM as DOC, the binding constant, K_{doc}, is defined as

$$PAH + NOM \rightleftarrows PAH\text{-}NOM$$

$$K_{doc} = [PAH\text{-}NOM]/[PAH][NOM]$$

For the first technique tried, fluorescence quenching (FQ), we used the procedure of Gauthier et al.[10] This technique is based on the observation that PAH fluorescence in aqueous solution is quenched upon association with humic material. For phenanthrene, an excitation wavelength of 251 nm and an emission wavelength of 366 nm were used along with a bandpass of 5 nm.

For the second technique, dialysis, the method of Carter and Suffet was employed.[11] Dialysis bags with a molecular-weight cutoff of 500 were used. Known concentrations of phenanthrene are placed in dialysis bags containing a known concentration of NOM at a pH of 7 for an equilibration time of 48 hr. From the original phenanthrene concentration and the concentration of free phenanthrene after the equilibration period, the amount of phenanthrene bound to the NOM is determined. Concerns about adsorption of PAH onto the dialysis material suggested that control experiments be performed. In our work, we have observed sorptive losses of up to 40%.

The reverse-phase separation technique of Landrum et al. was employed as the third technique.[12] After equilibration, a known volume of phenanthrene solution in NOM is slowly passed through the C-18 column. The free phenanthrene adsorbs onto the nonpolar column packing. The NOM-bound phenanthrene passes through the column due to ionization of humic/fulvic acid functional groups at a pH of 7. The free phenanthrene is subsequently eluted with methanol for measurement. Minimal losses of PAH (on the order of 5%) have been measured using this technique.

The FQ method provided the lowest values of the K_{doc} for the BA source, while the converse is true for the OC source (Table 7.2). Different trends can be observed for the dialysis results. The lowest recovery of PAH was observed for the dialysis technique, which indicates that these results are the most questionable. Better recovery was obtained with the Sep-Pak method; however, some losses were observed. Clearly, the selection of an analytical technique can influence binding results. As the FQ and Sep-Pak methods agreed within a factor of two, we selected the FQ method for subsequent experiments. FQ has an advantage over the other techniques in the short time required for analysis. However, there are some potential concerns about quantitative comparisons between binding constants for different NOM sources: when a PAH binds to a substrate, the nature of the substrate affects the degree of fluorescence quenching.[7]

RESULTS AND DISCUSSION

We have conducted batch experiments to study the three binary systems (PAH-NOM, PAH-α-Al$_2$O$_3$, and NOM-α-Al$_2$O$_3$), and the ternary system (PAH-NOM-α-Al$_2$O$_3$). The binding experiments reported herein have been based on the following set of experimental conditions: a temperature of 23°C (room temperature), a pH of approximately 8, and an ionic strength of approximately 0.01. Both bulk sources of aquatic NOM exhibited these approximate pH and ionic strength conditions. We have performed limited, related work indicating little difference in binding constants over a pH range of 6 to 8 and an ionic strength range of 0.01 to 0.1. Experimental evaluation of a wider range of pH conditions would likely reveal differences.[11] It is important to note that the work reported herein is based on an "intermediate" PAH in terms of water solubility and K$_{ow}$.

PAH-NOM Binary System

Binding constants have been measured for different molecular weight and humic/nonhumic fractions of the two aqueous NOM sources. Results for various NOM fractions are summarized in Table 7.3, obtained with the fluorescence quenching method. The following observations can be made:

1. Larger binding constants were found for the OC source than the BA source.
2. The humic fraction for the OC and BA sources exhibited a higher binding constant than the bulk NOM for these sources.
3. The higher-MW fraction of each source exhibited a higher binding constant than the lower-MW fraction.

Overall, the humic/high-MW fraction of each NOM source exhibits the greatest binding properties. Experiments performed for nonhumic fractions were done at different ionic strengths/conductivities than for bulk and humic fractions. While the magnitude of FQ may vary with the organic composition of the NOM source, we believe that it is predominantly the humic fraction of the NOM which binds PAHs. Therefore, the primary concerns are the humic content of a NOM source and its fluorescence properties. Moreover, it is likely that observed MW differences in binding are due to the humic composition of each MW fraction.

Table 7.3. Binding Constants, K_{doc}, for Bulk NOM and NOM Fractions (Based on FQ)

Fraction	DOC (mg/L)	K_{doc} ($\times 10^4$)
OC Source:		
Bulk	5.2	9.0 ± 0.5
Humic	4.0	11 ± 0.2
Nonhumic	1.2	0
<5K MW	3.3	8.3 ± 0.3
>5K MW	2.0	16 ± 0.2
BA Source:		
Bulk	5.7	5.7 ± 0.3
Humic	3.6	9.0 ± 0.5
Nonhumic	2.1	0
<5K MW	4.0	3.9
>5K MW	1.7	17
<1K MW	1.9	3.5 ± 0.5
>1K MW	3.4	7.7 ± 0.5

Note: Specification of MW fractions of less than or greater than some indicated value implies ranges of 0 to "indicated value" and "indicated value" to 0.45 μm, respectively.

NOM–α-Al$_2$O$_3$ Binary System

Batch experiments have been performed to measure equilibrium isotherms for the sorption of NOM onto α-Al$_2$O$_3$ under bulk NOM source conditions (pH ≈ 8). Isotherms for bulk NOM are portrayed in Figure 7.1. The effects of NOM coating on surface chemistry are highlighted by the change in electrophoretic mobility of the particles. The EPM values of NOM-coated alumina at pH 8.0 were –1.2 and –1.3 μm/sec per volt/cm for the OC and BA sources with surface coverages of 0.1 and 0.2 mg DOC/m^2, respectively. As expected, the NOM coating increased the negative surface charge of the alumina particles. In related work, Davis and Gloor found that higher-molecular-weight NOM preferentially adsorbs to the alumina surface.[13] Also, both Leenheer[14] and Jardine et al.[15] found that hydrophobic (i.e., humic) components of NOM

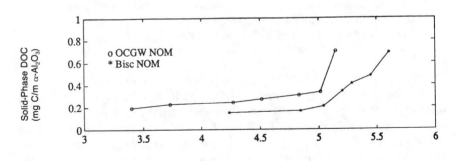

Figure 7.1. Equilibrium sorption isotherms for aqueous NOM on α-Al$_2$O$_3$.

have a greater affinity for soils than hydrophilic (i.e., nonhumic) components. Our binary system is intended to serve as a model of mineral-bound NOM; however, since we have selected alumina as our model mineral, the term mineral-bound more accurately reflects alumina-bound conditions. It is important to make a distinction between partitioning and adsorption within the context of our work: the former involves a "solvent extraction" phenomenon with NOM molecules behaving as a solventlike phase; the latter involves complexation between certain NOM molecules with ligand features (e.g., humic/fulvic acid) and the alumina surface.

PAH–α-Al$_2$O$_3$ Binary System

A single point isotherm experiment was run to determine the potential for PAH binding directly to the mineral surface in the complete absence of NOM. The addition of 5 g/L of α-Al$_2$O$_3$ to a 0.5 μM (89 μg/L) solution of phenanthrene in Milli-Q water resulted in an equilibrium concentration of 0.376 μM (66 μg/L), indicating a solid-phase concentration of 4.6 μg/g α-Al$_2$O$_3$ after 72 hr. As an approximation, this corresponds to an equivalent linear partition coefficient of 69.7 cm^3/g.

PAH–NOM–α-Al$_2$O$_3$ Ternary System

Relatively little is known about the chemical differences between aquatic NOM and soil NOM (i.e., soil organic matter) in a groundwater system. Intriguing questions exist about their genesis and interrelationships. Soil fulvic acid is defined as material extractable from soil under high pH (i.e., 13) conditions and remaining in solution under low pH conditions (i.e., 1). Aquatic NOM can enter the subsurface via recharge or by in situ leaching of organic-rich strata. As it follows advective flow, aquatic NOM comes into intimate contact with mineral matter, and sorption onto the mineral surface may occur if localized equilibrium does not exist. Thus, soil organic matter may have its origins as either aquatic NOM or as organic-rich strata created through sedimentary processes. Recognizing that NOM origin can affect chemical properties, we have attempted to study both of these scenarios.

Experiments have been conducted to study the partitioning of a PAH into α-Al$_2$O$_3$-bound NOM, using the OC* and BA* NOM samples. In these experiments, a known mass of α-Al$_2$O$_3$ (5 g/L) was contacted with a known volume of a NOM solution for an equilibration time of greater than 3 hr at each of three pH conditions (4.5, 6.0, and 8.0). The mixture was centrifuged and rinsed with a NaHCO$_3$/NaCl solution (I = 0.01) at each of the three pH conditions; while some desorption of NOM was observed at the higher pH (8), virtually no desorption was seen at the lower pH (4.5). The NOM-coated solids were then contacted with a phenanthrene solution/NaHCO$_3$/NaCl solution (I = 0.01) at pH levels of 4.5, 6.0, and 8.0. The mixture was centrifuged after an equilibration time of 72 hr. Based on measurements of initial and final phenanthrene, estimates of PAH partitioning into the mineral-bound NOM were made.

Thus, the overall protocol involved first contacting bulk NOM with alumina to produce residual NOM and mineral-bound NOM, and then separately contacting residual NOM and mineral-bound NOM with PAH to evaluate binding. The overall process is described below:

$$NOM + \alpha\text{-}Al_2O_3 \rightleftarrows NOM + NOM\text{-}\alpha\text{-}Al_2O_3$$

$$NOM\text{-}\alpha\text{-}Al_2O_3 + PAH \rightleftarrows PAH\text{-}NOM\text{-}\alpha\text{-}Al_2O_3$$

$$NOM + PAH \rightleftarrows PAH\text{-}NOM$$

The results are presented in Table 7.4. This tabular summary describes binding constants, K_{doc}, for various NOM sources/fractions as a function of pH. Also shown is other information including (1) source/fraction DOC, (2) calculated values of equivalent fraction organic carbon, f_{oc}, for NOM-coated alumina, (3) estimates of the fractional NOM surface coverage on alumina, and (4) calculated values of the PAH solid-phase (alumina) concentration. From Table 7.4 we can observe

1. No binding occurred between the PAH and mineral-bound NOM for both aquatic NOM sources (OC* and BA*).
2. Less binding occurred between the PAH and residual NOM than between the PAH and the original bulk NOM
3. Less binding occurred between the PAH and mineral-bound NOM than between the PAH and bulk NOM for the soil NOM source (Soil-FA).

It is noteworthy that binding of a PAH with mineral-bound NOM depends on the source of NOM; surface coverage of NOM on the mineral surface, and the configuration of NOM on the mineral surface (i.e., steric orientation, monolayer vs multilayer coverage, etc.). The observation of no binding between the PAH and mineral-bound NOM from both aquatic sources (OC* and BA*) may be explained by the following:

1. NOM coverage on the mineral surface is very low, and the PAH binding properties of the bare mineral surface control are weak.
2. Both aquatic sources of NOM actually represent "residual" NOM which was already in equilibrium with aquifer media at the actual field sampling site(s), thus suggesting that aquatic NOM may be a poor model for soil organic matter.

These results compare somewhat favorably with observations by Spurlock and Biggar.[16]

While the bulk NOM solution produced from the soil fulvic acid exhibited the largest binding constant, its binding properties were significantly reduced after its sorption onto the alumina surface. The observation of less binding of PAH with mineral-bound soil fulvic acid may be due to changes in the configu-

Table 7.4. Binding Constants for Mineral-Bound vs Dissolved NOM

Source	Bulk NOM	Residual NOM pH =			Mineral-Bound NOM pH =		
		4.5	6.0	8.0	4.5	6.0	8.0
DOC^a (mg/L)							
OC*	3.7	0	0		3.7	3.7	
BA*	7.8	1.6	2.3		6.2	5.5	
Soil-FA	24.9	2.5	9.8	15.0	22.4	15.1	9.9
f_{oc}^b (× 10^{-3} g OC/g α–Al_2O_3)							
OC*	—	—	—	—	0.74	0.74	
BA*	—	—	—	—	1.24	1.10	
Soil-FA	—	—	—	—	4.48	3.03	2.00
NOM Fractional Surface Coverage^c(m^2 OC/m α–Al_2O_3)							
OC*	—	—	—	—	0.20	0.20	
BA*	—	—	—	—	0.32	0.28	
Soil-FA	—	—	—	—	1.20	0.80	0.50
K_{doc}^d (× 10^4)							
OC*	9.5	0	0		0	0	
BA*	2.4	0	0		0	0	
Soil-FA	21.3	0	10.1	13.1	1.7	3.1	0
	(±0.2)		(±0.2)	(±0.2)	(±0.7)	(±0.1)	
PAH Solid-Phase Conc. (μg PAH/g α–Al_2O_3)							
α–Al_2O_3 w/o NOM Coating	—	—	—	—	n/a	4.6	n/a
α–Al_2O_3 w/ Soil-FA Coating	—	—	—	—	50.9	47.5	~4.6

^{a}For mineral-bound NOM: equivalent aqueous-phase concentration = bulk NOM – residual NOM.
^{b}Calculated value based on equivalent aqueous-phase concentration of mineral-bound NOM and mineral mass.
^{c}Assumptions: molecular size of NOM molecule = 2 angstroms = 2×10^{-10} m; surface area of mineral = 30 m^2/L.
^{d}Binding constants for mineral-bound NOM expressed in terms of equivalent aqueous-phase concentration.

ration of the NOM when it adsorbs to the mineral surface. This explanation was corroborated by results of a control experiment involving an NOM solution preequilibrated with PAH, subsequent addition of α-Al_2O_3 (5 g/L), and an observed 60% decrease in bound PAH. This behavior suggests that the subsequent adsorption of the NOM to the alumina surface changed its configuration and lowered its PAH binding properties, although more work is needed to verify this hypothesis. Our observation of no PAH binding by mineral-bound NOM at a pH of 8.0 may simply be due to the low NOM surface coverage of the mineral at high pH conditions unfavorable to NOM adsorption onto alumina. Differences in PAH binding at pH 4.5 vs pH 6.0 may be due to some direct binding of the PAH to the mineral surface itself.

The observation that residual NOM exhibits less PAH binding than the corresponding bulk NOM is consistent with the premise that higher-molecular-weight NOM is preferentially adsorbed to the alumina surface.[13]

Murphy et al. found that the binding capacity of humic acid sorbed onto a mineral surface for anthracene was different than that for the same humic acid in the aqueous phase.[17] This behavior was attributable to a difference in the structure of the sorbed humic phase: the orientation and structure of the humic acid on the mineral surface may affect the effective surface area of the mineral-bound organic phase, thus reducing the accessibility of hydrophobic domains able to bind anthracene. Moreover, sorption isotherms for anthracene on humic-coated mineral surfaces were nonlinear, suggesting that anthracene may have adsorbed onto the mineral surface as opposed to partitioning into the surface organic phase. Our estimates of surface coverage of the mineral surface by NOM help elucidate the relative importance of partitioning into the NOM coating as opposed to sorption onto the "bare" mineral surface.

A comparison of binding to actual soil organic matter is useful. The value of K_{oc} reported by Karickhoff for phenanthrene on soil is 1.2×10^4.[5] This value for soil NOM is lower than the values reported for bulk aquatic NOM from both the OC and BA sources (9.4 and 5.7×10^4, respectively), thus suggesting a weaker binding ability for soil organic matter as opposed to aquatic NOM, although more data is needed to further strengthen this point. The PAH solid-phase loadings (Table 7.4) indicate that, despite the lower binding properties of mineral-bound vs solution-phase NOM, the NOM-amended surface exhibits greater binding than the "bare" mineral (alumina) surface.

Adsorbed aquatic NOM appears to be a poor model of soil organic matter. While more work is needed, it also appears the specific soil fulvic acid evaluated did not behave in conformance with our original expectations of it functioning as a model for soil organic matter. However, before conclusive evidence can be put forth, other soil fulvic acids should be evaluated within a similar experimental context. Chiou et al. found that the binding of nonpolar organic compounds with soil organic matter is greater than with soil humic acid, presumably due to the more nonpolar constituents of the former, such as humin.[18] It is important to recognize that the relative abundance of dissolved vs mineral-bound NOM in a groundwater aquifer will influence the subsurface transport of PAHs.

CONCLUSIONS

Groundwater NOM sources vary in their ability to bind PAHs. Moreover, various fractions of NOM exhibit different binding properties: our results indicate that stronger binding is observed between PAH and higher-MW and humic fractions of aquatic NOM than lower-MW and nonhumic fractions. Fluorescence quenching has emerged as the best analytical approach to distinguishing between free vs NOM-bound PAH.

Mineral-bound NOM has been simulated by sorption of aqueous NOM onto a mineral (alumina) surface; we have observed weaker binding to mineral-bound NOM as opposed to dissolved NOM. The relative abundance of dissolved NOM (facilitated transport) versus mineral-bound NOM (attenuation/retardation) in a subsurface system can be expected to influence PAH transport.

ACKNOWLEDGMENT

This research has been supported by the Subsurface Science Program of the Ecological Research Division, Office of Health and Environmental Research, U.S. Department of Energy, Washington, DC (Program Director: Dr. F. J. Wobber).

REFERENCES

1. Chiou, C., R. Malcolm, T. Brinton, and D. Kile. "Water Solubility Enhancement of Some Organic Pollutants and Pesticides by Dissolved Humic and Fulvic Acids," *Environ. Sci. Technol.* 20:502–508 (1986).
2. McCarthy, J., and J. Zachara. "Subsurface Transport of Contaminants," *Environ. Sci. Technol.* 23:496–502 (1989).
3. Wershaw, R. "A New Model for Humic Materials and Their Interactions with Hydrophobic Organic Chemical in Soil-Water or Sediment-Water Systems," *J. Contam. Hydrol.* 1:29–45 (1986).
4. Karickhoff, S., D. Brown, and T. Scott. "Sorption of Hydrophobic Pollutants on Natural Sediments," *Water Resour. Res.* 13:241–248 (1979).
5. Karickhoff, S. "Semi-Empirical Estimation of Sorption of Hydrophobic Pollutants on Natural Sediments and Soils," *Chemosphere* 10:833–846 (1981).
6. Chiou, C., P. Porter, and D. Schmedding. "Partitioning Equilibria of Nonionic Organic Compounds between Soil Organic Matter and Water," *Environ. Sci. Technol.* 17:227 (1983).
7. Backhus, D., and P. Gschwend. "Use of Fluorescent Polycyclic Aromatic Hydrocarbon Probes to Study the Impact of Colloids on Pollutant Transport in Groundwater," *Environ. Sci. Technol.* (in press).
8. Amy, G., M. Collins, and P. King. "Comparing Gel Permeation Chromatography and Ultrafiltration for the Molecular Weight Characterization of Aquatic Organic Matter," *Journal AWWA* 79(1):43–50 (1987).
9. Thurman, E., and R. Malcolm. "Preparative Isolation of Aquatic Humic Substances," *Environ. Sci. Technol.* 15:463 (1981).
10. Gauthier, T., E. Shane, W. Guerin, W. Seltz, and C. Grant. "Fluorescence Quenching Method for Determining Equilibrium Constants for Polycyclic Aromatic Hydrocarbons Binding to Dissolved Humic Materials," *Environ. Sci. Technol.* 20:1162–1166 (1986).
11. Carter, C., and I. Suffet. "Binding of DDT to Dissolved Humic Materials," *Environ. Sci. Technol.* 16:735–740 (1982).
12. Landrum, P., S. Nihart, B. Eadle, and W. Gardner. "Reverse-Phase Separation Method for Determining Pollutant Binding to Aldrich Humic Acid and Dissolved Organic Carbon in Natural Waters," *Environ. Sci. Technol.* 18:187–192 (1984).

13. Davis, J., and R. Gloor. "Adsorption of Dissolved Organics in Lake Water by Aluminum Oxide: Effect of Molecular Weight," *Environ. Sci. Technol.* 15:1223–1229 (1981).

14. Leenheer, J. "Study of Sorption of Complex Organic Solute Mixtures on Sediments by Dissolved Organic Carbon Fractionation Analysis," in *Contaminants and Sediments,* Vol. 2, R. Baker, Ed. (Ann Arbor, MI: Ann Arbor Science, 1980).

15. Jardine, P., N. Weber, and J. McCarthy. "Mechanisms of Dissolved Organic Carbon Adsorption on Soil," *Soil Sci. Soc. Am. J.* 53:1378–1385 (1989).

16. Spurlock, F., and J. Biggar. "Effect of Naturally Occurring Soluble Organic Matter on the Adsorption and Movement of Simazine in Hanford Sandy Loam," *Environ. Sci. Technol.* 24:736–741 (1990).

17. Murphy, E., J. Zachara, and S. Smith. "Influence of Mineral-Bound Humic Substances on the Sorption of Hydrophobic Organic Compounds," *Environ. Sci. Technol.* (submitted for publication).

18. Chiou, C., D. Kile, and R. Malcolm. "Sorption of Vapors of Some Organic Liquids on Soil Humic Acid and Its Relation to Partitioning of Organic Compounds in Soil Organic Matter," *Environ. Sci. Technol.* 22:298–303 (1988).

CHAPTER 8

Calorimetric Acid-Base Titrations of Fulvic Acid

Mike Machesky

INTRODUCTION

The fate and transport of trace contaminants in surficial environments is influenced by interactions with naturally occurring organic compounds. A significant fraction of this organic matter can be classified as humic substances.[1] Humic substances possess appreciable quantities of acidic functional groups, and consequently, ambient pH values determine the fraction of these groups that exist in protonated or ionized forms. Moreover, the extent to which these acidic functional groups are protonated or ionized can significantly influence humic substance–trace contaminant interactions. Thus, a thorough understanding of pH effects on humic substances is necessary to assess the influence of humic substances on trace contaminant fate and transport.

Many studies have demonstrated that trace metal binding by humic substances is influenced by pH.[2,3] In general, trace metal binding increases as pH increases until metal ion hydrolysis reactions become important. These hydrolysis reactions decrease the net positive charge of the metal cation and probably decrease the favorable steric interactions (i.e., chelate structures) believed to be responsible for the high affinity binding observed between many trace metals and humic substances. The decrease in binding as pH decreases can be rationalized in terms of H^+ competition for available binding sites on humic substances. The kinetics of trace metal binding by humic substances are also influenced by solution pH. In one study, Ni(II) dissociated from the most labile fraction of available binding sites more slowly as pH was increased from 7.4 to 8.4.[4] Protons are also released upon binding of trace metals to humic substances, presumably due to the dissociation of weakly acidic protons to accommodate the formation of trace metal–humic substance bonds.[5] Thus, the interpretation of trace metal binding studies is aided by a thorough understanding of the acid-base properties of humic substances.

Ionic strength may also influence the equilibrium and kinetic aspects of trace metal binding by humic substances. Major cations can displace bound trace metals as ionic strength increases. Increasing Ca^{2+} concentrations from 10^{-5} to 10^{-3} M decreased Cu(II) binding by Suwannee River fulvic acid (SRFA).[6] A further increase in Ca^{2+} to 10^{-2} M, however, had no additional effect. In contrast, another study found no difference between 0 and 10^{-2} M Ca^{2+} in the extent of Cu(II) binding by Suwannee River humic acid.[7] In addition, the rate of Ni(II) dissociation from the most labile fraction of SRFA binding sites increased as ionic strength was increased from 2 to 100 mM.[4]

Fewer studies have examined the influence of pH and ionic strength on organic contaminant–humic substance interactions. In one such study, DDT bound by a humic acid increased as pH decreased from 9.2 to 6.0 and as Ca^{2+} increased.[8] Similarly, solubility enhancements by various DDT and PCB congeners in the presence of SRFA decreased as pH was increased from between 4 and 6 to 8.5.[9] This influence can be qualitatively rationalized by the concept that lower pH (or higher ionic strength) results in a net decrease in the negative charge of humic substances. Consequently, the hydrophobicity of humic substances is enhanced, and this allows hydrophobic trace organic association to increase.

Titrations with strong acid or base are frequently used to characterize the acidic functional group distribution of humic substances. However, several factors limit the utility of this technique. The buffer capacity of water below pH 4 and above pH 10 makes accurate determination of acidic functional group concentrations much more difficult in these pH regions. In addition, hysteresis is often observed between acid and base titration curves, particularly if solutions are aged under alkaline (pH > 8) conditions.[10,11] In alkaline solutions, pH has been observed to decrease with time, probably as a result of ester hydrolysis reactions.[12] Despite these difficulties, strong acid and base titrations are frequently used to estimate the concentrations and types of acidic functional groups present in humic substances.[13] Generally, total acidic functional group concentrations determined by titration range between 4 and 12 meq/g, and carboxylic and phenolic groups are the dominant types of groups present.[2]

A much less utilized variation of the titration technique is titration calorimetry. With this method, temperature changes accompanying titrations are measured, and modifying the basic instrumentation to include pH electrodes permits temperature and pH changes to be monitored simultaneously during titrations. Reaction free energies and enthalpies can both be directly determined with these data, and consequently, reaction free energies can be partitioned into energetic (enthalpic) and statistical (entropic) components. Applications of titration calorimetry have primarily dealt with homogeneous solution reactions of simple compounds.[14] Much less studied have been more complicated reactions, such as adsorption processes[15] or polyelectrolyte titrations,[16] and only a few of these studies have been concerned with humic substances.

A principal reason titration calorimetry is useful for investigating the acid-base functionality of humic substances is that simple carboxylic and phenolic acids possess widely different ionization enthalpies. Ionization enthalpies for simple carboxylic acids generally range between -1.5 and 0.8 kcal/mole, and those for simple phenols are 4.5 to 6.0 kcal/mole.[17] Phenolic hydrogens can also participate in hydrogen bond formation with adjacent groups, and this can result in more endothermic ionization enthalpies. For example, the ionization enthalpy for the first phenolic hydrogen of 1,2-dihydroxybenzene is 8.2 kcal/mole, and that for the phenolic hydrogen of salicylic acid is 8.6 kcal/mole.[17]

The most extensive previous calorimetric studies of humic substances are those of Choppin and Kullberg,[18] and Perdue.[19,20] In these studies, proton ionization enthalpies between pH 4 and 6 fell between -0.2 and 2.0 kcal/mole, which is within the range observed for simple carboxylic acids. At higher pH values (> 8), however, Perdue observed exothermic ionization enthalpies of -0.4 to -8 kcal/mole,[19] while Choppin and Kullberg found endothermic values of 6 to 7.5 kcal/mole.[18] The latter values are consistent with the ionization enthalpies expected for phenolic hydrogens, while the former values could not be satisfactorily explained.

The impetus for the present study was to address some of the experimental shortcomings of these previous calorimetric studies, which included large titration volumes[20] and the need to conduct separate pH and calorimetric titrations.[18] In addition, advantage was taken of recent progress in humic substance isolation and purification procedures and the existence of well-characterized humic substance standards. Finally, the effects of ionic strength and titration direction on enthalpy values were investigated.

METHODS

Two fulvic acids were the focus of the present study. Suwannee River fulvic acid was purchased from the International Humic Substances Society (IHSS). This fulvic acid has been extensively characterized by a variety of techniques and a recent publication summarizes what is known about its physical and chemical properties.[21] The second fulvic acid was isolated from a Brazilian peat deposit located 30 km south of Rio de Janeiro (BFA). Briefly, the isolation procedure was as follows: Repeated 12- to 16-hr extractions with 0.1 N NaOH were made in 50 mL polypropylene centrifuge tubes under $N_2(g)$ until extracts remained colorless. Extract solutions were centrifuged at 12,000 rpm for 20 min and filtered under $N_2(g)$ through 0.45-μm filters. Combined extracts were then acidified to pH 1.0 with 6 N HCl, and after 12–16 hr, centrifugation at 12,000 rpm for 20 min was used to separate the humic and fulvic acids. The fulvic acid extract was passed through an XAD-8 resin column, precleaned according to the Thurman and Malcolm procedure.[22] The column was rinsed with 0.65 column volumes of water and then back eluted with 0.1 N NaOH at 3 mL/min until the eluant was clear. The eluant was reacidified to pH 1.0 and again passed through the XAD-8 column, rinsed with water, and back-eluted

Table 8.1. Relative Peak Area Comparison of Major ^{13}C NMR Bands for SRFA and BFA

Band (ppm)	Assignment[a]	SRFA(%)[b]	BFA(%)
0–50	paraffinic	21	28
50–95	alcohol, methoxy, acetals, ethers	15	13
110–145	aromatic, olefinic	24	24
145–165	phenolic	9	6
165–185	carboxyl, ester	18	20
190–200	aldehyde, ketone	6	3
	% of Total Area Assigned	94	93

[a]Based on Hatcher et al.[23]
[b]Based on values from Leenheer et al.[24]

with NaOH. The eluant was next passed through an H^+-saturated cation exchange resin column (Bio-Rad AG-50W-X8, 20–50 mesh) twice. The resin was resaturated if the pH began to increase. Finally, the H^+-saturated BFA was freeze-dried and stored in a desiccator before use.

The BFA was characterized by solid-state CPMAS ^{13}C NMR (courtesy of Dr. Pat Hatcher, Penn State University). Table 8.1 lists relative areas of predominant bands for both SRFA and BFA.[23,24] Given that the relative errors in these areas are 5–10%,[23] the carbon distribution in these two fulvic acids is similar, except for the paraffinic carbon fraction which may be slightly greater in BFA.

A TRONAC (Orem, UT) isoperibol/isothermal calorimeter was used for the calorimetric determinations. All experiments were performed in the isoperibol mode with an initial fulvic acid solution volume of 25 mL. Several modifications were made to the basic instrument. A second high-precision Gilmont buret (2.5 mL capacity) was attached and a pH header (available from TRONAC) was added to accommodate micro glass and reference pH electrodes (Models MI-405 and MI-402 from Microelectrodes, Inc., Londonderry, NH). In addition, most instrument control and data acquisition functions were computer automated. An ADALAB-PC system (Interactive Microware, State College, PA) interfaced to an IBM-PC was used to control calibration heater and buret activation. This interface was also used to acquire digitized pH readings (12 bit A/D conversion) using the recorder channel output of an Orion 701A pH meter. Thermistor voltages were input to a Keithley Instruments Model 192 digital voltmeter and transferred over an IEEE-488 Bus (using Keithley software) to the IBM-PC. Thermistor and pH readings were collected each second and later transferred to diskettes for data reduction. Preliminary buret and thermistor calibrations were performed according to established procedures,[25] and data acquisition hardware and software were calibrated by determining the heat of neutralization of 0.1 M THAM with 0.1 M HCl.

A typical fulvic acid titration proceeded as follows: The pH electrodes were calibrated with 1:1 phosphate, phthalate, and carbonate buffers prepared fresh every 4–6 weeks according to standard procedures.[26] Next, 25.0 mL of 500 mg/L fulvic acid solution was transferred to the calorimeter dewar reaction vessel, and solid NaCl (99.999% pure) added to attain an ionic strength of 0.001 M. These fulvic acid solutions were prepared by dissolving the freeze-dried powders in distilled-deionized water and passing these solutions through H^+-saturated cation exchange resin (Bio-Rad AG-50W-X8, 20–50 mesh). Standardized base (0.1 N NaOH prepared from CO_2-free stock solutions) was added from one of the previously filled burets to increase the pH to 10.4 ± 0.1, and 20 min were allowed for thermal equilibration. Standardized 0.1 N HCl was then dispensed from the second buret in fifteen to twenty 30–40 μL volume increments with 90 sec between successive additions. The final pH ranged from 3.5 to 3.7. The temperature of the dewar reaction vessel was then lowered to the usual starting value (24.6°C), and the titration repeated with standard base (0.1 N NaOH) to pH 10.4 ± 0.1. An identical titration sequence was repeated in 0.01 and 0.1 M NaCl on the same FA solution, and all titrations were duplicated. Atmospheric CO_2 contamination was minimized during a titration by purging the top of the calorimeter insert around the stirring motor (the only connection between the dewar reaction vessel and the atmosphere) continuously with $N_2(g)$.

Data reduction procedures were performed using LOTUS-123 spreadsheets. The basic strategy was to correct for all conceivable extraneous reactions, both in terms of reaction enthalpies and H^+ consumption or release, and to ascribe the remainder of these quantities to the protonation (acid addition) or ionization (base addition) enthalpies of the fulvic acids (kcal/mole H^+). Nonchemical heat corrections included heat loss through the walls of the dewar reaction vessel and the heat of titrant dilution. Standard formulas were used to correct for these effects.[25] The principal data reduction formula used to estimate the quantity of added acid or base that reacted with the fulvic acid solutions was a rearranged form of the general titration curve equation in which the volume of strong acid or base titrant required to reach a specified pH value is the dependent variable.[27] For strong acid as the titrant this formula is

$$X = [C_B^\circ - C_T^{H2CO3} (\alpha_1 + 2\alpha_2) - K_w^\circ/(a_H + '\cdot\gamma_{-1}) + a_H + '/\gamma_{+1}] \quad (8.1)$$
$$\cdot V_o / [M_A + K_w^\circ/(a_H + '\cdot\gamma_{-1}) - a_H + '/\gamma_{+1}]$$

where X = volume of strong acid titrant required to reach the observed pH value

C_B° = initial concentration of strong base in the solution

C_T^{H2CO3} = total carbonate concentration assumed to be present (2 x 10^{-5} M in this study)

α_1, α_2 = fraction of total carbonate present as HCO_3^- and CO_3^{2-}, respectively, at a particular pH

K_w° = ion product of H_2O

$a_H + ' =$ measured hydrogen ion activity (10^{-pH})

$\gamma \pm 1 =$ univalent activity coefficient calculated using the Davies equation

$V_o =$ initial volume of the solution titrated

$M_A =$ molarity of the acid titrant

The total carbonate concentration was assumed to be 2×10^{-5} M (twice the concentration present in air equilibrated water below pH 6) since some additional CO_2 contamination may have occurred during the brief time the reaction vessel was open to the atmosphere at high pH (< 30 sec) to increase the ionic strength.

The initial strong base concentration was estimated to be

$$C_B^o = [K_w^o/(a_H + \cdot \gamma_-)] - [a_H + /\gamma_{+1}] + C_T^{H_2CO_3} (\alpha_1 + 2\alpha_2) \qquad (8.2)$$

where $a_H + =$ hydrogen ion activity at the start of the titration. For strong base titrations, the formulas are identical to Equations 8.1 and 8.2 except all signs are reversed, and C_A^o and M_B, the initial strong acid concentration and base molarity, respectively, are used in place of C_B^o and M_A. For the fulvic acid titrations, the volume of strong acid or base titrant added per interval was greater than that predicted with Equation 8.1, the excess being taken to equal the quantity of H^+ bound or released by the fulvic acid solutions. This theoretical correction procedure for solution blank effects assumes water formation and dissolved carbonate equilibria are responsible for all blank corrections— probably a reasonable assumption for the present titrations. However, the possibility that other extraneous reactions influenced the present titrations cannot be ruled out. In any event, the most significant extraneous chemical reaction in terms of both proton consumption and enthalpy effects is probably the water formation reaction:

$$H^+ + OH^- = H_2O \quad \Delta H^o{}_r = -13.34 \text{ kcal/mole} \qquad (8.3)$$

This correction was especially important at pH values < 5 during base titrations and > 9 during acid titrations.

RESULTS

Solution pH, bound and released H^+ (μmoles), and residual heat values (mcals) for a representative SRFA titration are listed in Table 8.2. The first pH values listed are those at the start of the titrations with acid or base, and subsequent values represent the pH at the end of a particular 90-sec titration interval. Similarly, the H^+ bound and H^+ released columns refer to the quantities of H^+ or OH^- that reacted with SRFA per interval as determined from the difference between the total H^+ or OH^- added and that calculated with the aid

Table 8.2. pH, H⁺ Bound or Released (μmoles), and Heat (mcal) Values per Titration Interval for Representative SRFA Acid and Base Titrations at 0.01 M Ionic Strength

Acid Titration			Base Titration				
pH	H⁺ Bound	Final Heat	pH	H⁺ Released	Net Heat	H₂O Diss. Correction	Final Heat
10.39	0	0	3.48	0	0	0	0
10.24	1.76	−10.26	3.60	1.58	−14.96	21.08	6.12
10.00	1.84	−10.95	3.76	1.89	−19.78	25.21	5.43
9.67	2.30	−11.91	3.93	2.47	−27.68	32.95	5.27
9.12	2.90	−12.70	4.14	2.75	−34.91	36.69	1.78
8.19	3.58	−12.75	4.35	3.25	−36.20	43.36	7.16
7.23	3.86	−0.90	4.59	3.50	−43.24	46.69	3.45
6.48	3.79	−0.21	4.86	3.71	−45.26	49.49	4.23
6.04	3.81	0.04	5.13	3.86	−45.95	51.49	5.54
5.67	3.84	0.09	5.47	3.91	−50.58	52.16	1.58
5.36	3.84	0.24	5.84	3.94	−45.14	52.56	7.42
5.07	3.81	0.31	6.31	3.91	−45.39	52.16	6.77
4.80	3.74	0.13	7.02	3.87	−46.27	51.63	5.36
4.56	3.60	0.36	8.00	3.96	−38.85	52.83	13.98
4.33	3.40	0.20	9.05	3.73	−32.23	49.76	17.53
4.13	3.15	0.18	9.61	3.14	−21.01	41.89	20.88
3.93	2.70	0.27	9.97	2.44	−13.66	32.55	18.89
3.77	2.44	0.18	10.21	1.95	−0.93	26.01	25.08
3.62	1.98	0.20	10.37	2.00	−0.58	26.68	26.10
			10.49	1.83	−0.53	24.41	23.88

of Equations 8.1 and 8.2. The corresponding final heat values represent the heat change associated with H^+ binding or release from SRFA. Final heat values for acid titrations are exothermic above pH 6.5 and slightly endothermic below pH 6.5. Dividing these final heat values (mcals) by μmoles H^+ bound is equivalent to kcal/mole H^+ reacted per interval. This reaction can be represented by

$$A^- + H^+ = HA \qquad (8.4)$$

where A^- and HA represent ionized and protonated forms of fulvic acid. Net residual heat values for base titrations (third column under base titrations) are always exothermic; to monitor the reversibility of the protonation reaction (Equation 8.4), these net heat values must be corrected for the endothermic heat of water dissociation. Thus, base titrations are represented as

$$HA + OH^- = A^- + H_2O \qquad \text{net reaction (exothermic)} \qquad (8.5)$$

$$\underline{H_2O = H^+ + OH^-} \qquad \text{water dissociation (13.34 kcal/mole)} \qquad (8.6)$$

$$HA = H^+ + A^- \qquad \text{overall reaction (endothermic)} \qquad (8.7)$$

Table 8.2 lists these contributions to the base titration results separately. The water dissociation enthalpy correction values (fourth column under base additions) are greater in absolute magnitude than the net heat values throughout the entire pH range, and especially above pH 9. This dominance of the water dissociation enthalpy correction has also been noted in conjunction with base titrations of polyelectrolytes.[28]

Representative acid and base titration curves for SRFA at the three ionic strengths investigated are presented in Figures 8.1 and 8.2, respectively. Figures 8.3 and 8.4 are corresponding titration curves for BFA. The titration curves for both SRFA and BFA are not significantly influenced by ionic strength or titration direction. In addition, total acid and base consumption between pH 10.4 and 3.6 is about 30% less for BFA than for SRFA.

Enthalpy data for SRFA acid and base titrations at the three ionic strengths investigated are presented in Figures 8.5 and 8.6, respectively. Figures 8.7 and 8.8 present corresponding data for BFA. Enthalpy values from two titrations for each ionic strength and titration direction were averaged over 0.3 pH

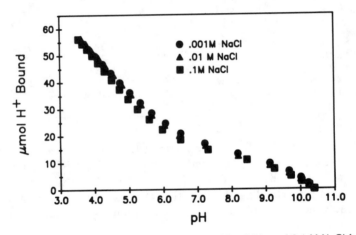

Figure 8.1.   Representative acid titration curves in 0.001, 0.01, and 0.1 M NaCl for SRFA.

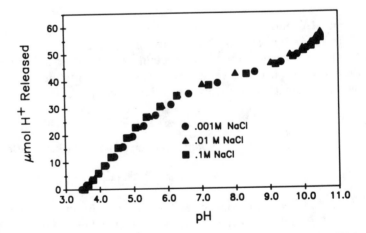

Figure 8.2.   Representative base titration curves in 0.001, 0.01, and 0.1 M NaCl for SRFA.

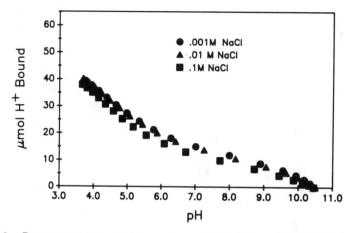

Figure 8.3. Representative acid titration curves in 0.001, 0.01, and 0.1 M NaCl for BFA.

intervals from 3.6 to 10.5 for inclusion in these figures. Corresponding pH values are the midpoint pH values for each interval (3.75, 4.05, . . .,10.35). This averaging and smoothing procedure allows precision estimates (± 1 SD) to be calculated for these enthalpy data.

The following are some general trends and features of these data:

1. Acid addition (or protonation) enthalpies are slightly endothermic (0 to $+1$ kcal/mole) below pH 6.5 for both fulvic acids.
2. Above pH 6.5, protonation enthalpies for both fulvic acids become increasingly more exothermic. The increase appears nearly linear for SRFA until pH 10, where values are about -7 ± 1 kcal irrespective of ionic strength. Above pH 10, enthalpy values appear to remain near -7 kcal/mole, although only two values are present in this region. Protonation enthalpies for BFA increase more rapidly above pH 6.5 and reach -5 to -6 kcal/mole between pH 8 and 9.

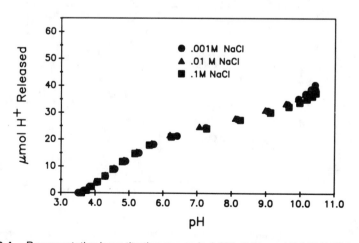

Figure 8.4. Representative base titration curves in 0.001, 0.01, and 0.1 M NaCl for BFA.

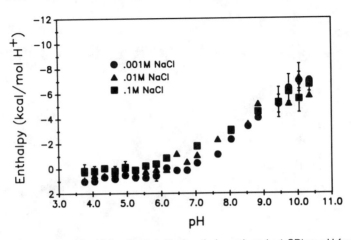

Figure 8.5. Average acid addition (protonation) enthalpy values (± 1 SD) vs pH for SRFA in 0.001, 0.01, and 0.1 M NaCl.

Above pH 9, enthalpies appear to exhibit an ionic strength dependence such that enthalpies at lower ionic strength are more exothermic. However, this may be an experimental artifact since data precision is poor.

3. Below pH 6, base addition (or ionization) enthalpies are greater in absolute magnitude and much less precise than corresponding acid addition enthalpies. Moreover, values appear to become more endothermic below pH 4.5.

4. Ionization enthalpies increase linearly above pH 6 for SRFA, with values above pH 10 reaching 9 to 10.5 kcal/mole. This is 3 to 4 kcal/mole greater in absolute value than acid addition enthalpies in the same pH region. Ionization enthalpy values for BFA also increase above pH 6 and reach 10 to 11.5 kcal/mole above pH 10.

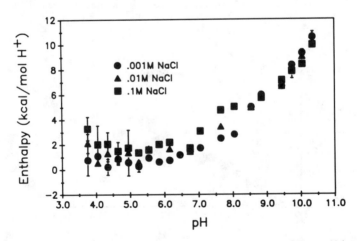

Figure 8.6. Average base addition (ionization) enthalpy values (± 1 SD) vs pH for SRFA in 0.001, 0.01, and 0.1 M NaCl.

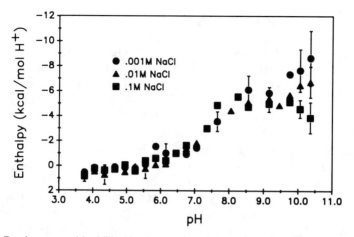

Figure 8.7. Average acid addition (protonation) enthalpy values (± 1 SD) vs pH for BFA in 0.001, 0.01, and 0.1 M NaCl.

DISCUSSION

The titration curves (Figures 8.1 to 8.4) are quantitatively similar to those previously documented in the literature. An inflection point near pH 7.5 is evident, as noted in other studies; in other respects also the curves are similar to those previously documented for various fulvic acids.[29,30] The buffer capacity of these solutions is also at a minimum in the pH 7 to 8 range since the spacing between pH values attained during the equal-volume titrant additions is greatest in this region.

The curves are reproducible, and acid-base hysteresis is much less than noted in several previous studies.[10-12] For example, comparison of the acid and

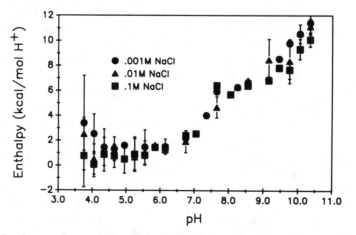

Figure 8.8. Average base addition (ionization) enthalpy values (± 1 SD) vs pH for BFA in 0.001, 0.01, and 0.1 M NaCl.

base titrations for both fulvic acids (Figures 8.1 to 8.4) reveals H^+ bound and released are similar over the same pH range. For SRFA, 47 μmoles H^+ are bound during acid titrations (average for all ionic strengths), and 45 μmoles H^+ are released during base titrations between pH 4 and 10. Corresponding values for BFA are 33 μmoles H^+ bound during acid titrations and 32 μmoles H^+ released during base titrations. The similarity between acid and base titrations in this study is most likely due to the relatively short duration of the titrations ($\sim$ 2 hr to complete acid and base titrations at each ionic strength) compared to previous studies.[12] In addition, ionic strength (0.001 to 0.1 M) has little effect on the quantity of H^+ bound or released by either fulvic acid—also noted previously.[5,29]

Titration data for SRFA in the present study are also quantitatively similar to those obtained previously. Proton consumption from pH 6 to 4 is equivalent to 2.0 meq H^+/g for SRFA compared to 2.1 meq H^+/g from the data cited in a previous study.[12] Similarly, H^+ consumption is about 0.65 meq H^+/g from pH 10 to 8 compared to 0.60 meq H^+/g observed previously. Proton consumption by BFA is equivalent to 1.4 meq/g from pH 6 to 4 and 0.62 meq/g from pH 10 to 8. Consumption of H^+ from pH 6 to 4 is indicative of protonation of carboxylic acid functional groups, while phenolic groups probably dominate from pH 10 to 8. Thus, H^+ consumption over these intervals can be used to estimate the titratable carboxyl and phenolic group densities of humic substances. Previous studies have suggested titratable carboxyl and phenolic group densities equal approximately twice the H^+ consumption between pH 6 and 4 and 10 and 8, respectively.[12] Thus, titratable carboxyl and phenolic group densities for SRFA are 4.0 and 1.3 meq/g, respectively, which are in good agreement with previous estimates.[12] Corresponding values for BFA are 2.8 meq/g for carboxylic and 1.2 meq/g for phenolic groups, similar to previous estimates for other peat-derived fulvic acids.[31] The good agreement between previous estimates and those of this study for SRFA suggests the theoretical procedure used to correct for solution blank effects in the present study (Equations 8.1 and 8.2) is satisfactory, at least over a limited pH range (3.6 to 10.4).

Ideally, acid (protonation) and base (ionization) addition enthalpies should be equal in absolute magnitude. In this study, however, acid addition enthalpies are generally 1–4 kcal/mole less in absolute magnitude, with the largest differences occurring at the extremes of the titration range (pH < 4.5; pH > 9.5). Net residual heat values for base additions decrease rapidly above pH 9 (Table 8.2). Consequently, any random or systematic sources of error which could affect base addition enthalpy values are of greater consequence at high pH. In addition, protonation enthalpies for simple and H-bonded phenolic groups do not exceed -9 kcal/mole, as illustrated in Table 8.3. However, ionization enthalpy values for both fulvic acids exceed 10 kcal/mole above pH 10 and do not appear to level off as do the acid addition enthalpies. A probable explanation for this discrepancy is that the apparent base addition enthalpies are simply approaching the water dissociation enthalpy value (+ 13.34

Table 8.3. pKa, Free Energy, and Protonation Enthalpy Values for Selected Mono-, Di-, and Polycarboxylic as Well as Phenolic Acids

	pKa	ΔG (kcal)	ΔH (kcal)
Monocarboxylic			
Acetic	4.76	6.49	0.1
Propanoic	4.87	6.64	0.2
Butanoic	4.81	6.56	0.6
Benzoic	4.20	5.73	−0.1
Salicylic	2.97	4.05	−0.8
Dicarboxylic			
Oxalic	$pK_1 = 1.30$	1.77	0.8
	$pK_2 = 4.27$	5.82	1.5
Malonic	$pK_1 = 2.83$	3.86	−0.3
	$pK_2 = 5.70$	7.77	0.9
Succinic	$pK_1 = 4.21$	5.74	−0.8
	$pK_2 = 5.64$	7.69	−0.1
Phthalic	$pK_1 = 2.94$	4.01	0.6
	$pK_2 = 5.40$	7.37	0.5
Polycarboxylic			
Polyacrylic	$\alpha = 0$	4.40	6.00	0.0
	$\alpha = 0.3$	5.70	7.77	0.9
	$\alpha = 0.8$	6.64	9.07	1.0
Phenolic			
Phenol	10.0	13.64	−5.6
1,2-Dihydroxybenzene	$pK_1 = 9.2$	12.55	−8.2
	$pK_2 = 13.0$	17.73	−5.0
Salicylic	13.1	17.87	−8.6
4-Hydroxybenzaldehyde	7.6	10.37	−4.3

Sources: Christensen et al.[17] and Gunnarsson et al.[33]

kcal/mole) as net residual heat values become progressively smaller above pH 9 (Table 8.2). Acid addition (protonation) enthalpy values, however, never exceed −9 kcal/mole; this is more consistent with the idea that phenolic groups dominate the acid-base equilibria of these fulvic acids above pH 8.

Differences between acid and base addition enthalpies also increase below pH 4.5, and in this region another factor may be contributing to the apparent irreversibility. It is conceivable that disaggregation or unfolding of fulvic acid molecules as pH increases from below 4.5 upon base addition may be contributing to the enthalpy irreversibility. Disaggregation or unfolding would probably be a net endothermic process since it would involve the disruption of aggregate stabilizing bonds such as van der Waals or H bonds. These endothermic processes would decrease the net exothermic heat values and, after correction for water dissociation, result in the more endothermic final enthalpy values observed (Figures 8.6 and 8.8). Acid addition enthalpies, however, do not suggest aggregation or folding is occurring below pH 4.5 since enthalpies are constant and more precise in this region. This could be due to the short time between successive acid additions (90 sec) compared to the holding time at low pH (20–30 min) before base additions could commence. The holding time was necessary to lower the temperature of the dewar reaction vessel to the desired starting value (24.6°C) and could have provided sufficient time for aggregation to occur. The precision of the base addition enthalpies in this region is also poor, which may be an indication that the extent of aggregation is different for the various titrations. Clearly, additional and more careful work in the low pH region would be necessary to confirm this hypothesis. However, H^+ consumption or release is not the major cause of the enthalpy discrepancy or poor precision since acid and base titration curves are largely reversible below pH 4.5 (Figures 8.1 to 8.4). In any case, acid addition enthalpy values are believed to more accurately reflect the enthalpies associated with proton binding or release at low pH.

In previous studies, protonation enthalpies observed for humic and fulvic acids between pH 4 and 6 varied from 0.2 to –1.7 kcal/mole.[18,19] In the present study, corresponding values range from 1.0 to 0 kcal/mole. These slightly endothermic protonation enthalpies are more consistent with the values found for most carboxylic acids (Table 8.3). Between pH 6 and 7, a rather distinct break toward more exothermic enthalpy values occurs, probably reflecting the increasing importance of phenolic groups above pH 7. This inflection is much more distinct than that apparent for the titration data alone (Figures 8.1 to 8.4). Moreover, a useful application of combined titration and enthalpy data may be to fit these data simultaneously in order to obtain functional group distributions, average pKa values, etc., for these or other humic substances.

Acid addition enthalpies rapidly increase above pH 7 to values of –5 to –9 kcal/mole near pH 10. Protonation enthalpy values previously reported for humic substances near pH 10 are 0.4 to 8 kcal/mole,[19] and –6 to –7.5 kcal/mole.[18] The latter values agree closely with those of this study and are consistent with those expected from the protonation of phenolic groups (Table 8.3). Moreover, phenolic groups can participate in H-bond formation with neighboring groups, and this can result in slightly more exothermic protonation enthalpies. For example, the first protonation enthalpy for 1,2-dihydroxybenzene is –8.2 kcal/mole, compared to –5.6 kcal/mole for phenol. Protonation enthalpies for the fulvic acids exceed –6 kcal/mole above pH 10; consequently, it is possible that H-bond formation is augmenting the enthalpies in this pH region, supporting the hypothesis that H bonds exist between phenolic and adjacent ketone groups in SRFA molecules.[24]

Enthalpy values for BFA increase more rapidly than those for SRFA above pH 6.5. For example, near pH 8 protonation enthalpies for BFA average about –5 kcal/mole, whereas those for SRFA are about –3 kcal/mole. This difference probably arises because the contribution of phenolic hydrogens to the total enthalpy values becomes more important for BFA at lower pH values. This is consistent with the ratio of titratable phenolic to carboxylic groups being 0.2 for SRFA and 0.3 for BFA.

Below pH 6.5, however, protonation enthalpies for BFA and SRFA are similar and constant, implying that the carboxyl group ionization enthalpies of these fulvic acids are not noticeably perturbed by electrostatic effects, believed to significantly affect the free energy values for simple[32] and polycarboxylic[33] acids. For example, the free energy difference between the second and first dissociation constants for succinic acid is 1.95 kcal/mole, and the apparent free energy value of polyacrylic acid increases from 6.0 to 9.1 as the degree of ionization increases from 0 to 0.8 (Table 8.3). Corresponding enthalpy differences are 0.7 and 1.0 kcal/mole for succinic and polyacrylic acid, respectively. Thus, free energy values increase with increasing ionization and enthalpy values become more endothermic. Furthermore, since the magnitude of the enthalpy change is only about one-third of the corresponding free energy change, experimentally observing a contribution from electrostatic effects for humic substances would be difficult. Consequently, enthalpy values are of little use in helping to define the importance of electrostatic effects to carboxyl group ionization of humic substances.

CONCLUSIONS

The fulvic acid protonation enthalpy data can be divided into three distinct regions on the basis of pH. Below pH 6.5, enthalpies are slightly endothermic (1 to 0 kcal/mole), which is consistent with carboxylic groups dominating the acid-base behavior of fulvic acids in this pH regime. Electrostatic or other factors do not appear to be influencing the protonation of carboxyl groups in this region since enthalpy values remain constant in this pH range. Above pH 8, protonation enthalpies are –4 to –9 kcal/mole, which suggests phenolic groups are dominating the acid-base properties in this pH region. Some of these phenolic groups may be H-bonded to adjacent groups, such as ketones, since enthalpy values exceed those generally observed for simple phenols (–4 to –6 kcal/mole). The region between pH 6.5 and 8 can be classified as a transition region between carboxylic and phenolic group dominance.

Ionic strength changes (0.001 to 0.1 M in NaCl) had no significant influence on these enthalpy data, suggesting binding or condensation of Na^+ counterions over the limited pH range investigated results in no significant enthalpy effects. Base addition enthalpies, however, were always 1 to 4 kcal/mole greater in absolute magnitude than acid addition enthalpies. This irreversibility was due primarily to a difference in residual heat values rather than the H^+ consumption or release since acid and base titration curves were largely reversible for both fulvic acids over the entire pH range. Above pH 9, base addition enthalpy values are much larger (> 10 kcal/mole) than those observed for simple phenolic acids. This is most likely an artifact of the small net residual heat values observed and suggests base addition enthalpy values are less accurate than corresponding acid addition enthalpy values above pH 9. Also, disaggregation or unfolding of fulvic acid molecules may be contributing to the more endothermic ionization enthalpies observed below pH 4.5. Consequently, acid addition enthalpy values are believed to more accurately reflect proton binding and release from fulvic acids over the entire pH range investigated.

Further applications of titration calorimetry to study humic substances can be envisioned. Additional experiments below pH 4.5 would help to sort out the possible heat effects associated with disaggregation or unfolding. In addition, studies to investigate trace metal binding enthalpies may well reveal inherent binding-site distributions of humic substances for trace metals. In this respect, the utility of titration calorimetry to reveal binding-site heterogeneity has already been demonstrated in adsorption studies.[15]

REFERENCES

1. Aiken, G. R. "Isolation and Concentration Techniques for Aquatic Humic Substances," in *Humic Substances in Soil, Sediment and Water*, G. R. Aiken, D. M. McKnight, R. L. Wershaw, and P. MacCarthy, Eds. (New York: John Wiley and Sons, 1985), pp. 363–385.
2. Buffle, J. *Complexation Reactions in Aquatic Systems* (Chichester, Eng.: Ellis Horwood, 1988), pp. 314–322.
3. Cabaniss, S. E., and M. S. Shuman. "Copper Binding by Dissolved Organic Matter: I. Suwannee River Fulvic Acid Equilibria," *Geochim. Cosmochim. Acta* 52(1):185–193 (1988).
4. Cabaniss, S. E. "pH and Ionic Strength Effects on Nickel-Fulvic Acid Dissociation Kinetics," *Environ. Sci. Technol.* 24(4):583–588 (1990).
5. Ephraim, J., S. Alegret, A. Mathuthu, M. Bicking, R. L. Malcolm, and J. A. Marinsky. "A Unified Physicochemical Description of the Protonation and Metal Ion Complexation Equilibria of Natural Organic Acids (Humic and Fulvic Acids). 2. Influence of Polyelectrolyte Properties and Functional Group Heterogeneity on the Protonation Equilibria of Fulvic Acid," *Environ. Sci. Technol.* 20(4):354–366 (1986).
6. McKnight, D. M., and R. L. Wershaw. "Complexation of Copper by Fulvic Acid from the Suwannee River—Effect of Counterion Concentration," U.S. Geological Survey Open-File Report 87–557 (1989), pp. 63–79.
7. Hering, J. G., and F. M. M. Morel. "Humic Acid Complexation of Calcium and Copper," *Environ. Sci. Technol.* 22(10):1234–1237 (1988).
8. Carter, C. W., and I. H. Suffet. "Binding of DDT to Dissolved Humic Materials," *Environ. Sci. Technol.* 16(11):735–740 (1982).
9. Kile, D. E., C. T. Chiou, and T. I. Brinton. "Interactions of Organic Contaminants with Fulvic and Humic Acids from the Suwannee River and Other Humic Substances in Aqueous Systems, with Inferences to the Structures of Humic Molecules," U.S. Geological Survey Open-File Report 87–557 (1989), pp. 41–57.
10. Paxeus, N., and M. Wedborg. "Acid-Base Properties of Aquatic Fulvic Acid," *Anal. Chim. Acta* 169:87–98 (1985).
11. Gregor, J. E., and H. K. J. Powell. "Effects of Extraction Procedures on Fulvic Acid Properties," *Sci. Total Environ.* 62:3–12 (1987).
12. Bowles, E. C., R. C. Antweiler, and P. MacCarthy. "Acid-Base Titration and Hydrolysis of Fulvic Acid from the Suwannee River," U.S. Geological Survey Open-File Report 87–557 (1989), pp. 205–229.
13. Perdue, E. M. "Acidic Functional Groups of Humic Substances," in *Humic Substances in Soil, Sediment and Water*, G. R. Aiken, D. M. McKnight, R. L.

Wershaw, and P. MacCarthy, Eds. (New York: John Wiley and Sons, 1985), pp. 493–526.

14. Grime, J. K. "General Analytical Applications of Solution Calorimetry," in *Analytical Solution Calorimetry*, J. K. Grime, Ed. (New York: John Wiley and Sons, 1985), pp. 163–298.

15. Machesky, M. L., B. L. Bischoff, and M. A. Anderson. "Calorimetric Investigation of Anion Adsorption onto Goethite," *Environ. Sci. Technol.* 23(5):580–587 (1989).

16. Lewis, E. A., T. J. Barkley, R. R. Reams, L. D. Hansen, and T. St. Pierre. "Thermodynamics of Proton Ionization from Poly(vinylammonium salts)," *Macromolecules* 17:2874–2881 (1984).

17. Christensen, J. J., L. D. Hansen, and R. M. Izatt. *Handbook of Proton Ionization Heats and Related Thermodynamic Quantities* (New York: John Wiley and Sons, 1976).

18. Choppin, G. R., and L. Kullberg. "Protonation Thermodynamics of Humic Acid," *J. Inor. Nucl. Chem.* 40:651–654 (1978).

19. Perdue, E. M. "Solution Thermochemistry of Humic Substances: Acid-Base Equilibria of River Water Humic Substances," in *Chemical Modeling in Aqueous Systems*, American Chemical Society Symposium Series 93, E. A. Jenne, Ed. (Washington, DC: American Chemical Society, 1979), pp. 99–114.

20. Perdue, E. M. "Solution Thermochemistry of Humic Substances—I. Acid-Base Equilibria of Humic Acid," *Geochim. Cosmochim. Acta* 42:1351–1358 (1978).

21. Averett, R. C., J. A. Leenheer, D. M. McKnight, and K. A. Thorn, Eds. *Humic Substances in the Suwannee River, Georgia: Interactions, Properties and Proposed Structures*, U.S. Geological Survey Open-File Report 87–557 (1989).

22. Thurman, E. M., and R. L. Malcolm. "Preparative Isolation of Aquatic Humic Substances," *Environ. Sci. Technol.* 15:463–466 (1981).

23. Hatcher, P. G., I. A. Breger, L. W. Dennis, and G. E. Maciel. "Solid-State ¹³C NMR of Sedimentary Humic Substances: New Revelations on Their Chemical Composition," in *Aquatic and Terrestrial Humic Materials*, R. F. Christman and E. T. Gjessing, Eds. (Ann Arbor, MI: Ann Arbor Science, 1983), pp. 37–82.

24. Leenheer, J. A., D. M. McKnight, E. M. Thurman, and P. MacCarthy. "Structural Components and Proposed Structural Models of Fulvic Acid from the Suwannee River," U.S. Geological Survey Open-File Report 87–557 (1989), pp. 335–359.

25. Eatough, D. J., J. J. Christensen, and R. M. Izatt. *Experiments in Thermometric Titrimetry and Titration Calorimetry* (Provo, UT: Brigham Young University Press, 1974), pp. 1–45.

26. Durst, R. A., W. F. Koch, and Y. C. Wu. "pH Theory and Measurement," *Ion-Selective Elec. Rev.* 9:173–196 (1987).

27. Brewer, S. *Solving Problems in Analytical Chemistry* (New York: John Wiley and Sons, 1980), pp. 179–199.

28. Morcellett, M. "Thermodynamics of the Protonation of Some Polyelectrolytes," *Thermochim. Acta* 142:165–173 (1989).

29. Dempsey, B. A., and C. R. O'Melia. "Proton and Calcium Complexation of Four Fulvic Acid Fractions," in *Aquatic and Terrestrial Humic Materials*, R. F. Christman and E. T. Gjessing, Eds. (Ann Arbor, MI: Ann Arbor Science, 1983), pp. 239–273.

30. Varney, M. S., R. F. C. Mantoura, M. Whitfield, D. R. Turner, and J. P. Riley. "Potentiometric and Conformational Studies of the Acid-Base Properties of Fulvic

Acid from Natural Waters," in *Trace Metals in Sea Water*, C. S. Wong, E. Boyle, K. W. Bruland, J. D. Burton, and E. D. Goldberg, Eds. (New York: Plenum Press, 1983), pp. 751–772.

31. Mathur, S. P., and R. S. Farnham. "Geochemistry of Humic Substances in Natural and Cultivated Peatlands," in *Humic Substances in Soil, Sediment and Water*, G. R. Aiken, D. M. McKnight, R. L. Wershaw, and P. MacCarthy, Eds. (New York: John Wiley and Sons, 1985), pp. 53–85.

32. Christensen, J. J., R. M. Izatt, and L. D. Hansen. "Thermodynamics of Proton Ionization in Dilute Aqueous Solution. VII. $\Delta H°$ and $\Delta S°$ Values for Proton Ionization from Carboxylic Acids at 25°C," *J. Amer. Chem. Soc.* 89(2):213–222 (1967).

33. Gunnarsson, G., H. Wennerstrom, G. Olofsson, and A. Zacharov. "Enthalpies of Proton Dissociation from Poly(acrylic acid): Comparison between Experiment and Theory for a Polyelectrolyte System," *J. Chem. Soc. Faraday Trans. I.* 76:1287–1295 (1980).

The Transport and Composition
of Humic Substances in Estuaries

Lewis E. Fox

INTRODUCTION

The transport of terrestrial organic carbon to the oceans of the world, as well as the distribution, transport, and biologic availability of trace elements and organics, can be significantly affected by the behavior of humic substances in many estuaries. The reason for this lies in the abundance and chemical nature of these substances. Humics comprise 40–90% of terrestrially derived dissolved organic carbon (DOC) in rivers.[1-3] They are natural sorbants and chelators, thereby affecting the distribution and fate of some trace elements[4-10] and organics[4,11-13] through chemical bonding. The high-molecular-weight portion of riverine humics precipitates in estuaries, thereby providing a mechanism for sedimentary concentration of organics and elements which could include toxic substances.

The term "humic substance" defines an enormously complex class of organic compounds generally attributed to breakdown products of biogenic organic matter through chemical mechanisms which are largely unknown. Humics constitute a large proportion of the organic matter in soils and, thus, have received considerable attention from soil chemists. However, concentrated study has revealed no absolute chemical structure for humics, leaving their identification dependent on general chemical characteristics. As a result, they are vaguely defined as yellow-to-black-colored, moderately high molecular weight, alkali-soluble polyelectrolytes,[14-17] believed resistant to microbial degradation.[16] To facilitate investigation, humic substances are subdivided according to their solubility into two major categories: humic and fulvic acids. Humic acid is defined as the acid insoluble portion of an alkali soil extract after washing with ethyl alcohol.[15] The definition is often modified to exclude the alcohol wash. Soluble material remaining after acidification is called fulvic acid.[15,16] Modern extraction is almost exclusively done with XAD sorbent resins.[18] However, sorbed organics are still separated into humic and fulvic acids by acidification.

These definitions can be traced back to the 1920s and represent only slight modifications to definitions developed by German chemists in the early nineteenth century. Modern analytical techniques have not altered these definitions but have added more specific chemical criteria. For instance, the elemental composition of humics varies only slightly, regardless of source.[16] Spectral analysis gives a generally similar profile for all humics as well. Unfortunately, the heterogeneity of humics causes peak broadening, thereby denying the possibility of absolute organic identification. Nevertheless, spectral analyses yield important general information when used in conjunction with other chemical analyses. From infrared spectral analyses coupled with various titration methods, carboxyl and phenolic hydroxyl groups are identified as the major functional groups associated with humics, although quinone, hydroxy-quinone, lactone, ether, and phenolic hydroxyl groups also occur.[15,17,19-32]

A variety of techniques have been employed to measure average weights of humic substances. Number-averaged methods indicate humic acid weights of 1000-2000 daltons,[16,33] and fulvic acid weights of 600-1000 daltons.[16,34-36] Weight-averaged methods such as membrane filtration, gel filtration, and ultracentrifugation yield much higher values (700-10[6] daltons),[16,22,37] but these methods generate solute-membrane and solute-gel interactions as well as molecular aggregation, suggesting the high values are artifacts and that number-averaged values are better.[38]

Nuclear magnetic resonance studies indicate humics have major aliphatic and alicyclic components,[29,39-41] which seem to dominate the aromatic components previously identified by degradation analyses.[15,29,42-45] Generally, aquatic humics seem to contain a larger aliphatic and smaller aromatic fraction than soil humics,[31,46] although the aromatic fraction is a basic structural component consisting of polycarboxyl-phenolic aromatics,[44] which are probably lignin residues.[3,47,48]

Precipitation in estuarine and coastal waters of some form of terrestrially derived humic substance can be inferred from a variety of indirect evidence. Brown interpreted changes in ultraviolet characteristics of waters along an estuarine salinity gradient as indicative of the aggregation of high-molecular-weight organics.[49] Kalle noted removal of yellow organic matter (humic substances) in coastal waters.[50] Sieburth and Jensen found that terrestrially derived humic compounds, which were identified with paper chromatography, precipitated rapidly in seawater.[51] Likewise, Swanson and Palacas found humic substances extracted from beach sands precipitated in seawater after resuspension in freshwater.[52] The distribution of lignins and $^{13}C:^{12}C$ isotope ratios in organic matter extracted from coastal sediments is reported to be indicative of the precipitation of significant amounts of terrestrial organic matter.[47,53,54] Substantial amounts of humic substances reported in Delaware Bay[55] and the New York Bight[46] are also suggestive of the precipitation of terrestrial humics.

Estuarine precipitation of terrestrial carbon is also evident from direct analyses of aquatic humic acid. Hair and Bassett reported a decrease in humic acid concentrations with increased salinity in a tributary of Long Island Sound.[56] They extracted humic acid with glacial acetic acid and isoamyl alcohol and quantified it spectrophotometrically against a soil humic extract.[57] Humic acid precipitation was also measured directly in seawater mixes of water from several Scottish rivers by the same analytical methods,[58] as well as from humic acid directly extracted from the Amazon Estuary.[59] Estuarine removal of humic acid was also observed in Delaware Bay[60] and several salt marsh tributaries to Delaware Bay, as well as the Mullica River, New Jersey, by direct carbon analysis (high-temperature oxidation) after acid extraction.[61,62]

Dissolved organic matter is arbitrarily separated from suspended organics by its ability to pass a 1-μm glass-fiber filter[63,64] (sometimes 0.8-μm[58]). Because of this definition, a considerable amount of colloidal-sized, suspended carbon is artificially classified as "dissolved." Therefore, it should be noted that much of the "dissolved" organic matter removed in estuaries is actually colloidal-sized, suspended organic material, as demonstrated by the observation that filtrates of river water, filtered with progressively finer filters, yield smaller amounts of reactive humic acid when mixed with seawater.[59]

The classical mechanics of colloidal aggregation and dispersion appear to explain humic acid removal in estuaries. Humic colloids undergo surface charge reduction through complexation with alkali earth cations in waters of increasing salinity. Charge reduction results in conformational shrinking, elimination of hydrated water, and subsequent hydrophonic colloidal aggregation through Van der Waals attraction. Evidence supporting this mechanism includes observation of less negative electrophoretic mobilities[65] and reduction of free binding sites on humic molecules in solutions of increasing ionic strength,[66] as well as the superior aggregation efficiency of calcium over magnesium and sodium.[67]

However, it appears that aquatic humic acid is not universally removed from estuarine waters. Eastman and Church report relatively conservative (unreactive) humic acid behavior in a small salt marsh creek.[68] More significantly, enormous inputs of humic acid to estuarine waters of the Orinoco River are evident in Figures 9.1 to 9.3. Figure 9.1 illustrates humic acid concentrations along the salinity gradient of the Orinoco Estuary. Clearly, some removal is evident, but the estuary is dominated by large inputs beginning at 6 ppt salinity. Estuarine concentrations of humic acid carbon exceed 3 mg C/L, which amounts to three-fourths of the concentration of all dissolved organic carbon in the Severn Estuary, U.K., for instance.[69] To put this into global perspective, the humic acid carbon released into the Orinoco Estuary, if annually consistent, would produce an average annual flux of 3.2×10^{12} g C/year using the discharge data reported by Lopez.[70] This is 63 times the average annual DOC discharge of the Severn Estuary[69] and over 0.5% of the average global DOC flux as estimated by Meybeck.[71]

The mechanism for humic acid release into Orinoco waters is unknown.

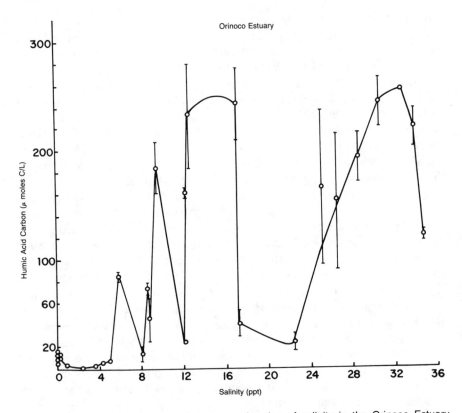

Figure 9.1. Humic acid carbon plotted as a function of salinity in the Orinoco Estuary, Venezuela. Error bars represent the standard deviation of duplicate samples.

From Figures 9.2 and 9.3, it is evident that the carbon is emanating from the bottom, but from Figure 9.3 it is clear that humic carbon is not directly related to sediment resuspension. These observations are supported by optical absorption measurements made by Blough et al.,[72] who also concluded that the Orinoco Estuary contained a strong source of colored organic material which is not produced by resuspension of sediments or through release from suspended particulate matter.

Humic acid comprises only 10–20% of total humic substances, so it is not clear from humic acid determinations alone the extent to which the majority of riverborne organic matter reacts in estuaries. The more indirect evidence of carbon precipitation, cited above, is also of little use in resolving the issue since the fraction of total estuarine carbon represented by sediment extracts or deduced from spectroscopic measurements cannot be directly ascertained. Measurement of estuarine DOC provides an indication of the extent of carbon removal in some estuaries, although inputs of DOC can totally obscure removal of humic acid carbon.[61] Studies of the transport of DOC in estuaries report input,[69,73] conservative (unreactive) behavior,[68,73] and removal.[58,59,61]

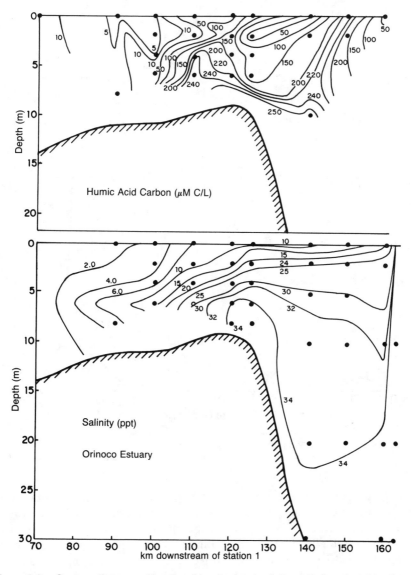

Figure 9.2. Contour diagrams of humic acid carbon and salinity plotted against river distance and water depth in the Orinoco Estuary.

Analysis of DOC data from a variety of estuaries is given in Table 9.1. In estuaries without significant DOC inputs, carbon removal ranges from 4 to 60%, with a mean of 19%, of the initial DOC. However, removal percentages below 5% cannot be distinguished from conservative behavior with certainty. Interestingly, Mantoura and Woodward conclude conservative DOC behavior for the Severn Estuary, U.K., from a very nice study that includes a large data

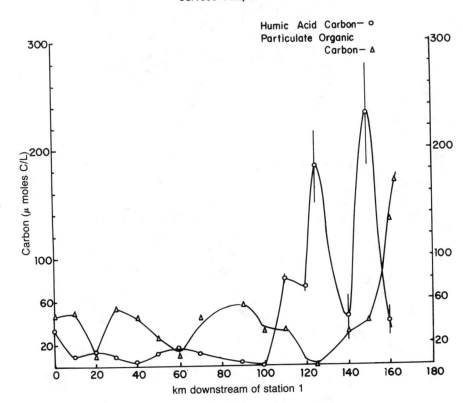

Figure 9.3.   Humic acid and particulate organic carbon plotted as functions of river distance in the Orinoco Estuary. Data consist exclusively of surface samples.

set taken over two years.[69] However, when data illustrated in the publication are analyzed after Officer,[74] removal is evident and seems to follow a distinct seasonal trend (Table 9.1). This could be an artifact from changes in river discharge (as is implied by the authors) since the Severn has a residence time of 3–6 months and DOC levels are reported to be linear functions of runoff.[69] Indeed, when riverine DOC levels increase from increased discharge, DOC-salinity diagrams artificially suggest estuarine removal. The reverse happens when discharge decreases. The problem is that only removal or near conservative behavior is evident in the Severn with no commensurate input, which would be expected when river discharge decreases.

It could also be argued that the maximum observed removals in June (Table 9.1) are artifacts resulting from higher carbon values at the seaward end-member due to high spring productivity. However, removal is still apparent when the seaward endmembers are moved to lower salinities. The point of this argument is not to fault the study, but to suggest the possibility that some

Table 9.1. Dynamics of Estuarine DOC and Humic Acid

River Estuary	% Nonconservative DOC[a] (+ Input - Removal)	River End Member DOC (mg C/L)	Carbon Added (+) or Removed (-) (mg C/L)	Humic Acid Carbon Removed (mg C/L)	Humic Acid Carbon Removed as % DOC River End Member	Reference
Severn						
25 April 1977	-15	4.4	-0.7	—	—	69
5-9 May 1977	-12[b]	4.7	-0.6	—	—	69
16-20 June 1977	-40(-16)[c]	5.8	-0.9	—	—	69
25 Oct.-5 Nov. 1977	-5	4.2	-0.2	—	—	69
9 Feb. 1978	-7	5.5	-0.4	—	—	69
14 Mar. 1978	-14(+21)[d]	4.4	-0.6	—	—	69
23 June-5 July 1978	-48(-17)[c]	4.8	-0.8	—	—	69
MEAN	-12 ± 4	4.8 ± 0.6	-0.6 ± 0.2			
River Stincher[e]	-4	7.6	-0.3	-0.58	8	58
River Cree[e]	-10	4.4	-0.4	-0.30	7	58
Glen Burn[e]	-22	13.5	-3.0	-2.75	20	58
Water of Luce[e]	-20	6.7	-1.4	-1.30	19	58
Amazon 1974	+68	2.5	+1.7	-0.13[f]	5	59
Amazon 1976				-0.15[f]	5	59
Amazon[e]				-0.20[f]	8[g]	59
Murderkill	+18	5.9	+0.3	-0.54	9	61
St. Jones	+20	5.4	+1.3	-0.55	10	61
Broadkill 10/78	+30	6.0	+1.8	-0.74	12	61
Mullica 1979	-37	8.5	-3.1	-1.80	21	61
Mullica 1980	-59	7.6	-5.0	-1.74	23	61
Delaware	-13	7.6	-5.0	-1.74	23	61
River Beaulieu 1976	-4	4.8	-0.2	—	—	73
River Beaulieu 1977	-4	5.4	-0.2	—	—	73

[a] Percentage input and removal is calculated from $(C_0 - C/C_0) \times 100$, where C_0 is the carbon value at the y-intercept of a line extended from the seawater end member, tangential to the concentration data on a plot of carbon versus salinity.[74]
[b] Only data with the freshwater end member is used.
[c] Value obtained when high salinity removal is ignored by redefining the seawater end member to approximately 29-ppt salinity.
[d] One data point indicates input.
[e] Data taken from mixing experiments.
[f] Humic acid carbon is assumed to be 1/2 the mass of humic acid.
[g] It is assumed that the Amazon river water used in this mixing experiment was taken in 1974 and contained DOC values specified for that sample.

small, average removal of DOC occurs in the Severn Estuary at approximately the level of humic acid removal reported for other estuaries (Table 9.1).

Humic acid removal, where it is measured in conjunction with DOC, is also illustrated in Table 9.1. The data is meager because DOC and humic acid are rarely measured together. Unfortunately, no estuaries with aberrant humic acid behavior, such as the Orinoco, are reported. Nevertheless, the data do indicate that humic acid is removed even in estuaries with large inputs of DOC. Clearly, different fractions of estuarine carbon can be doing different things at the same time. When not obscured by other processes, it is evident that the DOC removed generally equals humic acid removed—except in estuaries like the Mullica Estuary, where DOC removal exceeds humic removal by factors of 2 to 3.

The Mullica is an example of a "black water" river, that is, one with water of low pH, due to elevated concentrations of organic acids, and suspended iron (Appendix). The large pool of reactive carbon in such rivers may be related to unusually low divalent cation concentrations. The Mullica watershed, for example, contains average Mg^{2+} and Ca^{2+} concentrations of 24 and 26 μM, respectively,[75] which are 6 and 12 times below the Mg^{2+} and Ca^{2+} concentrations of an average of the world's rivers.[76] Since estuarine removal of humic material results from aggregation through cation sorption, watersheds that contain abnormally low cation concentrations, yet drain carbon-rich areas, could be expected to contain higher than average amounts of reactive DOC. Many such rivers exist throughout the world, the Rio Negro being a notable example. If the observed behavior of organic carbon in the Mullica can be extended to other black water estuaries, then they probably are regions of high organic transport and deposition.

COMPOSITION OF REACTIVE ORGANIC CARBON

A more detailed breakdown of Mullica DOC is given in Figures 9.4 and 9.5 and Table 9.2. Analytical methods for the illustrated measurements are given in the Appendix. From the illustrations and Table 9.2, it is evident that all fractions of organic carbon measured show a significant degree of removal. On a percentage basis the fluorescently active, undifferentiated carbon, which is referred to here as fulvic acid, appears to be the most unreactive portion of the DOC but precipitates the largest absolute amount of organic carbon. Humic acid and fulvic acid comprise 87% of the organic carbon removed, even though 73 and 80% of total carbohydrate and total amino acid are also removed, respectively, because, of course, humic and fulvic acids comprise 89% of the sum of organic carbon measured and 97% of the DOC. The discrepancy between the sum of organic carbon and DOC results from the fact that the mean DOC, determined by persulfate oxidation, was slightly less than the average sum of all organic fractions on a carbon basis. The difference, however, is barely significant at the 5% level.

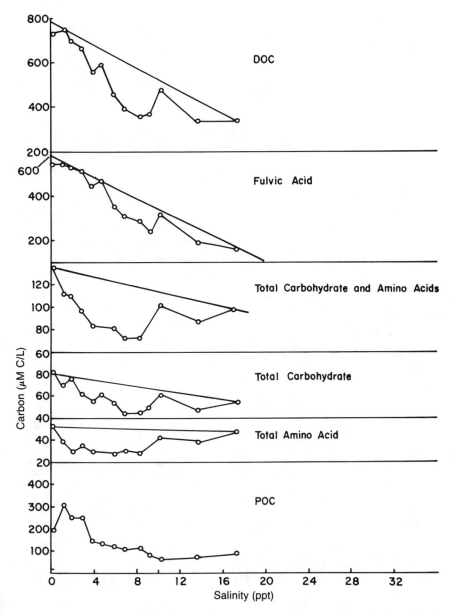

Figure 9.4.   Various fractions of organic carbon extracted from the Mullica Estuary plotted as a function of salinity. Individual fractions of organic carbon are labeled on the figure. The solid line represents the conservative mixing slope.

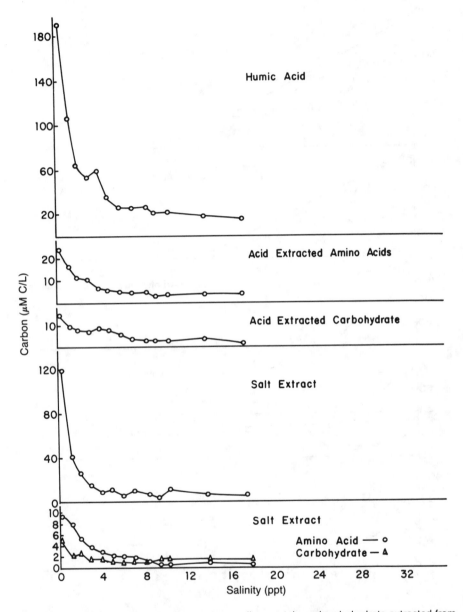

Figure 9.5.   Humic acid and salt extract, as well as protein and carbohydrate extracted from humic and salt extract from the Mullica Estuary, plotted as a function of salinity.

Table 9.2. Organic Carbon in the Mullica Estuary

Organic Classification	Max. Carbon (µM C/L)	Amount Removed (µM C/L)	% Removed	Carbon Removed As % Initial DOC	Carbon Removed As % DOC Removed	Carbon As % Humic Acid Removed	As % Total Carbohydrate Removed	As % Total Amino Acid Removed
Fulvic acid	570	240	44	30	52	—	—	—
Hydrolyzable carbohydrate	81	59	73	7	13	—	100	—
Hydrolyzable amino acid	51	41	80	5	9	—	—	100
Humic acid	193 ± 39	163 ± 39	84	21	35	100	—	—
Amino acid extracted from humic acid	25	23	92	3	5	14	—	56
Carbohydrate extracted from humic acid	14	10	72	1	2	6	17	—
TOTAL CARBON[a]	856 ± 39	470 ± 39	55	59	100	20	17	56
DOC	790 ± 40	465	59	59	100	—	—	—
POC	325	315	97	40	—	—	—	—
TOTAL CARBON	1115	780	70	—	—	—	—	—

Notes: Methods are given in Appendix.
[a]Humic-acid-derived carbohydrate and amino acid carbon was subtracted from total to prevent counting it twice, only as carbohydrate and amino acid and again as humic acid.

Table 9.3. Components of Aquatic Extracts on a Carbon Basis

Extract	% Protein[a]	% Carbohydrate	% Aromatic[b]	% Unaccounted Carbon
Mullica HA	8.6 ± 1.4	8.2 ± 1.4	17	66
Mullica salt extract	4.6 ± 0.9	2.3 ± 0.6	21	72
Mullica UFR	3.2 ± 1.4	1.8 ± 0.3	17	78
Broadkill HA	13.2 ± 2.2	10.2 ± 1.1	18	59
Broadkill salt extract	B.D.[c]	B.D.	8	92
Broadkill UFR	12.3 ± 1.4	9.1 ± 1.4	17	62

[a]Protein carbon was calculated from an average amino acid molar C:N ratio of 3.37 after direct determination of primary amine.
[b]Aromatic carbon was calculated from H NMR, assuming an average aromatic H:C ratio of 1.
[c]B.D. = below detection.

Determination of DOC in all of the estuaries cited in this chapter has been performed by the persulfate oxidation method.[63,77] Recently, the method has been shown to significantly underestimate carbon concentrations in seawaters, apparently because of its inability to oxidize ambient, high-polymer organics.[78,79] Because of these findings, direct comparisons of the efficiency of persulfate oxidation versus high-temperature combustion were performed on humic acid, salt-extracted organic matter, and particulate organic carbon from the Mullica and Broadkill estuaries. No systematic variation was detected. In addition, comparison of DOC values from the Delaware Bay, determined by persulfate oxidation and by the method of Sugimura and Suzuki[79] showed no systematic differences.[80] These findings indicate that persulfate oxidation seems an adequate method to measure high-molecular-weight organics in these estuaries.

Since errors in measurements of seawater DOC appear to result from the inability of persulfate to completely oxidize polymeric organic matter, the change in the oxidation efficiency of persulfate between estuarine and ocean waters may reflect a change in the composition of DOC toward a larger abundance of a specific type of high-polymer organics. One compositional difference may be in the lipid content. Colloidal-sized organics extracted from Chesapeake Bay waters contain <1% lipid[81] while sea water humics, which comprise 10–50% of the oceanic DOC, seem to be mainly composed of aliphatic organic acids formed by the autoxidative cross-linking of unsaturated lipids.[82] These polymeric substances, which are absent from estuarine waters, could comprise that portion of the oceanic DOC resistant to persulfate oxidation.

Comparison of the reactive organic pools from freshwaters of the Mullica and Broadkill Rivers are illustrated in Tables 9.3 and 9.4. The Broadkill was used as a comparison watershed because its drainage basin characteristics differ considerably from those of the Mullica. The Broadkill drains mostly land under cultivation, in contrast to the Mullica which drains a unique pine-forest biome. The reactive organic pools from these watersheds were extracted in three separate ways. First, humic acid was extracted because it is shown to be a reactive fraction of DOC in both watersheds.[61] Second, organic matter defined as salt filter residue (SFR) was extracted by mixing seawater with filtered river water and subsequently refiltering the mixture. This organic fraction was chosen because humic materials are thought to be coagulated by complexation with major seawater cations.[67,83] The third extract is called ultra-filter residue (UFR) and was obtained by filtering previously filtered (Whatman GFC) river water through a 10^5 molecular-weight exclusion ultrafiltration membrane.[84,85] This method was chosen because the reactive carbon pool is reported to be size dependent. Filtrates of river water filtered with progressively finer filters yield smaller amounts of humic acid.[59]

Examination of Table 9.3 reveals that most of the carbon recovered from all the extracts of both rivers is not acid-hydrolyzable protein, carbohydrate, or aromatic. The aromatic fraction was deduced from proton NMR analyses.[85] Only 5 to 25% of the extracted carbon was recovered as protein and carbohydrate. And the proportion of aromatic carbon was almost a constant 18%, except for the Broadkill SFR, which generally yielded very little identifiable carbon. Proton NMR analyses suggest that most of the unidentified carbon from these extracts is aliphatic. Significant peaks indicative of methyl, methylene, and hydroxyl protons are evident in all extracts.[84,85] An additional peak, evident only in the Broadkill extracts, is indicative of protons adjacent to any combination of carbonyl, alkene, or N-acetyl groups.[84] Another noticeable difference between Mullica and Broadkill extracts evident in Table 9.3 is that Mullica SFR and UFR extracts have similar proportions of protein and carbohydrate carbon, which amounts to 30–40% of the humic acid, while Broadkill humic acid and UFR extracts contain equal portions of protein and carbohydrate carbon, which is significantly larger relative to DOC than the Mullica. However, the differences in extracted protein and carbohydrate are only small percentages of the total carbon. Generally, data listed in Table 9.3 and cited HNMR analyses indicate that the organic composition of the extracts is similar with, perhaps, the exception of Broadkill SFR.

Table 9.4 reveals the differences among extracts on a physicochemical and size basis. Only 26 and 7% of the Mullica and Broadkill SFR extracts, respectively, are composed of humic acid carbon, which amounts to only 19 and 4% of the total humic acid carbon in the two rivers, respectively. Clearly, less than a quarter of the SFR pool is humic acid. On the other hand, not only is a much larger portion of the Mullica and Broadkill UFR humic acid, but all of the humic acid is extracted in the high-molecular-weight pool. In other words, nearly 3 and 1.5 times the humic acid carbon in the Mullica and Broadkill, respectively, could be precipitated in the estuaries if all the high-molecular-weight carbon is precipitated.

The portion of high-molecular-weight carbon that is also salt-extracted carbon (SFR) was not directly measured. However, from Table 9.4, 19% of the

Mullica DOC is humic acid and 16% is SFR. Since 25% of the SFR is humic acid, then 12% of the DOC is uniquely SFR, and since 19% of the humic acid is SFR, then about 16% is uniquely humic acid. Thus, 29% of the Mullica DOC is extracted as humic acid and SFR without overlap. This is close to the total percentage of high-molecular-weight carbon extracted and indirectly suggests (1) that humic acid and salt extract, together, compose the high-molecular-weight portion of the Mullica DOC, but (2) that this only accounts for half the DOC removed in the Mullica Estuary as determined from field measurements (Table 9.1). A similar analysis can be done for the Broadkill. Table 9.4 indicates humic acid carbon (HAC) and SFR total 20% of the Broadkill DOC. This agrees with the 25% UFR directly extracted. Unfortunately, DOC inputs in the Broadkill obscure total DOC removed. However, 20% removal is near the mean of estuaries cited in Table 9.1.

Table 9.5 illustrates the proportion of organic nitrogen from the Mullica and Broadkill in each extract. One major difference between organic nitrogen and carbon in the extracts is that practically all of the SFR nitrogen, from both rivers, can also be extracted as humic acid nitrogen (HAN). In addition, almost twice the sum of HAN and SFR N from the Mullica is extracted as high-molecular-weight nitrogen.

Tables 9.6 and 9.7 list amino acid concentrations from hydrolyzates of Mullica River water, humic acid, and SFR extracts. With the exceptions of glycine, alanine, and valine, the amino acids in the acid and salt extracts are identical. This is consistent with the large overlap of SFR and humic acid nitrogen evident in Table 9.5. A comparison of the hierarchy of amino acids between the Mullica River water and its extracts shows them to be practically identical and very similar to other black water rivers (Table 9.7). High abundances of aspartic acid, glutamic acid, and serine suggest the original, unhydrolyzed organic matter contained large densities of carboxyl and hydroxyl functionalities, consistent with the general characteristics of humic acids.

High abundances of organic acids in the Mullica River are consistent with low observed pH values (5.5) and low divalent cation concentrations[75] and tend to suggest a general reason for the larger organic removal observed in the estuary relative to other watersheds cited. As previously mentioned, humic acid removal appears to be the result of coagulation caused by complexation of acidic functional groups with major seawater ions, specifically Ca^{2+} and Mg^{2+}.[67,83,86] If the major portion of DOC is composed of large organic acids, as in the Mullica, judging from the low riverine pH values, then a major portion of the DOC will be complexed and coagulated during estuarine mixing with seawater. As a result, one could expect to find removal of larger portions of riverine DOC in black water estuaries because of the high abundances of organic acids and low divalent cation concentrations.

One of the puzzling aspects of the results of the extraction experiments is why a much larger portion of humic acid and high-molecular-weight organic matter was not extracted as SFR. Since salt-induced coagulation seems to be the major process responsible for removal of suspended humics, one would

HUMIC SUBSTANCES IN ESTUARIES   143

Table 9.4. Organic Carbon in Mullica and Broadkill Extracts

River Extract	Organic C as % DOC	% Humic Acid Carbon (HAC) in Extract	% SFR-C in Extract	HAC in Extract as % Total HAC	Organic C as % DOC after Correction for Overlap
Mullica HA	19.2 ± 2.5	100	18.5 ± 2.2	100	16.3 ± 2.5
Broadkill HA	18.0 ± 3.2	100	4.0 ± 2.0	100	16.2 ± 3.2
Mullica SFR	16.4 ± 3.9	25.8 ± 4.0	100	18.5 ± 2.2	12.4 ± 3.9
Broadkill SFR	2.8 ± 0.3	6.9 ± 3.0	100	4.0 ± 2.0	2.6 ± 0.3
Mullica UFR	33.2 ± 4.2	33.4 ± 2.3	—	108.5 ± 13.8	—
Broadkill UFR	25.1 ± 3.9	58.8 ± 8.1	—	118.0 ± 50.0	—
Mullica HA + SFR	—	—	—	—	28.7 ± 4.0
Broadkill HA + SFR	—	—	—	—	19.2 ± 0.5

Table 9.5. Organic Nitrogen in Mullica and Broadkill Extracts

River Extract	Organic N as % DON	% HAN in Extract	% SFR-N in Extract	HAN in Extract as % Total HAN	Organic N as % DON Corrected for Overlap
Mullica HA	24.1 ± 5.6	100	32.2 ± 4.3	100	21.2
Broadkill HA	19.8 ± 2.5	100	19.1 ± 4.1	100	19.8
Mullica SFR	8.0 ± 5.4	70.6 ± 15.4	100	32.2 ± 4.3	2.4 ± 5.4
Broadkill SFR	3.6 ± 4.5	102.7 ± 19.3	100	19.1 ± 4.1	0
Mullica UFR	51.7 ± 14.7	49.3 ± 5.1	—	107.5 ± 22.1	—
Broadkill UFR	26.2 ± 7.3	85.4 ± 15.9	—	129.0 ± 24.0	—
Mullica HA + SFR	—	—	—	—	23.6
Broadkill HA + SFR	—	—	—	—	19.8

expect salt extract (SFR) recovery equal to observed estuarine DOC removal, which was at least 60 and 12% in the Mullica and Broadkill, respectively. Instead, only a quarter of that amount was recovered (Table 9.4).

It is probable that by removing suspended material prior to mixing with seawater, the effectiveness of humic extraction was significantly reduced. Preston and Riley demonstrate humic adsorption to suspended clays increases with increasing salinity primarily through ion exchange.[87] Although increased humic removal was not quantified, it was clearly enhanced. The upper half of Figure 9.6 illustrates mean humic acid carbon values at different salinities taken from mixes of Mullica river water and seawater. Particulates were not removed prior to mixing. Clearly, carbon is removed as particulate organic carbon (POC) increases. The increase in suspended carbon equals approximately 34% of the DOC (Figure 9.6). This is approximately double the humic acid carbon removed (Figure 9.6) and is consistent with the proportion of high-molecular-weight organic carbon extracted from the Mullica (Table 9.4). Clearly, other organic carbon besides humic acid is removed in the Mullica. The amount of DOC removed in the mixing experiment, however, only accounts for half the organic carbon removed from the Mullica Estuary observed in the August 1980 sampling. It is possible some other, as yet undetermined, removal mechanism is operating in the Mullica Estuary, but considering the results of the extraction experiments and field observations from November 1979 (Table 9.1), it seems more probable that the large removal observed in 1980 was anomalous, and that removal of organic carbon in the Mullica is generally between 30 and 40% DOC.

KINETICS

Figures 9.6, 9.7, and 9.8 illustrate the results of kinetic experiments using waters from two salt marsh rivers, the Broadkill and St. Jones, and the Mullica River. River water and seawater were mixed to desired salinities, and both POC and humic acid carbon were measured periodically for 3 hours. From Figure 9.6, it is clear that humic acid carbon, averaged over 3 hours, is transferred to the POC pool in amounts which increase with salinity to 12 ppt. However, kinetic order was difficult to precisely determine because humic acid was not continuously removed with time. Instead, it oscillated between input and removal. This was especially pronounced in the higher salinity samples of the Mullica (Figure 9.7). Despite scatter, the Broadkill and St. Jones Rivers generally followed linear humic acid removal with time (Figure 9.8). The zero-order rate constants maximized at approximately 5 ppt salinity. Rate constants are illustrated as a function of salinity in Figure 9.9, where they are also compared to those calculated from data given by Sholkovitz.[58] Apparently, removal maximizes at low salinities (4–8 ppt), although the specific salinity at maximum varies among estuaries. The kinetic order of Mullica removal could not be determined because of scatter (Figure 9.7).

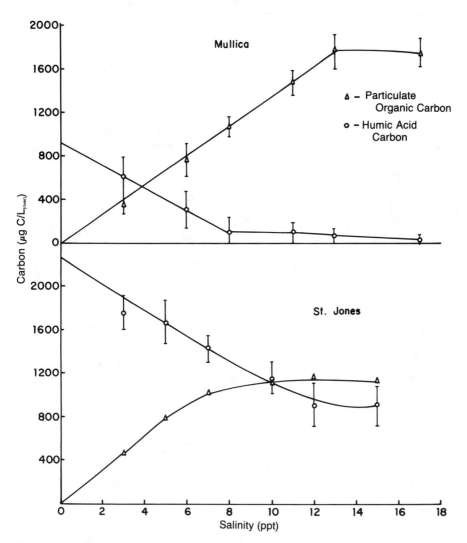

Figure 9.6. Particulate organic carbon and humic acid carbon from river water–seawater mixes plotted as a function of salinity. Each symbol represents the mean of eight samples taken over 3 hours. POC values were normalized to the POC concentration of the freshwater sample. Error bars represent standard deviations.

It appears that the size distribution of humic acid dispersions oscillates for hours after the dispersion has been perturbed. The results of exclusion chromatography indicate that aquatic humic size distributions are at least biomodal and tend to fluctuate with pH and salinity.[88,89] Thus, equilibrium between the various sizes of colloidal humic dispersions is easily perturbed and slow to reestablish. Recurrent initial removal over the first 15–30 minutes after seawater is added in the kinetic experiments (especially at the higher salinities in the

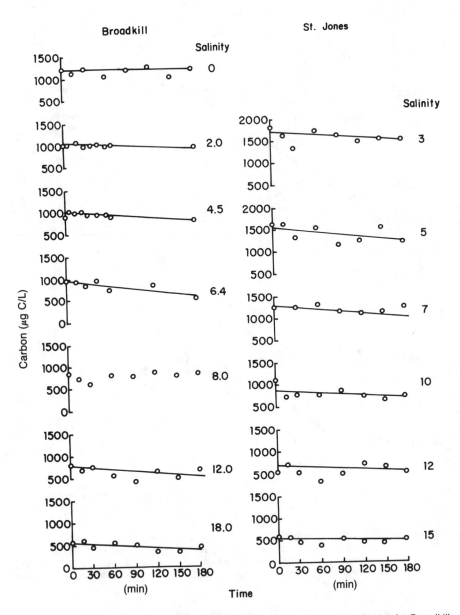

Figure 9.7. Humic acid carbon plotted as a function of time at various salinities for Broadkill and St. Jones river water–seawater mixes. Solid lines represent the least-squares linear fit to the data. The 8-ppt salinity sample from the Broadkill was not given a solid line because its slope was zero.

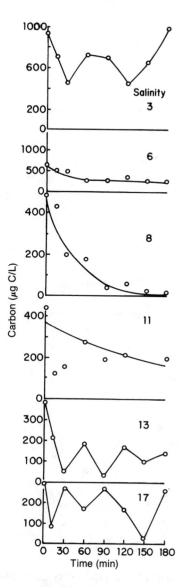

Figure 9.8.   Humic acid carbon plotted as a function of time at various salinities for the Mullica river water–seawater mix. Solid lines in panels of 6, 8, and 11 ppt salinity demonstrate the least-squares exponential best fit to the data. Solid lines in panels of 3, 13, and 17 ppt salinity merely connect data points. Zero-to second-order kinetics give an equally poor fit to the data.

Mullica — see Figure 9.7)) suggests that there may never be sufficient time for establishment of equilibrium in rapidly mixing estuaries, and that removal is kinetically controlled.

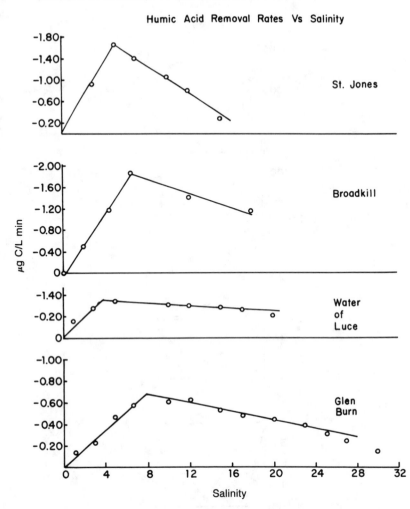

Figure 9.9.   The zero-order humic acid removal rate constant is plotted as a function of salinity. Rate constants were determined from Figure 9.8 and Sholkovitz.[58]

METALS

Complexation or sorption of a variety of trace elements to aquatic humic substances has been studied.[90] Specifically, Fe, Co, Pb, Cd, Cu, Zn, Ni, Hg, UO_2, Ag, Al, Sb, and even Au are reported to be complexed with some form of natural humic substance.[5-10] The behavior of humic-metal complexes in estuaries can have a direct influence on the distribution and deposition of the elements in question. Elements strongly associated with humics will follow the fate of the organic matter to which they are bound and will remain unreactive in estuarine waters. However, complexation with aquatic humics increases the mobility of many elements, thereby enhancing their exposure to the food chain after they are released in estuarine waters or sediments. The speciation of elements in estuaries is difficult to measure directly because of vanishingly small concentrations — and equally hard to calculate, with practical certainty, because of changing pH and ionic strength levels as well as possibly significant influences from uncharacterized organic matter. Thus, most of the information regarding the chemical state of trace elements in estuaries is limited to their physicochemical behavior — specifically, whether or not they are removed and whether the removal is commensurate with humic material. Powell et al. report Fe, Al, and Cu to be organically bound in waters of the Ochlockonee Estuary, while Mn, Zn, Cd, and Ni were not.[91] Fe and Al were rapidly removed before 10 ppt salinity, but Mn, Zn, and Cd were conserved. Organic sorption was deduced from extraction with anionic exchange resin.[91] Sholkovitz observed removal of humic acid, Fe, Mn, and Al from river water–seawater mixes of waters from four Scottish rivers, although chemical complexation between the metals and humic acid could not be deduced from the experiment.[58] Preston and Riley report that fairly unstable Al-humate complexes form in freshwater, but none are observed at salinities above 3 ppt.[87]

Amorphous iron concentrations are far in excess of amorphous ferric hydroxide solubility in oxic waters due to formation of colloidal suspensions which are stabilized by sorption with organic matter. These suspensions aggregate in salt solutions[67,92] following classical coagulation theory,[83] thus producing the estuarine removal of suspended ferric iron so often reported.[58,59,73,93-99]

Aquatic humic substances have long been associated with dispersed iron.[89,90,100-102] Humic extracts are reported to complex suspended iron through coordination of carboxyl functionalities,[103] resulting in more negative electrophoretic mobilities[65,104,105] and competitive inhibition of phosphate complexation.[106,107] Thus, we would expect good correlation between dissolved humic carbon and suspended iron in estuaries. However, humic acid carbon does not correlate well with estuarine iron.[62] In addition, the kinetics of humic acid aggregation clearly differ from those of suspended iron, which follows a second-order path with rates that increase as a continuous function of salinity.[62] This, in conjunction with other evidence, suggests that humic acid is not chemically bound with iron.[62]

However, organic nitrogen coextracted with humic acid and salt extract correlates well ($r = .98$) with suspended iron in the Mullica Estuary (Figure 9.10). The organic matter represented by these fractions is designated acid- and salt-extracted amino acid and total extracted nitrogen. However, amino acid is deduced from fluorometric analysis of primary amine after acid hydrolysis (Appendix). Although cleavage of amide linkages is the most likely source, some caution must be used when interpreting these results as coming exclusively from protein hydrolysis. Nonproteinaceous nitrogen compounds, such as aminoglucans and aromatic amines, are found in humic materials.[43,108,111] Nevertheless, primary amine represents 70% of the organic nitrogen in both extracts.

Tables 9.6 and 9.7 show that 47 and 51% of the fluorometrically determined

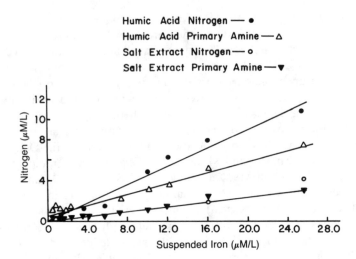

Figure 9.10. Organic nitrogen coextracted with humic acid and salt extract from the Mullica Estuary is plotted as a function of suspended iron. Solid lines represent the linear least-squares fit to the data. Correlation coefficients were .98.

amine in the Mullica river was directly measured as amino acid. Thus, greater than one-third of the extract nitrogen can be directly attributed to protein hydrolysis.

Proteineous organics coexist in humic substances extracted from both soils and river waters. Haworth identified polypeptides that could be extracted from soil humics with warm water, as well as those so strongly bound that they could only be separated by acid hydrolysis.[108] He also noted commensurate separation of iron and aluminium. Biederbeck and Paul report 53% of soil humic nitrogen is hydrolyzable, with 36% as recoverable amino acid.[109] Lytle and Perdue report greater than 96% of the hydrolyzable amino acids in the Williamson River, Oregon, were coextracted with humic materials on XAD-7 macroreticular resin.[112] Furthermore, the compositional hierarchy of individual amino acids from this river and the Satilla River, Georgia,[19] are practically identical to those of the Mullica.

The amino acid composition of humics extracted from the rivers listed in Table 9.7 is surprisingly constant considering the variation in geographic region—and hence source of humic material—represented by these watersheds. It is possible that organic sorption with suspended iron is mediated by iron-oxidizing bacteria. Under these conditions, the nature of sorbed organics would be expected to be more uniform than if sorption occurred by random reaction with DOC.

The apparent ubiquitous presence of acidic protein moieties in humic extracts, as well as the correlation between acid- and salt-extracted amine and suspended iron, suggests that a large portion of organic matter sorbed to suspended ferric hydroxide in estuaries is composed of acidic polypeptides.

Table 9.6. The Distribution of Amino Acids in the Mullica River (Concentrations are Given in μmol N)

Amino Acid	Total Riverine Amino Acid	Acid Extract[a]	Salt Filter Residue
Cysteic acid[b]	0.22	0.07	0.07
Aspartic acid	0.36	0.42	0.35
Threonine[b]	0.19	B.D.[c]	B.D.
Serine	0.24	0.28	0.32
Glutamic acid	0.23	0.25	0.28
Glycine[d]	0.57	0.56	0.11
Alanine[d]	0.30	0.24	B.D.
Valine[d]	0.12	0.14	B.D.
Methionine	B.D.	B.D.	B.D.
Isoleucine	0.08	0.10	0.07
Leucine	0.12	0.14	0.14
Tyrosine	0.04	0.04	0.04
Phenylalanine	0.05	0.07	0.07
Histidine[b]	0.02	0.06	0.11
Lysine	0.09	0.10	0.11
Arginine	0.03	B.D.	B.D.
Proline[b]	0.15	B.D.	B.D.
Ammonia	6.81	1.42	0.87

[a]Humic acid.
[b]Amino acids in riverine sample with concentrations different from humic acid and salt filter residue.
[c]B.D. = below detection.
[d]Amino acids concentrations different in humic acid than salt filter residue.

The nature and degree of their chemical interaction with humic extracts are unclear. However, approximately 30% of the acid- and salt-extracted nitrogen from the Mullica River could not be identified as primary amine, despite fairly good correlation with suspended iron (Figure 9.10). This could be the result of hydrolysis inefficiency or an indication of other organic nitrogen compounds. Hydrolysis efficiencies were tested on serum albumin (Appendix), but some polypeptides may resist acid attack when intertwined with humic polymers. Therefore, the best estimate of the proportion of proteineous material found in the organics associated with suspended iron is between 40 and 70%.

When measured, the hydrolyzable amino acid portion of riverine DOC is small, about 5%. Nevertheless, a relatively small proportion of acidic proteineous organic matter may well be all that is required to stabilize suspended ferric hydroxide. From Figure 9.10, a mass ratio for ferric iron:amino acid

Table 9.7. Order of Abundance of Amino Acids Making Up >60% of the Total

Mullica River Sample	Acid Extract[a]	Salt Extract	Rivers from Beck et al.[19]	River from Lytle and Perdue[112]
Glycine	Glycine	Aspartic acid	Glycine	Glycine
Aspartic acid	Aspartic acid	Serine	Aspartic acid	Aspartic acid
Alanine	Serine	Glutamic acid	Alanine	Alanine
Serine	Glutamic acid		Serine	Serine
Glutamic acid			Threonine	Glutamic acid

[a]Humic acid.

carbon of 4.7 can be calculated. If it is assumed that the mass of total amino acid is double the amino acid carbon, then ~ 1.4 mg of ferric iron could be stabilized by 0.6 mg of humic protein.

SUMMARY

Humic substances are defined as moderately-high-molecular-weight, alkali-soluble polyelectrolytes. They are further subdivided into humic and fulvic acids according to their solubility in acid solution. Humic acids are insoluble at low pH. Humic substances comprise 40–90% of terrestrially derived dissolved organic carbon in rivers. The chemical composition of these substances include major aliphatic and alicyclic components with basic aromatic structural components which are probably lignin residues. Most functional groups of these substances are carboxyl and hydroxyl functionalities.

Aquatic humic substances are observed to precipitate in estuaries. Primarily, humic acid is the portion of DOC removed, although input and conservative behavior have been observed. Enormous inputs are observed in the Orinoco Estuary. On average, approximately 20% of riverine DOC is precipitated in estuaries, although the range is 4–60% depending on the characteristics of the watershed in question.

Classical mechanics of colloidal aggregation and dispersion appear to explain humic acid removal in estuaries. Humic colloids undergo surface charge reduction through complexation with alkali earth cations in waters of increasing salinity. Charge reduction results in conformational shrinking, elimination of hydrated water, and subsequent hydrophobic colloidal aggregation through Van der Waals attraction.

The kinetics of humic acid aggregation appear to be complex. Zero- to second-order kinetics could be assigned to removal reactions in mixtures of river water and seawater. Measurements are scattered due to oscillation in the size distribution of suspended humic acid after salinity is increased. However, averaged rate constants increase with salinity to a maximum of 4–12 ppt.

Compositional studies of humic material extracted from the Mullica and Broadkill Rivers by acidification, salting out, and size exclusion demonstrated that

1. protein and carbohydrate comprise 5–24% of the total extract carbon with the high-molecular-weight extract from the Mullica having the lowest percentage and humic acid extracted from the Broadkill, the highest
2. the percentage of aromatic carbon in all extracts was 18 ± 1 except in the salt extract from the Broadkill
3. the 60–92% uncharacterized carbon in each extract contained a large abundance of methyl, methylene, and hydroxyl protons and — except for a pronounced presence of carboxyl, alkene, and/or N-acetyl groups in the Broadkill — all extracts were generally similar in terms of organic composition

Physicochemical differentiation of each extract suggested humic acid and SFR carbon overlapped by less than 25% and that the sum of humic acid and SFR carbon equaled the high-molecular-weight carbon extracted. Indirect evidence suggested SFR extract recovery would be 4 times larger if suspended particulates were not removed prior to mixing. Compositional and kinetic studies suggested 1.5–3 times the carbon removed as humic acid is actually precipitated during estuarine mixing.

Many metals have been reported complexed to aquatic humics in freshwaters, although iron and perhaps aluminium and copper are the only ones reported to coprecipitate with humics in estuaries. Iron, although rapidly removed in estuarine waters (as is humic acid), does not correlate well with humic acid carbon along a salinity gradient, and the kinetics of iron removal are different from those of humic acid removal. Instead, suspended iron and organic nitrogen, coprecipitated with humic acid, correlate well. The organic nitrogen extract consists of 70% primary amine, suggesting that a large abundance of proteineous material is sorbed to suspended iron. High abundances of aspartic acid and glutamic acid in the nitrogeneous extract suggest iron sorption may be the result of complexation through carboxyl functionalities.

ACKNOWLEDGMENTS

The author thanks Jon Sharp for providing laboratory facilities and advice for this research as well as George Aiken for a critical review of the manuscript.

APPENDIX: METHODS OF ANALYSIS FOR VARIOUS ORGANICS EXTRACTED FROM WATERS OF THE MULLICA ESTUARY

Study Area

The Mullica drainage basin encompasses 1475 km² of the Atlantic coastal plain in New Jersey.[101] The river is characterized by high organic matter and suspended iron concentrations.[93,96,99,101] High suspended iron values are the result of abundant limonite deposits, produced by bacterially mediated oxidation of ferrous iron leached from subsoils by acidic groundwaters.[113] High humic–organic matter abundances are the result of pine forest vegetation.[61,82]

The Mullica Estuary begins in marshlands and progresses into a large, shallow embayment, Great Bay. The bay is bound on the ocean side by a barrier bar island, which focuses communication with the ocean through a narrow inlet. Sampling was confined to the lower-salinity two-thirds of the estuary, where terrestrial organic carbon and iron are removed.

Analysis

A flow diagram of the preparation and treatment of each water sample taken during a continuous transect of the Mullica Estuary is given in Figure 9.11. Onsite sample preparation consisted of filtration through prebaked (450–500°C) Whatman, GFC filters (nominal pore size, 1.2 μ). The filter residues were dessicated and returned to the laboratory for analysis of particu-

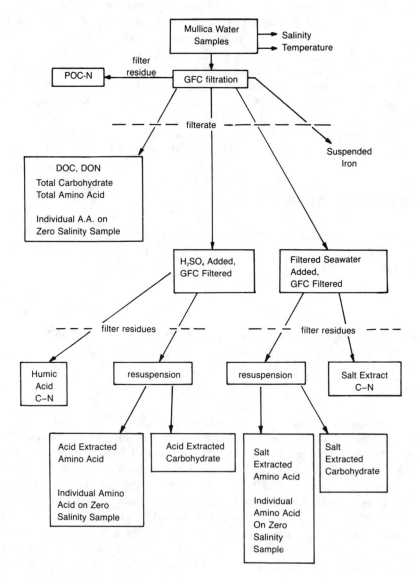

Figure 9.11. Flow diagram of sample treatment.

late organic carbon and nitrogen (POC-PON). The filtrates were returned to the laboratory for analysis of dissolved organic carbon, nitrogen, total carbohydrate, total amino acid, suspended ferrous and ferric iron. Likewise, carbon, nitrogen, amino acid, and carbohydrate from humic acid and carbon, nitrogen, amino acid, and carbohydrate from the salt extract, both taken from the filtrate, were measured.

Salt extracts were obtained by mixing the estuarine sample filtrate with filtered, offshore seawater in proportions to achieve a constant salinity of 20 ppt. The procedure was done to examine organic matter naturally aggregated by seawater during estuarine mixing, thus enabling comparison with humic acid (acid-extracted carbon) from the same waters (Figure 9.11).

Particulate organic carbon was analyzed both by high-temperature oxidation in a Hewlett Packard CHN analyzer[64] and by a modified persulfate oxidation method.[63,77] Dissolved organic carbon (DOC) was determined by infrared analysis after persulfate oxidation.[63,77] Dissolved organic nitrogen was determined after Solorzano and Sharp.[114]

Acid- (humic acid) and salt-extracted carbon was determined both by high-temperature oxidation[84] and by persulfate oxidation. Nitrogen was determined by high-temperature oxidation. Results of persulfate oxidation and high-temperature oxidation were within 20%, with no systematic variation.

Total hydrolyzable carbohydrate was determined spectrophotometrically by complexation of 3-methyl-2-benzothiazolinone hydrochloride (MBTH) with formaldehyde, which was produced by aldol cleavage with periodic acid after carbohydrate hydrolysis and reduction of resulting monosaccharides.[115,116]

Acid- and salt-extracted hydrolyzable carbohydrate was determined spectrophotometrically with MBTH after modified sample preparation. After sample acid or sample seawater mixes were filtered, filters were transferred to dry, prebaked ampules where one mL of 1 M HCl and 10 mL deionized water were added. The ampules were sealed and autoclaved for 4 h. Hydrolysis efficiency was measured and appears to be constant after 2 h. After cooling, ampules were opened, and 10 mL of sample were transferred to prebaked scintillation vials. Two mL of 0.45 M NaOH was added to neutralize the acid. Sample solutions were then reduced in the dark by addition of 0.1 mL 10% KBH_4, excess BH_4^- neutralized with HCl, and the resulting monosaccharides determined after centrifugation.[115,116]

Total amino acids were determined fluorometrically after acid hydrolysis and derivatization with fluorescamine.[117,118] Standards consisted both of a mixture of aspartic acid, alanine, and leucine, and of serum albumin. The method specifically measures primary amines. To arrive at a total amino acid carbon value, it is assumed each primary amine represents one amino acid with an average carbon ratio of 3:4. High proportions of basic amino acids would cause an overestimation of the total amino acid carbon values using this ratio. However, individual amino acid analysis indicates these samples contained a low proportion of basic amino acids.

Acid- (humic acid) and salt-extracted amino acid was determined fluorometrically with modified sample preparation. Extracts were collected on prebaked Whatman GFC filters as described above, and transferred to prebaked ampules with 5 mL deionized water and 5 mL concentrated HCl. The ampules were sealed and autoclaved for 4 h. After autoclaving, 10 mL of solution was transferred to a prebaked vial and evaporated. The residue was resuspended in 10 mL of 0.4 M borate buffer and centrifuged for 5 min. The supernatant was derivatized with fluorescamine, and the resulting fluorescence determined on an Aminoc-Bowman spectrophotofluorimeter at 390-nm excitation and 480-nm emission.

A single low salinity sample of Mullica river water was concentrated and acid-hydrolyzed, and amino acids were determined with ninhydrin after chromatographic separation; the low-salinity humic acid and salt extracts underwent the same process. Both fluorescamine and ninhydrin gave the same results on a nitrogen basis, provided ammonia was considered with the total ninhydrin amine. The ammonia was likely the result of deamination during evaporation. Ammonia was determined by endophenol complexation.[119]

Fulvic acid is defined here by fluorescence. It is given a carbon value by its linear relationship with the uncharacterized DOC in this watershed. Although fluorescence is a characteristic of humic substances, it varies with molecular weight[120] and pH value[121] and is, therefore, intended to be defined only for this watershed. In effect, fulvic acid is the uncharacterized dissolved carbon in the Mullica Estuary. Uncharacterized carbon is defined as the difference between DOC and the sum of total carbohydrate, amino acid, humic acid, and salt extract on a molar carbon basis.

Suspended iron was determined spectrophotometrically with ferrozine.[122] Salinity was determined conductiometrically with an induction salinometer.

REFERENCES

1. Reuter, J. H., and E. M. Perdue. "Importance of Heavy Metal Organic Matter Interactions in Natural Waters," *Geochim. Cosmochim. Acta* 41:325–334 (1977).
2. Thurman, E. M., and R. L. Malcolm. "Structural Study of Humic Substances: New Approaches and Methods," in *Aquatic and Terrestrial Humic Materials*, R. F. Christman and E. T. Gjessing, Eds. (Ann Arbor, MI: Ann Arbor Science, 1983), pp. 1–24.
3. Ertel, J. R., J. I. Hedges, A. H. Devol, and J. E. Rickey. "Dissolved Humic Substances of the Amazon River System," *Limnol. Oceanogr.* 31:739–754 (1986).
4. Mayer, L. M. "Geochemistry of Humic Substances in Estuarine Environments," in *Humic Substances in Soils, Sediments, and Water*, G. R. Aiken, D. M. McKnight, R. L. Wershaw, and P. MacCarthy, Eds. (New York: Wiley-Interscience, 1985), p. 211.
5. Matson, W. R., H. E. Allen, and P. Rekshan. "Trace-Metal-Organic Complexes in the Great Lakes," American Chemical Society, Div. Water, Air Waste Chem., Gen. Paper (1969), pp. 164–168.
6. Pauli, F. W. "Heavy-Metal Humates and Their Behavior against Hydrogen Sulphide," *Soil Sci.* 119:98–105 (1975).

7. Nissenbaum, A., M. J. Baedecker, and I. R. Kaplan. "Studies on Dissolved Organic Matter from Interstitial Waters of a Reducing Marine Fjord," in *Advances in Organic Geochemistry*, H. R. von Gaertner and H. Wehner, Eds. (Oxford: Pergamon Press, 1971), pp. 427–440.
8. Curtin, G. C., H. W. Lakin, and A. E. Hubert. "The Mobility of Gold in Mull (Forest Humus Layer)," U.S. Geological Survey Prof. Paper 700-C:127–129 (1970).
9. Ramamoorthy, S., and D. J. Kushner. "Heavy-Metal Binding Sites in River Water," *Nature* 256:399–401 (1975).
10. Boyle, R. W., W. M. Alexander, and G. E. M. Aslin. "Some Observations on the Solubility of Gold," Geological Survey Can. Paper 75-24 (1975).
11. Chiou, C. T., R. L. Malcolm, T. I. Brinton, and D. E. Kile. "Water Solubility Enhancement of Some Organic Pollutants and Pesticides by Dissolved Humic and Fulvic Acids," *Environ. Sci. Technol.* 20:502–508 (1986).
12. Means, J. C., and R. Wijayaratne. "Role of Natural Colloids in the Transport of Hydrophobic Pollutants," *Science* 215:968–970 (1982).
13. Wijayaratne, R., and J. C. Means. "Affinity of Hydrophobic Pollutants for Natural Estuarine Colloids in Aquatic Environments," *Environ. Sci. Technol.* 18:121–123 (1984).
14. Mukherjee, P. N., and A. Lahiri. "Rheological Properties of Humic Acid from Coal," *J. Colloid Sci.* 11:240–243 (1956).
15. Stevenson, F. J., and J. H. A. Butler. "Chemistry of Humic Acids and Related Pigments," in *Organic Geochemistry*, G. Eglinton and M. T. J. Murphy, Eds. (New York: Springer-Verlag, 1969), pp. 534–557.
16. Schnitzer, M., and S. U. Khan. *Humic Substances in the Environment* (New York: Marcel Dekker, 1972).
17. Stevenson, F. J., and K. M. Goh. "Infrared Spectra of Humic Acids and Related Substances," *Geochim Cosmochim. Acta* 35:471–483 (1971).
18. Thurman, E. M., and R. L. Malcolm. "Preparative Isolation of Aquatic Humic Substances," *Environ. Sci. Technol.* 15:463–466 (1981).
19. Beck, K. C., J. H. Reuter, and E. M. Perdue. "Organic and Inorganic Geochemistry of Some Coastal Plain Rivers of the Southeastern United States," *Geochim. Cosmochim. Acta* 38:341–364 (1974).
20. Black, A. P., and R. F. Christman. "Characteristics of Colored Surface Water," *Journal AWWA* 55:753–770 (1963).
21. Dubach, P., and N. C. Mehta. "The Chemistry of Soil Humic Substances," *Soils and Fertilizers* 26:293–300 (1963).
22. Gjessing, E. T. *Physical and Chemical Characteristics of Aquatic Humus* (Ann Arbor, MI: Ann Arbor Science, 1976).
23. Lamar, W. L. "Evaluation of Organic Color and Iron in Natural Surface Waters," U.S. Geological Survey Prof. Paper 600-D (1968), pp. D24-D29.
24. Martin, D. F., M. T. Doig, and R. H. Pierce. "Distribution of Naturally Occurring Chelators (Humic Acids) and Selected Trace Metals in West Coast Florida Streams," Florida Dept. Nat. Resources Mar. Res. Lab., St. Petersburg, Prof. Paper Series 12 (1971).
25. Otsuki, A., and T. Hanya. "Some Precursors of Humic Acid in Recent Lake Sediments Suggested by Infrared Spectra," *Geochim. Cosmochim. Acta* 31:1505–1515 (1967).
26. Otsuki, A., and T. Hanya. "Fractional Precipitation of Humic Acid from Recent

Sediment by Use of *N,N*-Dimethylformamide and Its Infrared Absorption Spectra," *Nature* 212:1467–1563 (1966).

27. Rashid, M. A. "Role of Humic Acids of Marine Origin and Their Different Molecular Weight Fractions in Complexing Di- and Trivalent Metals," *Soil Sci.* 111:298–306 (1971).

28. Rashid, M. A., and J. D. Leonard. "Modifications in the Solubility and Precipitation Behavior of Various Metal Salts as a Result of Their Interaction with Sedimentary Humic Acids," *Chem. Geol.* 11:89–97 (1973).

29. Schnitzer, M., and S. I. M. Skinner. "The Peracetic Acid Oxidation of Humic Substances," *Soil Sci.* 118:322–331 (1974).

30. Steelink, C. "What is Humic Acid?" *J. Chem. Ed.* 40:379–384 (1963).

31. Stuermer, D. H., and J. R. Payne. "Investigation of Seawater and Terrestrial Humic Substances with Carbon-13 and Proton Nuclear Magnetic Resonance," *Geochim. Cosmochim. Acta* 40:1109–1114 (1976).

32. Wagner, G. H., and F. J. Stevenson. "Structural Arrangement of Functional Groups in Soil Humic Acid as Revealed by Infrared Analysis," *Soil Sci. Soc. Proc. Am.* 29:43–48 (1965).

33. Visser, S. A., and H. Mendel. "X-Ray Diffraction Studies on the Crystallinity and Molecular Weight of Humic Acids," *Soil Biol. Biochem.* 3:259–265 (1971).

34. Wilson, S. A., and J. H. Weber. "A Comparative Study of Number-Averaged Dissociation Corrected Molecular Weights of Fulvic Acids Isolated from Water and Soil," *Chem. Geol.* 19:285–293 (1977).

35. Reuter, J. H., and E. M. Perdue. "Calculation of Molecular Weights of Humic Substances from Colligative Data: Application to Aquatic Humus and Its Molecular Size Fractions," *Geochim. Cosmochim. Acta* 45:2017–2022 (1981).

36. Aiken, G. R., and R. L. Malcolm. "Molecular Weight of Aquatic Fulvic Acids by Vapor Pressure Osmometry," *Geochim. Cosmochim. Acta* 51:2177–2184 (1987).

37. Flaig, W. A. J., and H. Beutelspacher. "Investigations of Humic Acids with the Analytical Ultracentrifuge," in *Isotopes and Radiation in Soil Organic-Matter Studies* (Vienna: International Atomic Energy Agency, 1968), pp. 20–30.

38. Aiken, G. R. "Evaluation of Ultrafiltration for Determining Molecular Weight of Fulvic Acid," *Environ. Sci. Technol.* 18:978–981 (1984).

39. Ruggiero, P., F. S. Interesse, L. Cassidei, and O. Sciacovelli. "H-NMR Spectra of Humic and Fulvic Acids and Their Peracetic Oxidation Products," *Geochim. Cosmochim. Acta* 44:603–609 (1980).

40. Grant, D. "Chemical Structure of Humic Acid," *Nature* 270:709–710 (1977).

41. Neyrand, J. A., and M. Schnitzer. "The Chemistry of High Molecular Weight Fulvic Acid Fractions," *Can. J. Chem.* 52:4123–4132 (1974).

42. Hansen, E. M., and M. Schnitzer. "The Alkaline Permanganate Oxidation of Danish Illuvial Organic Matter," *Soil Sci. Soc. Am. Proc.* 30:745–748 (1966).

43. Lawson, G. J., and J. W. Purdie. "Chemical Constitution of Coal. 4. Examination of Sub-Humic Acids Produced by Ozonization of Humic Acid," *Fuel* 45:115–129 (1966).

44. Liao, W., R. F. Christman, J. D. Johnson, D. S. Millington, and J. R. Hass. "Structural Characterization of Aquatic Humic Material," *Environ. Sci. Technol.* 16:403–410 (1982).

45. Wilson, M. A., R. P. Philip, A. H. Gillam, and K. R. Tate. "Comparison of the Structures of Humic Substances from Aquatic and Terrestrial Sources by Pyroly-

sis Gas Chromatography–Mass Spectrometry," *Geochim. Cosmochim. Acta* 47:497–502 (1983).

46. Hatcher, P. G., I. A. Breger, and M. A. Mattringly. "Structural Characteristics of Fulvic Acids from Continental Shelf Sediments," *Nature* 285:560–562 (1980).

47. Hedges, J. I., and P. L. Parker. "Land-Derived Organic Matter in Surface Sediments from the Gulf of Mexico," *Geochim. Cosmochim. Acta* 40:1019–1029 (1976).

48. Pempkowiak, J., and R. Pocklington. "Phenolic Aldehydes as Indicators of the Origin of Humic Substances in Marine Environments," in *Aquatic and Terrestrial Humic Materials*, R. F. Christman and E. T. Gjessing, Eds. (Ann Arbor, MI: Ann Arbor Science, 1983), pp. 371–386.

49. Brown, M. "Transmission Spectroscopy Examination of Natural Waters, Part C," *Estuar. Coast. Mar. Sci.* 5:309–317 (1977).

50. Kalle, K. "The Problem of Gelbstoff in the Sea," *Oceanogr. Mar. Biol. Ann. Rev.* 4:91–104 (1966).

51. Sieburth, J. M., and A. Jensen. "Studies on Algal Substances in the Sea. 1. Gelbstoff (Humic Material) in Terrestrial and Marine Waters," *J. Exp. Mar. Biol. Ecol.* 2:174–189 (1968).

52. Swanson, V. E., and J. G. Palacas. "Humates in Coastal Sands of Northwest Florida," U.S. Geological Survey Bull. 1214-B: B1-B29 (1965).

53. Gardner, W. S., and D. W. Menzel. "Phenolic Aldehydes as Indicators of Terrestrially Derived Organic Matter in the Sea," *Geochim. Cosmochim. Acta* 38:813–822 (1974).

54. Shultz, D. J., and J. A. Calder. "Organic $^{13}C/^{12}C$ Variations in Estuarine Sediments," *Geochim. Cosmochim. Acta* 40:381–385 (1976).

55. Swain, F. M. "Biogeochemistry of Humic Compounds," *Proc. Intern. Meeting Humic Substances* (Nieuwersluis, Pudoc, Wageningent, 1972).

56. Hair, M. E., and C. R. Bassett. "Dissolved and Particulate Humic Acid in an East Coast Estuary," *Estuar. Coast. Mar. Sci.* 1:107–111 (1973).

57. Martin, D. F., and R. H. Pierce, Jr. "A Convenient Method of Analysis of Humic Acid in Fresh Water," *Environ. Lett.* 1:49–52 (1971).

58. Sholkovitz, E. R. "Flocculation of Dissolved Organic and Inorganic Matter during the Mixing of River Water and Sea Water," *Geochim. Cosmochim. Acta* 40:831–845 (1976).

59. Sholkovitz, E. R., E. A. Boyle, and N. B. Price. "The Removal of Dissolved Humic Acids and Iron during Estuarine Mixing," *Earth Planet. Sci. Lett.* 40:130–136 (1978).

60. Sharp, J. H., C. H. Culberson, and T. M. Church. "The Chemistry of the Delaware Estuary. General Considerations," *Limnol. Oceanogr.* 27:1015–1028 (1982).

61. Fox, L. E. "The Removal of Dissolved Humic Acid during Estuarine Mixing," *Estuar. Coast. Shelf Sci.* 16:431–440 (1983).

62. Fox, L. E. "The Relationship between Dissolved Humic Acids and Soluble Iron in Estuaries," *Geochim. Cosmochim. Acta* 48:879–884 (1984).

63. Sharp, J. H. "The Total Organic Carbon in Seawater – Comparison of Measurement Using Persulfate Oxidation and High Temperature Combustion," *Mar. Chem.* 1:211–231 (1973).

64. Sharp, J. H. "Improved Analysis for 'Particulate' Organic Carbon and Nitrogen from Sea Water," *Limnol. Oceanogr.* 19:984–989 (1974).

65. Tipping, E., and D. Cooke. "The Effects of Adsorbed Humic Substances on the

Surface Charge of Goethite in Fresh Water," *Geochim. Cosmochim. Acta* 46:75–80 (1982).

66. Mantoura, R. F. C., A. Dickson, and J. P. Riley. "The Complexation of Metals with Humic Materials in Natural Waters," *Estuar. Coast. Mar. Sci.* 6:387–408 (1978).

67. Eckert, J. M., and E. R. Sholkovitz. "The Flocculation of Iron Aluminium and Humates from River Water by Electrolytes," *Geochim. Cosmochim. Acta* 40:847–848 (1976).

68. Eastman, K. W., and T. M. Church. "Behavior of Iron, Manganese, Phosphate and Humic Acid during Mixing in a Delaware Salt Marsh Creek," *Estuar. Coast. Mar. Sci.* 18:447–458 (1984).

69. Mantoura, R. F. C., and E. M. S. Woodward. "Conservative Behavior of Riverine Dissolved Organic Carbon in the Severn Estuary: Chemical and Geochemical Implications," *Geochim. Cosmochim. Acta* 47:1293–1309 (1983).

70. Lopez, J. C. S. "Condiciones Hidrogeoquimicas de la Region Estuarina– Deltaica del Orinoco Durante El Mes de Noviembre de 1985," PhD Thesis, Universidad de Oriente, Instituto Oceanografico de Venezuela, Cumaná, Venezuela (1989).

71. Meybeck, M. "Carbon, Nitrogen, and Phosphorus Transport by World Rivers," *Am. J. Sci.* 282:401–450 (1982).

72. Blough, N. Y., B. Dister, and O. C. Zafiriou. "Optical Absorption Spectra of Waters from the Orinoco River Outflow: Terrestrial Input of Colored Organic Matter to the Caribbean," *EOS* 71:171 (1990).

73. Moore, R. M., J. D. Burton, P. J. L. Williams, and M. L. Young. "The Behavior of Dissolved Organic Material, Iron, and Manganese in Estuarine Mixing," *Geochim. Cosmochim. Acta* 43:919–926 (1979).

74. Officer, C. B. "Discussion of the Behavior of Non-Conservative Dissolved Constituents in Estuaries," *Estuar. Coast. Mar. Sci.* 9:91–94 (1979).

75. Means, J. L., R. F. Yuretich, D. A. Crerar, D. J. Kinsman, and M. P. Borcsik. *Hydrogeochemistry of the New Jersey Pine Barrens*, New Jersey Geological Survey Bull. 76 (Trenton, NJ: Department of Environmental Protection, 1981).

76. Stallard, R. F., and J. M. Edmond. "Geochemistry of the Amazon. 2. The Influence of Geology and Weathering Environments on the Dissolved Load," *J. Geophys. Res.* 88:9671–9688 (1983).

77. Menzel, D. W., and R. F. Vaccaro. "The Measurement of Dissolved Organic Carbon and Particular Carbon in Sea Water," *Limnol. Oceanogr.* 3:138 (1964).

78. Suzuki, Y., and Y. Sugimura. "A Catalytic Oxidation Method for the Determination of Total Nitrogen Dissolved in Sea Water," *Mar. Chem.* 16:83–97 (1985).

79. Sugimura, Y., and Y. Suzuki. "A High-Temperature Catalytic Oxidation Method for the Determination of Nonvolatile Dissolved Organic Carbon in Sea Water by Direct Injection of a Liquid Sample," *Mar. Chem.* 24:105–131 (1988).

80. Sharp, J. H. Personal communication (1990).

81. Means, J. C., and R. D. Wijayaratne. "Chemical Characterization of Estuarine Colloidal Organic Matter: Implications for Adsorptive Processes," *Bull. Mar. Sci.* 35:449–461 (1984).

82. Harvey, G. R., D. A. Boran, S. R. Piotrowicz, and C. P. Weisel. "Synthesis of Marine Humic Substances from Unsaturated Lipids," *Nature* 309:244–246 (1984).

83. Ong, H. L., and R. E. Bisque. "Coagulation of Humic Colloids by Metal Ions," *Soil Sci.* 106:220-224 (1968).

84. Fox, L. E. "Geochemistry of Humic Acid during Estuarine Mixing," in *Aquatic and Terrestrial Humic Materials*, R. F. Christman and E. T. Gjessing, Eds. (Ann Arbor, MI: Ann Arbor Science, 1983), pp. 407-427.

85. Fox, L. E. "The Geochemistry of Humic Acid and Iron during Estuarine Mixing," PhD Thesis, University of Delaware, Newark, DE (1981).

86. Tipping, E., and M. Ohnstad. "Aggregation of Aquatic Humic Substance," *Chem. Geol.* 44:349-357 (1984).

87. Preston, M. R., and J. P. Riley. "The Interactions of Humic Compounds with Electrolytes and Three Clay Minerals under Simulated Estuarine Conditions," *Estuar. Coast. Mar. Sci.* 14:567-576 (1982).

88. Gjessing, E. T. "Use of Sephadex Gel for the Estimation of Molecular Weight of Humic Substances in Natural Waters," *Nature* 208:1091-1092 (1965).

89. Ghassemi, M., and R. F. Christman. "Properties of the Yellow Organic Acids of Natural Waters," *Limnol. Oceanogr.* 13:583-597 (1968).

90. Jackson, K. S., I. R. Jonasson, and G. B. Skippen. "The Nature of Metals-Sediment-Water Interactions in Freshwater Bodies, with Emphasis on the Role of Organic Matter," *Earth-Science Rev.* 14:97-146 (1978).

91. Powell, R. T., B. L. Lewis, and W. M. Landing. "Trace Metal Speciation in an Organic Rich Estuary," *EOS* 71:99 (1990).

92. Matsunaga, K., K. Igarashi, S. Fukase, and H. Tsubota. "Behavior of Organically Bound Iron in Sea Water of Estuaries," *Estuar. Coast. Shelf Sci.* 18:615-622 (1984).

93. Coonley, S. S., E. B. Baker, and H. D. Holland. "Iron in the Mullica River and in Great Bay, N.J.," *Chem. Geol.* 7:51-63 (1971).

94. Aston, S. R., and R. Chester. "The Influence of Suspended Particles on the Precipitation of Iron in Natural Waters," *Estuar. Coast. Mar. Sci.* 1:225-231 (1973).

95. Liss, P. S. "Conservative and Non-Conservative Behavior of Dissolved Constituents during Estuarine Mixing," in *Estuarine Chemistry*, J. D. Burton and P. S. Liss, Eds. (London: Academic Press, 1976), pp. 93-127.

96. Boyle, E. A., J. M. Edmond, and E. R. Sholkovitz. "The Mechanism of Iron Removal in Estuaries," *Geochim. Cosmochim. Acta* 41:1313-1324 (1977).

97. Santschi, P. H., H. Y. Li, and S. R. Casson. "The Fate of Trace Metals in Narragansett Bay, Rhode Island: Radio-Trace Experiments in Microcosms," *Estuar. Coast. Mar. Sci.* 10:635-654 (1980).

98. Mayer, L. H. "Aggregation of Colloidal Iron during Estuarine Mixing: Kinetics, Mechanism, and Seasonality," *Geochim. Cosmochim. Acta* 46:2527-2535 (1982).

99. Fox, L. E., and S. C. Wofsy. "Kinetics of Removal of Iron Colloids from Estuaries," *Geochim. Cosmochim. Acta* 47:211-216 (1983).

100. Shapiro, J. "Effect of Yellow Organic Acids on Iron and Other Metals in Water," *Journal AWWA* 56:1062-1082 (1964).

101. Crevar, D., J. L. Means, R. F. Yuretich, M. P. Borcsik, J. L. Amster, D. W. Hastings, G. W. Knox, K. E. Lyon, and R. F. Quiett."Hydrogeochemistry of the New Jersey Coastal Plain. 2. Transport and Deposition of Iron, Aluminium, Dissolved Organic Matter, and Selected Trace Elements in Steam, Ground, and Estuarine Water," *Chem. Geol.* 33:23-44 (1981).

102. Francois, R. "Sedimentary Humic Substances: Structure, Genesis, and Properties," *Rev. Aquat. Sci.* 3:41–80 (1990).

103. Senesi, N., S. M. Griffith, M. Schnitzer, and M. G. Townsend. "Binding of Fe^{+3} by Humic Materials," *Geochim. Cosmochim. Acta* 41:969–976 (1977).

104. Tipping, E. "The Adsorption of Aquatic Humic Substances by Iron Oxides," *Geochim. Cosmochim. Acta* 45:191–199 (1981).

105. Tipping, E. "Some Aspects of the Interactions between Particulate Oxides and Aquatic Humic Substances," *Mar. Chem.* 18:161–169 (1986).

106. Carpender, P. D., and J. D. Smith. "Effect of pH, Iron and Humic Acid on the Estuarine Behavior of Phosphate," *Environ. Tech. Lett.* 6:65–72 (1984).

107. Young, T. C., and W. G. Comstock. "Direct Effects and Interactions Involving Iron and Humic Acid during Formation of Colloidal Phosphorus," in *Sediments and Water Interactions*, P. G. Sly, Ed. (New York: Springer-Verlag, 1986), pp. 461–470.

108. Haworth, R. K. "The Chemical Nature of Humic Acid," *Soil Sci.* 111:71–79 (1971).

109. Biederbek, V. O., and E. A. Paul. "Fractionation of Soil Humate with Phenolic Solvents and Purification of Nitrogen-Rich Portion with Polyvinylpyrrolidone," *Soil Sci.* 115:357–366 (1973).

110. Ogner, G. "Fractionation of Humus Hydrolysates by Ion Exchange Resins," *Soil Sci.* 110:86–92 (1970).

111. Stepanov, V. V. "Reaction of Humic Acids with Some Nitrogen Containing Compounds," *Soviet Soil Sci.* 2:167–173 (1969).

112. Lytle, C. R., and E. M. Perdue. "Free, Proteineous, and Humic-Bound Amino Acids in River Water Containing High Concentrations of Aquatic Humus," *Environ. Sci. Tech.* 15:224–228 (1981).

113. Crerar, D., G. W. Knox, and J. L. Means. "Biogeochemistry of Bog Iron in the New Jersey Pine Barrons," *Chem. Geol.* 24:111–135 (1979).

114. Solorzano, L., and J. H. Sharp. "Determination of Dissolved Organic Nitrogen in Natural Waters," *Limnol. Oceanogr.* 25:751–754 (1980).

115. Burney, C. M., and J. M. Sieburth. "Dissolved Carbohydrates in Seawater. 2. A Spectrophotometric Procedure for Total Carbohydrate Analysis and Polysaccharide Estimation," *Mar. Chem.* 5:15–28 (1977).

116. Johnson, K. M., and J. M. Sieburth. "Dissolved Carbohydrates in Seawater. 1. A Precise Spectrophotometric Analysis from Monosaccharides," *Mar. Chem.* 5:1–13 (1977).

117. Bohlen, P., S. Stein, W. Parnian, and S. Udenfriend. "Fluorometric Assay of Protein in the Nanogram Range," *Arch. of Bioch. Biophys.* 155:213–220 (1973).

118. North, B. "Primary Amines in California Coastal Waters: Utilization by Phytoplankton," *Limnol. Oceanogr.* 20:20–27 (1975).

119. Strickland, J. D. H., and T. R. Parsons. *A Practical Handbook of Seawater Analysis* (Ottawa, Canada: Fisheries Research Board, 1972).

120. Levesque, M. "Fluorescence and Gel Filtration of Humic Compounds," *Soil Sci.* 113:346–353 (1972).

121. Christman, R. F., and M. Ghassemi. "Chemical Nature of Organic Color in Water," *Journal AWWA* 58:723–741 (1966).

122. Stookey, L. L. "Ferrozine – New Spectrophotometric Reagent for Iron," *Anal. Chem.* 42:779 (1970).

CHAPTER 10

The Hydrolysis of Suwannee River Fulvic Acid

Ronald C. Antweiler

INTRODUCTION

Fulvic and humic acids are operational terms for naturally occurring, complex organic substances that occur ubiquitously in nature, and whose chemical structures have not been determined. Although these materials have long been studied, little is adequately known about them, mostly because they defy separation into pure components,[1] even though intricate separation techniques have been devised.[2] Further, until about ten years ago, almost all of the research on humic materials was done on soil and sediment humic acids. In the early 1980s, with the advent of adequate techniques for isolating and concentrating aquatic humic substances,[3,4] an inevitable question arose concerning the similarity between these and humic materials extracted from soils or sediments. Chiefly because of the need to have consistency, the International Humic Substances Society (IHSS) established reference materials, not as "standards" in the sense of being representative, but as common materials on which various people worldwide could conduct experiments and have a common point of reference.

Because of these factors, the IHSS stream standard fulvic acid from the Suwannee River was chosen as the focus of study to redetermine such simple characteristics as titratable functionality.[5] Many previous authors using various humic materials noted that the pH of humic and fulvic acids tended to decrease upon standing in alkaline solution.[6-9] The reasons given for this decrease have been numerous. Bremner[10] and Swift and Posner[11] noted that oxidation occurred at high pH, but since these works many authors have taken meticulous care to ensure the absence of oxygen and CO_2, and still the consumption of hydroxide has been noted. Breger suggested that the consumption of hydroxide could be caused by the splitting of lignin fragments within humic molecules.[12] Mikita noted that in alkali, humic substances formed stable free radicals and that these could consume hydroxide.[13] Davis and Mott proposed that consumption was caused by the degradation of polysaccharide moieties within the fulvic acid, involving a "peeling" process.[14] Davis and Mott[14] and

163

Stevenson[15] suggested that fulvic acids may contain enolizable ketones, and these in turn could undergo a slow tautomerization, which could account for the consumption. Paxeus and Wedborg suggested the possibility that the slow hydroxide loss could be nothing more than acidic deprotonation, but that it was slow because of diffusion problems caused by the size and shape of the molecules.[16] Periodically researchers have failed to notice a pH drift and have ascribed its presence in others' work to artifacts of technique.[17,18] However, Bowles et al. suggested that the consumption was at least partially caused by the hydrolysis of ester moieties in the fulvic acid,[5] a hypothesis which has been suggested recently by others as well.[19,20] In this chapter, I reaffirm this hypothesis and present additional confirmatory evidence for it being at least partially responsible for the alkali loss.

MATERIALS AND METHODS

Materials

The fulvic acid used was the IHSS reference fulvic acid concentrated from the Suwannee River, Georgia. Details of its concentration, isolation, and purification can be found in Malcolm et al.[21] The aspirin (acetylsalicylic acid), coumarin, coumarin 3-carboxylic acid, glucuronolactone, chlorogenic acid, potassium acid phthalate (KHP), coumalic acid, 2-benzofuran carboxylic acid, sodium hydroxide (NaOH), and hydrochloric acid (HCl) were all Aldrich reagent grade.

Equipment

The titrations were carried out on a Radiometer system which included the following items: pH glass electrode model G2040C, reference electrode model K4040, pH meter model PHM84, and automatic titrator model TTT80. The system included the ABU80 autoburette, with accuracy of ± 0.25 μL; the REC80 servograph recorder; an REA260 derivation unit, which plots the first derivative of pH vs titrant volume; an REA270 pH stat unit, which plots volume of titrant consumed at a constant pH as a function of time; and a TTA80 titration assembly with built-in stir bar. This assembly allowed the sample to be isolated in a sample chamber where ultrapure N_2 gas could be bubbled through; the sample chamber was immersed inside a constant temperature bath (all experiments were conducted at $25 \pm 0.2°C$).

Infrared spectra were determined on a Perkin Elmer 580 infrared spectrophotometer, with resolutions of 3.7 cm^{-1} for a scan time of 16 min. Samples were cast films on AgCl plates, requiring only 0.25 mg of material for a spectrum, and were prepared by evaporation of the sample under N_2.

Methods

Samples were accurately weighed (± 0.2 mg) and transferred to a sample vessel where distilled deionized water was added. The sample was sealed from the atmosphere and allowed to stand for 20 min to equilibrate with both the constant temperature bath and the N_2 atmosphere before any experiments were begun. The NaOH and HCl were standardized against predried KHP by rapidly titrating 20 times to achieve accurate statistics. These standards were recalibrated daily. In addition, the flow of N_2 into the sample chamber was constantly regulated to the same value, as was the rate of stirring of the sample, and the temperature was checked daily. Also, the pH electrodes and instrument response time were rigorously examined every morning against buffered solutions at pH 6.86 ± 0.01 (at 25°C), 4.01 ± 0.01 (at 25°C) and 10.00 ± 0.02 (at 25°C). These solutions were also used to check the electrodes before and after every experiment.

The manual titration was accomplished at the overall rate of 2.5 μL of 0.1 N NaOH per minute. In the buffered pH regions (below pH 5 and above pH 11), 40 μL of base were added in an aliquot, and the solution was let stand for 16 min before the next aliquot of base was added (40 μL/16 min = 2.5 μL/min). In the pH range 5 to 10, the amount of base added was decreased to 20 μL (corresponding to an 8-min delay before a fresh aliquot was added), and finally to 10 μL (with a corresponding 4-min delay). The recorder was set to monitor the pH vs time. The entire procedure typically took about 6 hours per sample. This type of titration is referred to in the text as "discontinuous titration."

The pH stat experiments were accomplished as follows. After the sample had equilibrated with N_2 and with the constant temperature bath, it was rapidly titrated with base to the target pH (typically pH 10). Then, the automatic titrator was set to add only sufficient base to hold the solution there. The recorder was set to monitor the amount of base added as a function of time. pH stat experiments were always run for at least 15 hours.

For selected experiments, an aliquot of solution was taken before and after, and from these aliquots cast-film plates were made (by evaporating the aliquots under N_2). These cast-film plates were subsequently run on the infrared spectrophotometer.

RESULTS

Following earlier results of Bowles et. al,[5] two initial experiments were performed on the Suwannee River fulvic acid (SFA). In the first, the discontinuous titration experiment (Figure 10.1), SFA was titrated as described above: successive aliquots of base were added to achieve an overall titration rate of 2.5 μL/min (= 0.25 μM/min); the total time required to titrate the 10.1 mg of SFA was 352 min. The expected result was for the curve to display a stair-step

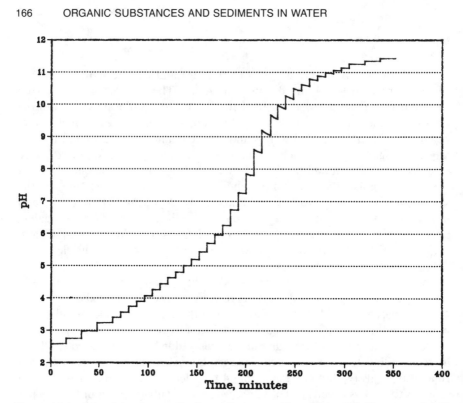

Figure 10.1. The discontinuous titration of Suwannee River fulvic acid. Titrant was 0.1 N NaOH; titration rate was 2.5 μL/min.

pattern throughout the entire experiment, as shown in Figure 10.2 for the discontinuous titration of potassium acid phthalate (KHP). We would expect the pH to remain constant in between additions of base, or, if anything, for the pH to rise slightly due to diffusion from the burette tip; this assumes that the kinetics of deprotonation are rapid, which is a reasonable assumption. But, as can be seen in Figure 10.1, from about pH 5.5 to pH 11.1 there is a decrease in the pH between successive additions of base, creating a sawtooth pattern, and implying that throughout this entire range, hydroxide is being consumed. The greatest drop between successive base additions occurs near pH 9.2 — in the 8 min between alkali additions the pH dropped from 9.20 to 9.05. Closer inspection shows that the decrease is not linear — the greatest rate of pH drop occurs initially, and the shape of the drop is concave upwards, implying that hydroxide is being consumed at a greater rate the higher the pH. To further study this phenomenon, we performed a second series of experiments on SFA, the pH stat experiments, to examine the rate at which base was being consumed at a given pH. In Figure 10.3 the results of pH 9 and 10 stat experiments are shown. Initially, hydroxide is consumed rapidly; with the passage of time, the rate of consumption decreases, approaching zero eventually. The pattern shown in this figure is typical, showing an exponential decrease in base con-

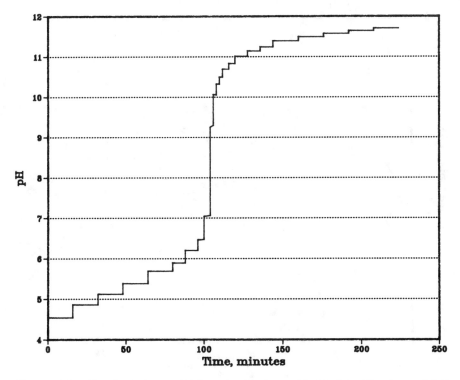

Figure 10.2. The discontinuous titration of potassium acid phthalate. Titrant was 0.1 N NaOH; titration rate was 2.5 μL/min.

sumption with respect to time. Again, the effect of pH is seen – at pH 10, the curve has leveled off by about 6 hours; at pH 9, the curve is not fully horizontal even after 20 hours.

We then conducted a series of pH stat experiments on various model compounds. In Figure 10.4, some of the results for glucuronolactone are shown for pH stat experiments at pH 8, 9, and 10. The shape of the curves are exponential and are strongly dependent upon the pH: each pH unit appears to speed up the rate of hydrolysis by a factor of about 10. In Figure 10.5, the results for pH 10 stat experiments on coumarin-3-carboxylic acid (C3C), aspirin, and 2-benzofuran carboxylic acid (2BF) are shown. Again, for aspirin and C3C, the shapes of the curves are exponential and represent the hydrolysis of the ester bond due to intramolecular catalysis.[22] 2-Benzofuran carboxylic acid does not consume any appreciable quantity of base, and therefore is not undergoing any hydrolysis at pH 10. Similar pH 10 stat experiments were conducted with coumarin, coumalic acid, chlorogenic acid, and 4-acetoxybenzoic acid. Of these, only chlorogenic acid does not consume hydroxide at pH 10; in all the others, the ester linkage is being hydrolyzed. It is clear that the shapes of the curves for SFA are similar to those for the hydrolyzing esters.

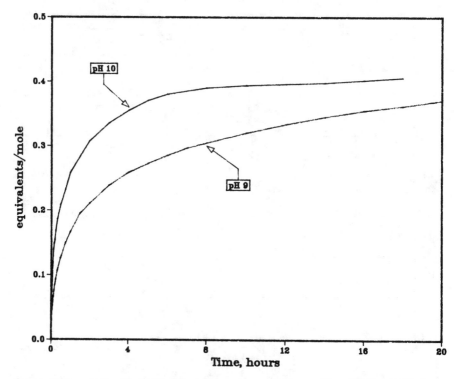

Figure 10.3.   pH 9 and pH 10 stat experiments on Suwannee River fulvic acid. Ordinate values were determined assuming a molecular weight of 800 for SFA. Titrant was 0.1 N NaOH.

This similarity strongly suggested that ester moieties within the SFA were undergoing hydrolysis. To substantiate this claim, we took aliquots of SFA before and after exposure to alkali, adjusting the pH of the latter to acid pH; these aliquots were evaporated under N_2 onto AgCl disks, and the resulting cast films were examined on an infrared spectrophotometer (IR). The IR absorbance spectra for the 2000 cm⁻¹ to 1200 cm⁻¹ region are shown in Figure 10.6, and as can be seen, the ester region around 1750 cm⁻¹ virtually disappears after exposure to base. We therefore believe that at least part of the hydroxide consumption is caused by ester hydrolysis. Further evidence for this has been offered by Wilson et. al.,[20] who showed via nuclear magnetic resonance spectroscopy (NMR) the progressive appearance with time of hydrolysis products as a sample of fulvic acid was exposed to base.

In an effort to further understand the process of ester hydrolysis, we conducted a battery of extended pH stat experiments, which for convenience we refer to as "hysteresis experiments," on the model esters listed in Table 10.1. In these, the following procedure was used:

1. The sample was titrated from its initial pH to pH 10 rapidly (approximately 2 min). In Tables 10.2 and 10.3, I stands for the number of equivalents (eq) per

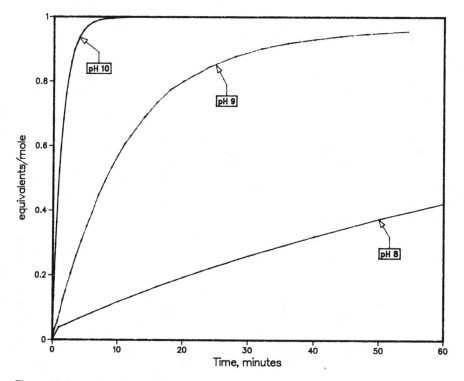

Figure 10.4.   pH 8, pH 9, and pH 10 stat experiments on glucuronolactone. Titrant was 0.1 N
NaOH.

mole used to titrate from a predetermined acidic pH (usually 4) to pH 10; it is
a measure of the number of acidic groups per molecule which deprotonate
between pH 4 and pH 10.

2. The sample was held at pH 10 for at least 15 hours. In Tables 10.2 and 10.3, S
   is the number of eq per mole of sample which were consumed during this
   entire time; it represents the number of esters per molecule which hydrolyze at
   pH 10. It must be noted that the electrodes showed a slow linear drift at pH 10
   which caused the addition of between 0.05 and 0.30 μeq/hr of base. S is
   therefore corrected for this drift.

3. The sample was back-titrated with acid to pH 4 and held there until no more
   acid was consumed (typically 1 hour). In Tables 10.2 and 10.3, B is the
   number of eq per mole of sample which were consumed in titrating from pH
   10 to pH 4 and holding the pH there; B represents the number of acidic groups
   (including phenols) per molecule which protonate between pH 10 and pH 4,
   plus the number of alcohol-acid groups per molecule which esterify (if any).

4. The sample was forward-titrated a second time to pH 10. In Tables 10.2 and
   10.3, F is the number of eq per mole of sample which were consumed in going
   from pH 4 to pH 10. F, like I, is a measure of the deprotonation of the acidic
   groups between pH 4 and 10.

5. The sample was held a second time at pH 10. In Tables 10.2 and 10.3, H is the
   number of eq per mole of sample which were consumed during this time. H,

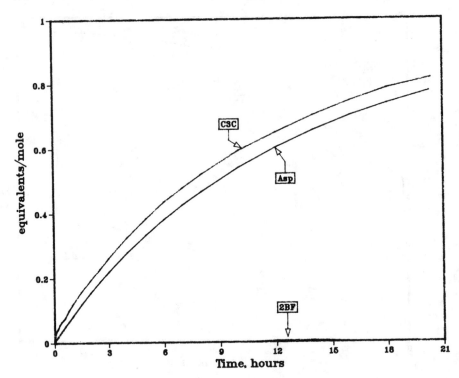

Figure 10.5.   pH 10 stat experiments on coumarin-3-carboxylic acid, aspirin, and 2-benzofuran carboxylic acid. Titrant was 0.1 N NaOH.

like S, is a gauge of hydrolysis. Again, its value is corrected for electrode drift.

The scheme developed in the hysteresis experiments therefore allows one to further characterize the hydrolysis which is occurring. There are several possibilities. First, if there were no hydrolysis at pH 10 (the case for 2-benzofuran carboxylic acid and chlorogenic acid—see Table 10.2), then the S and H columns read 0, and I = B = F (approximately). This is so because there is no slow consumption of base, and no irreversible reactions are occurring.

Second, if the hydrolysis were completely reversible, i.e., if the compound completely reesterified to the original ester upon back-titrating (the case for coumarin-3-carboxylic acid), then once again we have I = F (approximately)—the two forward titrations are identical. However, S and H do not equal zero because hydrolysis of the ester is slowly occurring. Further, B = I + S because the hydrolysis products completely reform to the original ester during the back titration.

Third, if the compound irreversibly hydrolyzes and there is no reesterification, then S > 0 (the hydrolysis of the ester consumes base during the pH stat experiment); H = 0, or is small compared to S, because the original ester has

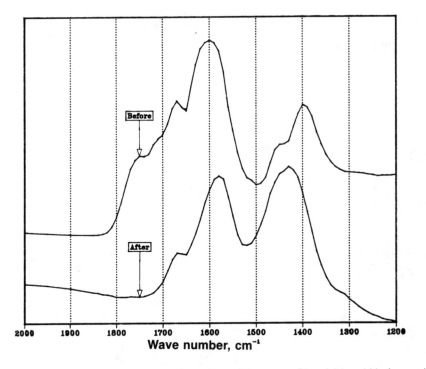

Figure 10.6.   IR absorbance spectra of cast films of Suwannee River fulvic acid before and after exposure to base. The pH of the solution of SFA after exposure to base was adjusted to acidic pH to minimize counterion effects. The ester stretching region is between 1710 cm^{-1} and 1780 cm^{-1}.

been destroyed leaving nothing to hydrolyze; and B = F approximately, because there are no irreversible reactions occurring between the end of the first pH 10 stat and the second forward titration to pH 10. This case is demonstrated by aspirin, coumarin, and glucuronolactone in Table 10.2.

Finally, if there is hydrolysis followed by esterification, but the new ester formed is not the same as the original ester, then S > 0 (hydrolysis occurs during the pH 10 stat); and B is not equal to F, because esterification during the back titration will consume acid in addition to the acid and phenol protonation. This case may be demonstrated by coumalic acid in Table 10.2.

In Table 10.3, the data from the hysteresis experiments on the SFA are presented, with the same column headings as in Table 10.2. Because S > 0, we have concluded that hydrolysis is occurring, but since B is virtually equal to F and H = 0, we believe that no irreversible reactions are occurring after the pH 10 stat, that is, no esterification; the reactions mimic the behavior of aspirin, coumarin, and glucuronolactone and indicate that irreversible hydrolysis is occurring. This analysis describes fulvic acid as if it were a discrete compound; however, the hysteresis experiments imply that at least the majority of hydrolyzable esters in fulvic acid are undergoing irreversible hydrolysis. Figure 10.7

Table 10.1. Model Compounds Used in the pH Stat Experiments

Compound	Elemental Formula	M.W.	Structural Formula	Approx. Sol. in H_2O (mg/L)
Acetyl salicylic acid (Asp)	$C_9H_8O_4$	180.2		3000
Coumalic acid (CA)	$C_5H_4O_4$	140.1		900
Coumarin-3-carboxylic acid (C3C)	$C_{10}H_6O_4$	190.2		350
2-Benzofuran-carboxylic acid (2BF)	$C_9H_6O_3$	162.1		570
Chlorogenic acid (Chl)	$C_{16}H_{18}O_9$	354.3		1700
Coumarin (Cou)	$C_9H_6O_2$	146.2		700
Glucuronolactone (Glu)	$C_6H_8O_6$	176.1		>2000

Note: The letters in parentheses are the shorthand notation followed in the text and in Table 10.2 to refer to these compounds.

shows for SFA one of the hysteresis experiments, graphically indicating the similarity between the back titration and the second forward titration, and the lack of similarity between these and the initial forward titration.

DISCUSSION

Another possible explanation for the data presented here is the keto-enol tautomerization hypothesis proposed by Davis and Mott;[14] not only would this hypothesis explain the slow base consumption, but it would also demonstrate

Table 10.2. Results of the Hysteresis Experiments on the Model Compounds Listed in
Table 10.1

Compound	I	S	B	F	H
2BF	0.14	0.00	0.13	0.18	0.00
Chl	1.30	0.00	1.30	1.28	0.00
C3C	1.11	0.91	1.96	1.16	0.32
	0.43	0.86	1.36	0.40	0.77
	0.35	0.89	1.28	0.35	0.64
Asp	0.34	0.76	0.87	—	—
	0.28	0.77	0.75	0.82	0.11
Glu	0.00	1.00	0.69	0.70	0.00
	0.00	0.99	0.18	0.17	0.00
Cou	0.04	1.75	1.53	1.53	0.00
CA	0.96	0.28	1.27	1.00	0.00
	0.72	1.12	1.71	0.75	0.03
	0.24	1.04	1.48	0.51	0.08

Note: The units are the number of equivalents of hydroxide per mole of material. I = initial
forward titration, S = pH 10 state, B = back titration, F = second forward titration, H =
second pH 10 stat. For a detailed explanation of these headings, refer to the text.

IR spectral changes similar to those seen in Figure 10.6. However, Leenheer et
al. have shown via sodium borohydride reduction that SFA contains no appreciable amounts of tautomerizable ketone moieties, indicating that this hypothesis cannot adequately explain the data.[23]

It is worthwhile to briefly reinterpret Figure 10.3 in terms of the hypothesis
of ester hydrolysis. In Figures 10.3 to 10.5, the ordinate is equivalents/mole,
which is independent of both the strength of the base and of the amount of
sample used. For model compounds, the ordinate value is easily calculated; for
SFA, however, an assumption of the average molecular weight is necessary,
and, following Aiken et al.,[24] we have assumed the molecular weight to be 800.
Once this number is decided, the ordinate value becomes a measure of the
number of moles of ester destroyed per mole of sample. In Figures 10.4 and
10.5, the curves level off at (or approach) an ordinate value of 1, correspond-

Table 10.3. Results of the Hysteresis Experiments on the Suwannee River Fulvic Acid

pH stat	I	S	B	F	H
9.0	2.43	0.42	—	—	—
10.0	2.96	0.45	3.08	2.81	0.0
10.0	2.86	0.45	3.18	2.90	0.0
10.0	3.00	0.51	3.19	2.98	0.0
10.0[a]	2.19	0.54	2.65	3.07	0.0
10.1	3.38	0.44	3.42	3.24	0.0

Notes: A molecular weight of 800 is assumed for SFA. The units are the number of equivalents
of hydroxide per mole of SFA. I = initial forward titration, S = pH 10 stat, B = back
titration, F = second forward titration, H = second pH 10 stat. For a detailed
explanation of the headings, refer to the text.
[a]Sample was re-freeze-dried prior to the experiment.

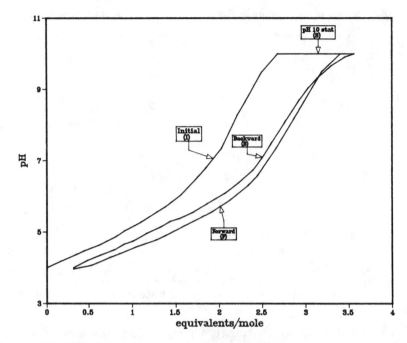

Figure 10.7. Hysteresis experiment on Suwannee River fulvic acid demonstrating irreversible hydrolysis. Forward titrant was 0.1 N NaOH; backward titrant was 0.1 N HCl. Abscissa values assume a molecular weight of SFA of 800.

ing to total hydrolysis of the esters. For SFA, the curves level off at about 0.4, which implies that up to about 0.4 ester moieties per molecule are hydrolyzable in the pH range of 9 to 10.

At the risk of being accused of unjustified speculation, we may theorize somewhat on the implications of ester hydrolysis on the nature of the functional groups in fulvic acid. In general, the ester bond is chemically stable and requires rather severe conditions to hydrolyze, much more severe than the experimental conditions used in this chapter. The type of ester which will hydrolyze under mildly basic conditions tends to be either phenolic or lactonic. Reversible hydrolysis almost always requires a cyclical structure and an incentive to reesterify, e.g., coumarin-3-carboxylic acid. Because SFA undergoes hydrolysis but does not appear to undergo appreciable reesterification, and because Wilson et al. noticed the appearance of hydrolysis products,[20] we may therefore speculate that the hydrolyzable ester moieties within fulvic acid are nonreversible phenolic in nature, similar to aspirin. It must be stressed that this remains conjecture—we have made several leaps of faith.

One of the more persistent arguments against the ester hydrolysis hypothesis is the observation that humic and fulvic acids are isolated by exposure to strong alkali, especially in the case of soil and sediment humic substances. If there were any esters originally present, the argument goes, they would have long ago hydrolyzed, thereby being absent to hydrolyze again. This certainly may be the case for soil and sediment humic substances; however, the procedure for the isolation of aquatic humic substances requires only very short exposure times to strong alkali, and realistically one might conclude that not all of the original esters were destroyed in the isolation procedure. In addition, there persists the possibility of esterification, even though it was not observed in the current set of experiments — Bowles et al. found some evidence of esterification by observing a slow consumption of acid at low pH values.[5] In our experiments, the low pH range was only 4; perhaps esterification requires pH 1 or 2 to occur. Finally, there is the possibility that the freeze-drying of the humic/fulvic solution (the last step in the aquatic isolation procedure) enhances the formation of esters. Regardless of these observations, the ester hypothesis suffers no more than most of the other proposals from the exposure-to-strong-base argument — if, in fact, the slow base consumption by humic substances is not an artifact of technique, and the process is not reversible, then the isolation procedure must not have had sufficient time to destroy whatever is causing the phenomenon, and it matters little whether that be stable free radicals, or lignin fragment degradation or amino acid hydrolysis.

The above discussion is indicative of the problems faced by the scientist: ultimately, the problem returns to our basic definitions of humic and fulvic acids. Because these definitions are operational, and because the method of isolation may alter the material, we need both to be conscious of the possible changes we have made by our isolation procedure, and at the same time to be searching for better, less-destructive isolation techniques — a point that has been made by others as well (e.g., Aiken[25]). One possible avenue of inquiry is to examine filtered whole water samples from locations where there is sufficient DOC, such as the Suwannee River. This clearly has the disadvantage of including all portions of the DOC together, but nevertheless allows one to study unaltered material. Another possibility for aquatic materials is to develop methods which never expose humic substances to concentrated acid or base — for example, adapting the method of Gregor and Powell.[26]

This chapter has demonstrated by IR spectroscopy and pH stat experiments the presence of esters in the IHSS fulvic acid; those who hope to extrapolate what is occurring in situ must bear in mind that the material they are using in the laboratory is not equivalent to what occurs "out there," though it may be similar enough. Of paramount importance for humic substances research is the necessity for detailing carefully the isolation and experimental procedures used, and developing a less-destructive, uniformly accepted procedure which all researchers can use.

DISCLAIMER

The use of trade names is for identification purposes only and does not constitute endorsement by the U.S. Geological Survey.

ACKNOWLEDGMENT

I am indebted to E. C. Bowles, P. MacCarthy, J. A. Leenheer, R. L. Wershaw, and G. R. Aiken for the many useful ideas and criticisms offered.

REFERENCES

1. MacCarthy, P., and J. A. Rice. "An Ecological Rationale for the Heterogeneity of Humic Substances: A Holistic Perspective of Humus," in *Proceedings of the Chapman Conference on the Gaia Hypothesis, San Diego, California, March 1988,* S. H. Schneider and P. J. Boston, Eds. (Cambridge, MA: MIT Press, 1990).
2. Leenheer, J. A., P. A. Brown, and T. I. Noyes. "Implications of Mixture Characteristics on Humic-Substance Chemistry," in *Aquatic Humic Substances: Influence on Fate and Treatment of Pollutants,* Advances in Chemistry Series 219, I. H. Suffet and P. MacCarthy, Eds. (Washington, DC: American Chemical Society, 1989), pp. 25-39.
3. Aiken, G. R., E. M. Thurman, R. L. Malcolm, and H. F. Walton. "Comparison of XAD Macroporous Resins for the Concentration of Fulvic Acid from Aqueous Solution," *Anal. Chem.* 51:1799-1803 (1979).
4. Thurman, E. M., and R. L. Malcolm. "Preparative Isolation of Aquatic Humic Substances," *Environ. Sci. Technol.* 15:463-466 (1981).
5. Bowles, E. C., R. C. Antweiler, and P. MacCarthy. "Acid-Base Titration and Hydrolysis of Fulvic Acid from the Suwannee River," in *Humic Substances in the Suwannee River, Georgia: Interactions, Properties and Proposed Structures,* R. C. Averett, J. A. Leenheer, D. M. McKnight, and K. A. Thorn, Eds., U.S. Geological Survey Open-File Report 87-557 (1989), pp. 205-229.
6. Martin, A. E., and R. Reeve. "Chemical Studies of Podzolic Illuvial Horizons: III. Titration Curves of Organic-Matter Suspensions," *J. Soil Sci.* 9:89-100 (1958).
7. Pommer, A. M., and I. A. Breger. "Equivalent Weight of Humic Acid from Peat," *Geochim. Cosmochim. Acta* 20:45-50 (1960).
8. Schnitzer, M., and J. G. Desjardins. "Molecular and Equivalent Weight of Organic Matter of a Podzol," *Soil Science Society of America Proceedings* 26:362-365 (1962).
9. Reuter, J. H., and E. M. Perdue. "Importance of Heavy Metal-Organic Matter Interactions in Natural Waters," *Geochim. Cosmochim. Acta* 41:325-334 (1977).
10. Bremner, J. M. "Some Observations on the Oxidation of Soil Organic Matter in the Presence of Alkali," *J. Soil Sci.* 1:198-204 (1950).
11. Swift, R. S., and A. M. Posner. "Autoxidation of Humic Acid under Alkaline Conditions," *J. Soil Sci.* 23:381-393 (1972).
12. Breger, I. A. "Chemical and Structural Relationship of Lignin to Humic Substances," *Fuel* 30:204-208 (1951).
13. Mikita, M. A. "Studies on the Structure and Reactions of Humic Substances," PhD Dissertation, University of Arizona, Tucson (1980).
14. Davis, H., and C. J. B. Mott. "Titrations of Fulvic Acid Fractions. II: Chemical Changes at High pH," *J. Soil Sci.* 32:393-397 (1981).
15. Stevenson, F. J. *Humus Chemistry: Genesis, Composition, Reactions* (New York: John Wiley and Sons, 1982).

16. Paxeus, N., and M. Wedborg. "Acid-Base Properties of Aquatic Fulvic Acid," *Anal. Chim. Acta* 169:87–98 (1985).
17. Borggaard, O. K. "Experimental Conditions Concerning Potentiometric Titration of Humic Acid," *J. Soil Sci.* 25:189–195 (1974).
18. Gregor, J. E., and H. K. J. Powell. "Protonation Reactions of Fulvic Acids," *J. Soil Sci.* 39:243–252 (1988).
19. Gregor, J. E., and H. K. J. Powell. "Effects of Extraction Processes on Fulvic Acid Properties," *Sci. Total Environ.* 62:3–12 (1987).
20. Wilson, M. A., P. J. Collin, R. L. Malcolm, E. M. Perdue, and P. Cresswell. "Low Molecular Weight Species in Humic and Fulvic Fractions," *Org. Geochem.* 12:7–12 (1988).
21. Malcolm, R. L., G. R. Aiken, E. C. Bowles, and J. D. Malcolm. "Isolation of Fulvic and Humic Acids from the Suwannee River," in *Humic Substances in the Suwannee River, Georgia: Interactions, Properties and Proposed Structures,* R. C. Averett, J. A. Leenheer, D. M. McKnight, and K. A. Thorn, Eds., U.S. Geological Survey Open-File Report 87-557 (1989), pp. 23–37.
22. Capon, B. "Intramolecular Catalysis," in *Essays in Chemistry,* Vol. 3, J. N. Bradley, R. D. Gillard, and R. F. Hudson, Eds. (London: Academic Press, 1972), pp. 127–156.
23. Leenheer, J. A., M. A. Wilson, and R. L. Malcolm. "Presence and Potential Significance of Aromatic Ketone Groups in Aquatic Humic Substances," *Org. Geochem.* 11:273–280 (1987).
24. Aiken, G. R., P. A. Brown, T. I. Noyes, and D. J. Pinkney. "Molecular Size and Weight of Fulvic and Humic Acids from the Suwannee River," in *Humic Substances in the Suwannee River, Georgia: Interactions, Properties and Proposed Structures,* R. C. Averett, J. A. Leenheer, D. M. McKnight, and K. A. Thorn, Eds., U.S. Geological Survey Open-File Report 87-557 (1989), pp. 163–179.
25. Aiken, G. R. "A Critical Evaluation of the Use of Macroporous Resins for the Isolation of Aquatic Humic Substances," in *Humic Substances and Their Role in the Environment,* F. H. Frimmel and R. F. Christman, Eds. (New York: John Wiley and Sons, 1988), pp. 15–31.
26. Gregor, J. E., and H. K. J. Powell. "Acid Pyrophosphate Extraction of Soil Fulvic Acids," *J. Soil Sci.* 37:577–585 (1986).

PART II

SORPTION INTERACTIONS WITH SOILS, SEDIMENTS, AND DISSOLVED ORGANIC MATTER

CHAPTER 11

Immobilization of Organic Contaminants by Organo-Clays: Application to Soil Restoration and Hazardous Waste Containment

Stephen A. Boyd, William F. Jaynes, and Brenda S. Ross

INTRODUCTION

The sorption of nonionic organic contaminants (NOCs) from water by soils and subsoils is controlled predominately by the organic matter fraction. The mineral phases of soils tend to be strongly hydrated in the presence of water. The preferential adsorption of water by minerals deactivates these surfaces as adsorbents of NOCs, which cannot effectively compete with water for the adsorption sites. As a result, NOCs in soil-water systems interact primarily with the organic matter phase, and the uptake of NOCs by soils is strongly correlated with the soil organic matter content.

The mechanistic function of soil organic matter in the sorption of NOCs from water can be described satisfactorily as a partition process.[1,2] This is a process of solubilization of the organic solute in the amorphous organic matter phase and is analogous to the partitioning of NOCs between water and an immiscible bulk organic solvent phase such as octanol or hexane. Although these processes are mechanistically similar, the solubility of NOCs in natural soil organic matter is approximately one-tenth as great as in typical hydrocarbon solvents such as octanol.

The sorptive uptake of NOCs from water by soils is characterized by the following properties that are consistent with the concept of solute partitioning:[3]

1. linear sorption isotherms extending to a high relative solute concentration, i.e., equilibrium concentration (C_e)/aqueous solubility
2. inverse dependence of the magnitude of the sorption coefficient on the water solubility of the solute
3. low and constant equilibrium heats of sorption
4. lack of evidence of competitive effects on sorption in multisolute systems

In contrast to solute partitioning, surface adsorptive mechanisms would be expected to display nonlinear isotherms, high exothermic heats of sorption, and adsorbate competition in multisolute systems.

The degree to which NOCs tend to associate with soil depends on

1. the amount of soil organic matter
2. the structure or solvency of the organic partition phase (which is relatively constant for natural soil organic matter)
3. the water solubility of the organic solute

As a result, NOCs of relatively high water solubility (ppm range) are relatively mobile in low organic matter soils or subsoils. The practical manifestation of this mobility is that these compounds are frequently found as groundwater contaminants and represent a serious threat to human health.

We have recently shown that the sorptive capabilities of naturally occurring clays for removing NOCs from water can be greatly improved by relatively simple chemical modifications that take advantage of the cation exchange properties of clays.[4-12] Most clay minerals possess a net negative charge due to isomorphous substitution in the aluminosilicate structure. In nature, this negative charge is neutralized by inorganic exchange ions such as Na^+ or Ca^{2+}. The presence of these strongly hydrated metal ions on the exchange sites of clays imparts a hydrophilic nature to the mineral surfaces. By simple ion exchange reactions, the naturally occurring inorganic exchange ions can be replaced by a variety of organic cations, thus changing the surface from hydrophilic to organophilic (as illustrated in Figure 11.1). As a result, the organo-clays formed may exhibit strong sorptive capabilities for removing NOCs from water.

In this chapter, we will summarize our previous research on the properties of organic cation-exchanged clays and soils as sorbents for NOCs that are commonly found as groundwater contaminants. This research has demonstrated that quaternary ammonium cations can be used to make stable organo-clay complexes, from soils or bentonite clays, that have high affinities for NOCs. The practical goal of this research has been to use organo-clays to prohibit the migration of NOCs through soils and subsoils as well as through clay containment barriers.

EXPERIMENTAL PROCEDURES

Details of the materials and methods have been given previously.[4-12] The general methods for preparation of the organo-clays and for obtaining the sorption data are summarized below.

Preparation of the Organo-Clays

The <2-μm clay size fractions were obtained by wet sedimentation, and subsequently Na^+, Mg^{2+}, or Ca^{2+} saturated, frozen, and freeze-dried. To

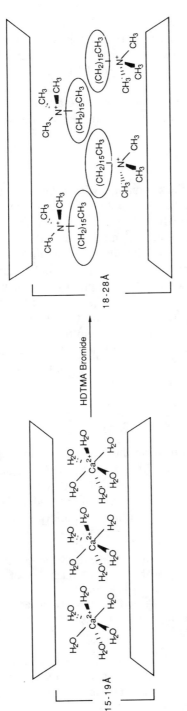

Figure 11.1. Schematic showing the ion exchange reaction of hexadecyltrimethylammonium (HDTMA) ions for Na$^+$ ions on smectite. The result is to change the nature of the clay surface from hydrophilic to organophilic.

prepare hexadecyltrimethylammonium (HDTMA) clays, aqueous HDTMA was added to an aqueous suspension of clay in an amount equivalent to the cation exchange capacity (CEC) of the clay. The clay was subsequently washed with distilled water, frozen, and freeze-dried. The tetramethylammonium (TMA) and trimethylphenylammonium (TMPA) clays were prepared as above except that the TMA and TMPA cations were added in a three- to five-fold excess of the CEC of the clay to ensure complete saturation of the CEC by the organic cations.

Sorption Isotherms

Sorption isotherms of the aromatic hydrocarbons were obtained by the batch equilibration technique. A quantity of organo-clay was weighed into screw-top Corex glass centrifuge tubes that were then filled with aqueous solutions of the test compound. A range of initial concentrations of each compound, up to about 70% of the aqueous solubility, was used. The tubes were closed with aluminum-foil-lined screw caps and agitated overnight on a reciprocating shaker. The solid and liquid phases were then separated by centrifugation, and a portion of the aqueous phase was extracted with hexane or carbon disulfide. The organic extractant was then analyzed using gas chromatography. Isotherms were constructed by plotting the amount sorbed vs the aqueous equilibrium concentration. The amount sorbed was calculated from the difference between the initial and equilibrium aqueous phase concentrations.

RESULTS AND DISCUSSION

The organic cations used to form the organo-clays were quaternary ammonium cations of the general form $[(CH_3)_3NR]^+$, where R is an alkyl or aromatic hydrocarbon. The structures of some of the quaternary ammonium cations (also referred to as QUATs) used are shown in Table 11.1. This class of organic cations offers several advantages that may be important in the eventual application of this technology:

1. They contain a permanent positive charge that is not pH dependent.
2. They are commercially available in large quantities at reasonable costs (approximately $0.50 to $1.00 per pound).
3. A wide variety of QUATs with different R groups are available, thereby offering synthetic versatility to produce organo-clays with different surface characteristics.

In addition, QUATs are already used widely in a variety of applications (e.g., detergents, fabric softeners, antistatic sprays, and swimming pool additives), which would, in general, suggest acceptable environmental compatibility.

In general, the organo-clays that we have studied (all derived from 2:1 clays) fall into two groups depending on the size or hydrophobicity of the so-called

Table 11.1. Quaternary Ammonium Cations Used to Prepare Organo-Clays

Name	Abbreviation	Structure
hexadecyltrimethyl ammonium	HDTMA	
tetramethyl ammonium	TMA	
trimethylphenyl ammonium	TMPA	
benzyltrimethyl ammonium	BTMA	
benzyltriethyl ammonium	BTEA	
benzyltributyl ammonium	BTBA	

hydrophobic tail, or R group, of the QUAT. The first class of modified clays is based on the use of QUATs with relatively large alkyl hydrocarbon R groups; these organo-clays are referred to as organophilic clays. Hexadecyltrimethylammonium (HDTMA) smectite (R is a C_{16} hydrocarbon) represents this class of organo-clays. Members of the second class of organo-clays are derived from smaller organic cations such as tetramethylammonium (TMA) or trimethylphenylammonium (TMPA) cations. These clays are functionally and mechanistically distinct from the former class of organo-clays, as will be described later in this chapter.

Organophilic Clays: HDTMA Smectite, Illite, and Vermiculite

Large organic cations such as HDTMA are very competitive exchange ions that will nearly stoichiometrically displace inorganic exchange ions such as Na^+ and Ca^{2+} from the exchange sites of clays. This is illustrated by the steep initial rise in the adsorption isotherm shown in Figure 11.2, which represents the ion exchange of HDTMA on a Ca-saturated smectite. This isotherm shows that the HDTMA is strongly adsorbed by the clay when added in an amount less than or equal to the cation exchange capacity of the clay.

The stoichiometric exchange of added HDTMA ions is a very important and beneficial characteristic for the commercial preparation of organo-clays, and in the application of this technology for soil modification (as discussed later). The most economical method for preparing organo-clays commercially is by

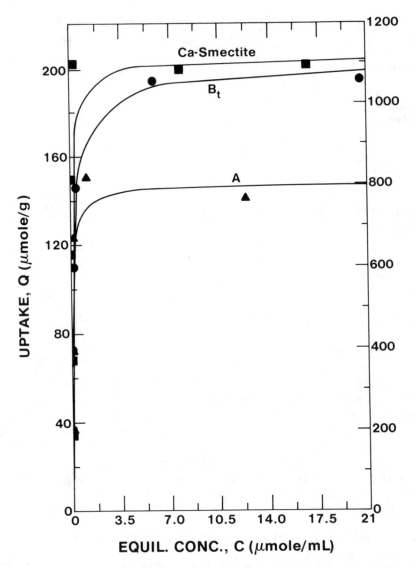

Figure 11.2. Adsorption isotherms showing the ion exchange of ^{14}C-labeled hexadecyltrimethylammonium cations on Ca^{2+} smectite (*right scale*), and on the Marlette soil A and B_t horizons (*left scale*). Reprinted from Boyd et al.[10] by permission from *Nature* 333:345–347. Copyright 1988, Macmillan Magazines, Ltd.

the so-called dry process, wherein the dry clay is mixed directly with the QUAT and extruded as a pastelike material that is subsequently dried and ground. In this process there is no separation of liquid and solid phases (which, if required, would be very costly) and thus no opportunity to remove any excess QUAT. A less competitive exchange ion (e.g., TMA) would need to be added

Table 11.2. Properties of Clay Minerals and Derived HDTMA Clays

Sample Designation	Mineralogy	Cation Exchange Capacity (meq/100 g)	Charge (mol⁻/unit cell)	d(001)[a] (Å)	Organic Matter (%)
VSC	Vermiculite with biotite and hydrobiotite, < 0.5 mm	80	1.32	28	20.48
SAz	High-charge smectite, < 2 μm	130	1.14	22.9	28.65
SWy	Low-charge smectite, < 2 μm	87	0.68	17.7	21.75
IMt	Illite with minor kaolinite and quartz, < 2 μm	24	1.44	10	10.54

Source: Jaynes and Boyd.[4]
[a]Refers to the HDTMA-exchanged clays.

in excess of the CEC to force the exchange reaction by mass action (otherwise the resulting organo-clay would be only partially saturated with the organic cation). If required, this would result in a dry-mixed product with nonexchanged (free) QUAT, which would obviously be undesirable in terms of cost and product quality.

A series of 2:1 clays were used to prepare homoionic HDTMA clays by adding HDTMA in an amount equivalent to the CEC of the clay.[4] Four of these clays are listed in Table 11.2, including a vermiculite (VSC), a high-charge smectite (SAz), a low-charge smectite (SWy), and an illite (IMt). The organic matter (HDTMA) contents of the HDTMA-exchanged clays ranged between 10 and 29% (w/w) and paralleled the CEC of the clays, which decreased in the order: SAz > SWy > VSC > IMt. The organic matter contents of these clays indicate complete adsorption of the added HDTMA, resulting in a complete saturation of the CEC by HDTMA cations. The basal spacings (distance between clay layers) of these HDTMA clays ranged from 10 Å (IMt) to 28 Å (VSC) and paralleled the unit charge of the expandable clays, which decreased in the order VSC > SAz > SWy. Illite is a nonexpandable clay; here HDTMA adsorption is limited to the external surfaces of the clay particle. Thus, the organic matter contents of HDTMA clays are determined by their CECs, and the interlayer distances by their unit charge densities.

Representative sorption isotherms of benzene, ethylbenzene, and naphthalene on an HDTMA smectite (SAz) are shown in Figures 11.3, 11.4, and 11.5. Sorption of these compounds by clay containing only inorganic exchange ions (e.g., Na^+, Ca^{2+}, Mg^{2+}) was negligible,[4] but for the HDTMA smectite very substantial sorption occurred. The isotherms shown in Figures 11.3, 11.4, and 11.5 for HDTMA smectite (SAz) are linear over a wide range of solute concentrations, suggesting that the solutes were partitioned from water into the HDTMA clay.

Highly linear isotherms were also observed for the sorption of several aro-

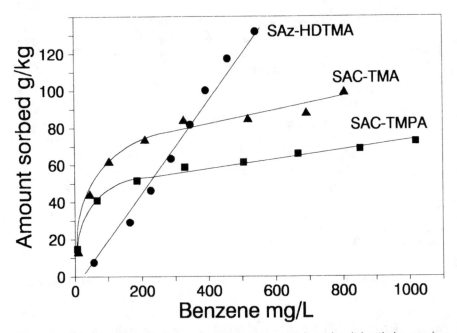

Figure 11.3. Sorption isotherms of benzene on hexadecyltrimethylammonium (HDTMA)-exchanged high-charge (SAz) smectite, tetramethylammonium (TMA)-exchanged low-charge (SAC) smectite, and trimethyl-phenylammonium (TMPA)-exchanged (SAC) smectite. *Source:* Jaynes and Boyd. [4,5]

matic hydrocarbons by the HDTMA-exchanged vermiculite, low-charge smectite, and illite.[4] The sorption coefficients (equal to the slope of the linear isotherms) are given in Table 11.3 for benzene, toluene, and ethylbenzene sorption by the HDTMA clays listed in Table 11.2. The linear sorption coefficient (K) is the ratio of the concentration of the (sorbed) solute in the HDTMA clay to its equilibrium solution concentration in water. Thus, higher K values correspond to a higher degree of sorption.

The K values listed in Table 11.3 show that there are significant differences in the efficacy of the HDTMA clays as sorbents for NOCs depending on the clay type, although all the HDTMA clays listed have impressive sorptive capabilities. The high-charge HDTMA smectite (SAz) was the most effective sorbent, showing a K value nearly five times as high as the low-charge HDTMA smectite (SWy). Interestingly, the HDTMA vermiculite (VSC) sorbed significantly more ethylbenzene than HDTMA smectite (SWy), even though the clays have nearly identical HDTMA contents. For benzene sorption, however, the differences in these clays were marginal. These results illustrate an important point: HDTMA clays with larger basal spacings (i.e., high-charge clays) can more effectively accommodate larger solute molecules such as alkyl-substituted benzenes. For smaller, more compact molecules such as benzene,

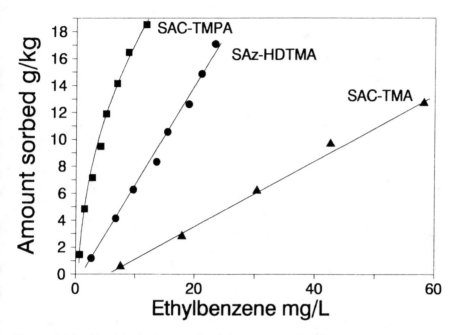

Figure 11.4. Sorption isotherms of ethylbenzene on hexadecyltrimethylammonium (HDTMA)-exchanged high-charge (SAz) smectite, tetramethylammonium (TMA)-exchanged low-charge (SAC) smectite, and trimethylphenylammonium (TMPA)-exchanged (SAC) smectite. *Source:* Jaynes and Boyd.[4,5]

steric restrictions are less evident among the 2:1 clays, and sorption depends primarily on the HDTMA content of the clay.

The greater basal spacings in higher-charge clays results from a closer packing of the organic cations on the clay surfaces, forcing the hydrophobic tails to adopt a more upright orientation (as shown in Figure 11.6). With lower unit charge density clays, there is a lower surface density of exchanged organic cations, which allows the hydrophobic tails to fold over on the mineral surface. This arrangement is commonly referred to as a "bilayer"[13] and results in a lower basal spacing of ≈ 18 Å. Thus, in general, when preparing organophilic clays, higher charge density clays are desirable to produce greater sorptive uptake of NOCs. Larger chain alkylammonium compounds (such as those with two alkyl-hydrocarbon groups, e.g., dimethyldioctadecylammonium) might also be used to induce greater expansion of low-charge clays, and hence cause greater retention of NOCs.

The mechanistic function of the HDTMA-derived organic phase of these clays can be described as a partition medium for NOCs.[9] The dissolution of NOCs into this phase is functionally and conceptually similar to the dissolution of NOCs in a bulk phase organic solvent such as octanol, except that here the organic partition phase is fixed on the clay surfaces. The log K_{om} values of these clays (Table 11.4; $K_{om} = 100\ K/\%$ organic matter) are similar to the

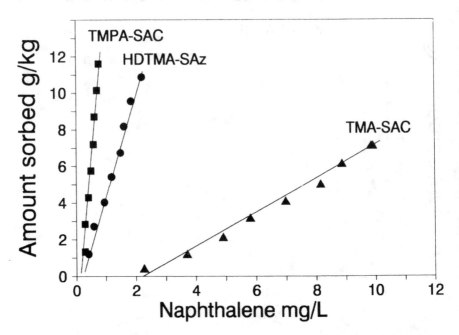

Figure 11.5. Sorption isotherms of naphthalene on hexadecyltrimethylammonium (HDTMA)-exchanged high-charge (SAz) smectite, tetramethylammonium (TMA)-exchanged low-charge (SAC) smectite, and trimethylphenylammonium (TMPA)-exchanged (SAC) smectite. *Source:* Jaynes and Boyd.[4,5]

corresponding log K_{ow} values. This property, along with the linearity of the sorption isotherms, provides strong mechanistic evidence for the concept that the HDTMA-derived organic phase functions as a powerful partition medium for NOCs.

Table 11.3. Linear Partition Coefficients (K) of HDTMA-Exchanged Soils and Clays

Sample	Benzene	Toluene	Ethylbenzene
Clays			
HDTMA-vermiculite (VSC)	68	169	448
HDTMA-smectite (SAz)	184	319	583
HDTMA-smectite (SWy)	53	74	127
HDTMA-illite (IMt)	39	77	156
Oshtemo soil (B$_t$ horizon)			
untreated	0	0	0.21
HDTMA-treated	3.83	6.59	12.84
Marlette soil (B$_t$ horizon)			
untreated	0	0	0.41
HDTMA-treated	16.50	26.90	62.51
St. Clair soil (B$_t$ horizon)			
untreated	0	0.17	0.81
HDTMA-treated	15.96	34.47	75.54

Sources: Data for clays from Jaynes and Boyd;[4] data for soils from Lee et al.[7]

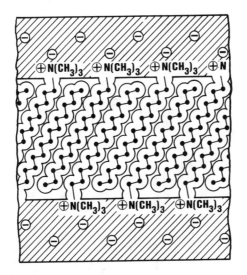

Figure 11.6. A schematic representing a high-charge smectite or vermiculite intercalated with hexadecyltrimethylammonium $[(CH_3)_3N(CH_2)_{15}CH_3]^+$ cations. Carbon-carbon bonds are shown as ● — ●. Galley region of the clay resembles a bulk organic solvent phase and functions as a highly effective partition medium for removing nonionic organic contaminants from water.

Soil Modification

The observation that vermiculite, illite, and several smectites exchanged with HDTMA were all highly effective sorbents for removing NOCs from water suggested the possibility of the in situ formation of organo-clays derived from soil or subsurface materials. These heterogeneous materials would usually contain a variety of clay mineral types. It is clear that the ability to greatly enhance the immobilization of organic contaminants in soils and subsoils would be extremely useful in limiting the extent of contamination and in protecting groundwater supplies.

Recently, we have examined the sorptive characteristics of whole soils that have been modified by the addition of HDTMA cations in an amount equivalent to the CEC of the soil.[7,10] Properties of the soils used in these studies are given in Table 11.5, and the organic matter contents of the HDTMA-treated soils are shown in Table 11.6. These studies have shown conclusively that the sorptive uptake of NOCs by HDTMA-exchanged soils was greatly enhanced, and this enhancement was observed for high and low clay content soils. These results are illustrated in Figure 11.7, which shows the sorptive uptake of benzene, toluene, and ethylbenzene by the untreated and HDTMA-treated St. Clair (41% clay) and Oshtemo (6%) soils. The dramatic improvement in the sorptive capabilities for removing NOCs from water of the HDTMA-soils is apparent from the sorption isotherms shown in Figure 11.7.

To quantify the effects of soil modification by HDTMA, the sorption coef-

Table 11.4. Organic Matter Normalized Partition Coefficients (K_{om}) of HDTMA-Exchanged Clays and Soils (B_t Horizon)

Solute	log K_{om}							log K_{ow}
	Vermiculite (VSC)	Illite (IMt)	Smectite (SAz)	Smectite (SWy)	Oshtemo Soil	Marlette Soil	St. Clair Soil	
Benzene	2.52	2.57	2.81	2.38	2.53(1.26)[a]	2.53	2.56	2.13
Toluene	2.92	2.86	3.05	2.53	2.76(1.94)	2.74	2.89	2.69
Ethylbenzene	3.34	3.17	3.31	2.77	3.06(1.98)	3.11	3.23	3.15

Sources: Data for soils from Lee et al.;[7] data for clays from Jaynes and Boyd.[4]

[a] Values in parentheses are log K_{om} values for natural (untreated) soils reported by Chiou.[3]

Table 11.5. Soil Properties

Property	Oshtemo B_t	Marlette B_t	St. Clair B_t
Cation exchange capacity (meq/100 g)	3.5	14.6	18.3
Particle size (%)			
sand	89.3	38.8	21.0
silt	4.4	31.6	34.9
clay[a]	6.3	29.6	41.4
illite	xxx	xxx	xxxx
vermiculite	xxx	xxx	xxx
kaolinite	xxx	xxx	xxx
chlorite	xx	xx	
quartz	xx	xx	xx
pH	5.84	5.40	6.72

Source: Data from Lee et al.[7]
[a]xx, small (5–15%); xxx, moderate (15–30%); xxxx, abundant (30–50%).

ficients of benzene, toluene, and ethylbenzene on the untreated and HDTMA-treated Oshtemo, Marlette, and St. Clair soils are shown in Table 11.3. These data show that the sorption coefficients were increased by a factor of 100 or more by this simple soil modification procedure.

It is also instructive to examine the organic matter normalized sorption coefficients (K_{om}) shown in Table 11.4. The K_{om} values are used to compare the effectiveness of HDTMA-derived organic phases vs natural soil organic matter on an equal weight basis. The K_{om} values of the HDTMA-treated (B_t horizon) soils, where organic matter is derived primarily from exchanged HDTMA (Table 11.6), were about 10 times higher than the K_{om} values of untreated soil, where organic matter is entirely natural. Thus, not only can the organic matter content of soils be increased by ion exchange reactions with HDTMA, but the added organic matter is 10–30 times more effective (as a sorbent for NOCs) on a unit weight basis than natural soil organic matter. Importantly, the K_{om} values of the HDTMA-treated Oshtemo soil (6% clay) were equal to those of the Marlette (30% clay) and St. Clair (41% clay) soils, showing that the effectiveness of exchanged HDTMA was not diminished in the low-clay content soil. The log K_{om} values of the HDTMA-treated soils (Table 11.4), which have clay mineralogy dominated by illite and vermiculite (Table 11.5), are very similar to the log K_{om} values of HDTMA-illite and HDTMA-vermiculite (Table 11.3).

It is also noteworthy that the log K_{om} values of the HDTMA-treated soils agree closely with the corresponding log K_{ow} values (Table 11.4). The similarity

Table 11.6. Organic Matter (OM) Content and Distribution in HDTMA-Treated Soils

	Oshtemo B_t	Marlette B_t	St. Clair B_t
Soil OM (%)	0.22	0.60	0.88
HDTMA-derived OM (%)	0.90	4.25	3.50
Total OM (%)	1.12	4.85	4.38

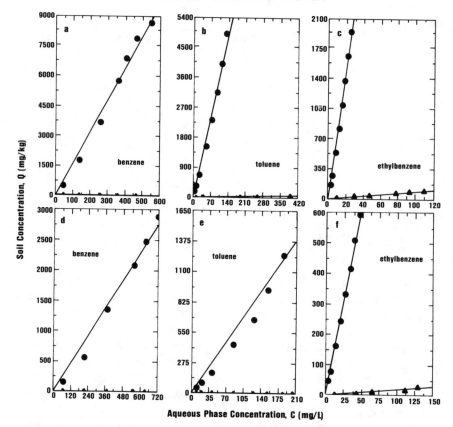

Figure 11.7. Sorption of benzene, toluene, and ethylbenzene on untreated (*triangles*) and hexadecyltrimethylammonium-treated (*circles*) St. Clair (*a-c*) and Oshtemo (*d-f*) B_t horizon soils. Reprinted from Lee et al.[7] Copyright 1989, American Chemical Society.

of the log K_{om} and log K_{ow} values and the high linearity of the isotherms provide strong mechanistic evidence that exchanged HDTMA in soils acts as a highly effective partition medium for NOCs.

In our recent studies of organic cation-exchanged soils,[7,10] we also showed that the ion exchange reaction of HDTMA for naturally occurring inorganic exchange ions was essentially stoichiometric. When HDTMA is added at levels less than or equal to the CEC of the soil, essentially all HDTMA was adsorbed via ion exchange. The result is illustrated in Figure 11.2 for HDTMA adsorption by the Marlette A and B_t horizon soils. Once adsorbed, HDTMA was held irreversibly, as shown by the inability of high-ionic-strength salt solutions to displace exchanged ^{14}C-HDTMA.[7] Thus, this soil modification technology offers the following properties that are important and desirable in its practical application:

1. complete exchange of the organic cation
2. irreversible sorption of the organic modifier
3. high sorption of NOCs by the modified soil

Organo-Clays as Surface Adsorbents: TMA and TMPA Smectites

A second class of organo-clays that we have studied are those in which small organic cations such as tetramethylammonium (TMA) or trimethylphenylammonium (TMPA) are used as exchange ions on smectites.[5,6,8,11,12] The mechanism of uptake and sorptive properties of these clays are distinctly different from clays exchanged with large organic cations such as HDTMA. In the latter case, sorption of NOCs occurs via partitioning into the HDTMA-derived organic phase that forms due to interactions between the large alkyl hydrocarbon moieties of the QUAT on the clay surface. When comparatively smaller QUATs are used (e.g., TMA or TMPA), these exchange ions exist as discrete entities on the smectite layers and, therefore, do not form partition phases. The surfaces of these clays can be viewed as containing isolated quaternary ammonium exchange cations that are separated by "free" (uncovered) planar aluminosilicate mineral surfaces.

The properties of low-charge and high-charge TMA and TMPA smectites are summarized in Table 11.7. These clays have N_2 BET surface areas of approximately 200 m^2/g, demonstrating the existence of significant free mineral surface area. The basal spacings of these clays are about 14 to 15 Å, which corresponds to an interlayer separation of about 4.7 to 5.7 Å. The organic carbon contents of 4 to 5% for the TMA smectites and 9 to 10% for the TMPA smectites indicate that the exchange sites are fully occupied by TMA or TMPA cations.

Sorption isotherms for the uptake of benzene from water by TMA smectite (SAC), TMPA smectite (SAC), and HDTMA smectite (SAz) are shown in Figure 11.3. The Langmuir-type isotherms of TMA and TMPA smectites sharply contrast with the highly linear benzene isotherm for HDTMA smectite and suggest a surface adsorptive mechanism occurs for TMA and TMPA smectites. These isotherms show that TMA smectite (SAC) is a very effective

Table 11.7. Properties of Low-Charge and High-Charge Tetramethylammonium (TMA) and Trimethylphenylammonium (TMPA) Smectites

	TMA Smectite		TMPA Smectite	
	SAC[a]	SAz	SAC	SWa[b]
Cation exchange capacity (meq/100g)	90	120	90	107
Charge (mol⁻/unit cell)	0.64	1.14	0.64	0.96
Organic carbon (g/100g)	4.0	5.0	9.15	9.66
d(001) (Å)	13.8	13.8	14.5	15.2
N_2 surface area (m^2/g)	206	187	n.a.	n.a.

Sources: Data for TMA smectite from Lee et al;[6] data for TMPA smectite from Jaynes and Boyd.[5]

[a]SAC was supplied by American Colloid Co. as Wyoming bentonite.
[b]SWa is a high-charge smectite from Grant County, Washington, supplied by the Clay Minerals Society Repository.

adsorbent of benzene, especially at equilibrium solution concentrations of less than 200 ppm, where it exhibits greater uptake of benzene than HDTMA smectite (SAz). However, TMA smectite (SAC) also exhibits strong shape/size selectivity in the uptake of aromatic hydrocarbons from water, resulting in progressively lower sorption of larger molecules.[8] This result can be seen by comparing the sorption isotherms of benzene, ethylbenzene, and naphthalene (Figures 11.3, 11.4, and 11.5) by TMA smectite (SAC). These results also show that the relation between water solubility and uptake is different for TMA smectite and for HDTMA smectite. For TMA smectite, sorption decreases in the same order as water solubility (benzene > toluene > naphthalene); for HDTMA smectite sorption increases as water solubility decreases, consistent with the concept of solute partitioning which involves hydrophobic (nonpolar) interactions with the HDTMA phase. Interestingly, the shape selectivity of TMA smectite (SAC) was not observed for the adsorption of organic vapors (benzene, toluene, xylene) by the dry clay.[6]

Lee et al. also observed strong competitive effects on sorption by TMA smectite in a binary solute system consisting of benzene and toluene: the presence of benzene reduced the uptake of toluene.[6] The nonlinear sorption isotherms, shape selectivity, and competitive effects exhibited by TMA smectite are manifestations of the surface adsorptive behavior of this clay in the sorption of NOCs from water.

The effect of surface charge density on the uptake of aromatic hydrocarbons by TMA smectite was also evaluated by Lee et al.[6] As illustrated in Figure 11.8, the high-charge TMA smectite (SAz) was considerably less effective than the low charge TMA smectite (SAC) in the sorption of benzene from water. Thus, TMA smectites derived from low-charge clays are more effective sorbents for NOCs than those derived from high-charge clays, but high-charge HDTMA clays are more effective than low-charge HDTMA clays.[4] Apparently, the closer packing of TMA ions in the high-charge smectite diminishes the number of adsorption sites of sufficient size to accommodate aromatic solutes such as benzene.

The high affinity of TMA smectite (SAC) for benzene suggests that such clays may be useful for immobilizing organic compounds with relatively high water solubility, for which organophilic clays (e.g., HDTMA smectite) are less effective. However, it would obviously be desirable to develop clays in which the high affinity for benzene, characteristic of TMA smectite, is extended to include water-soluble alkyl benzenes and polynuclear aromatic hydrocarbons, often present as co-contaminants with benzene. Such organo-clays could be useful as liners in petroleum tank farms, underground storage tanks, and in waste disposal reservoirs in general, as well as in the purification of petroleum-contaminated water.

Recently, we have reported that trimethylphenylammonium (SAC) smectite is an effective adsorbent of the most water-soluble aromatic hydrocarbon constituents of petroleum, including benzene, toluene, ethylbenzene, p-xylene, butylbenzene, and naphthalene.[5] These results are illustrated in Figures 11.3,

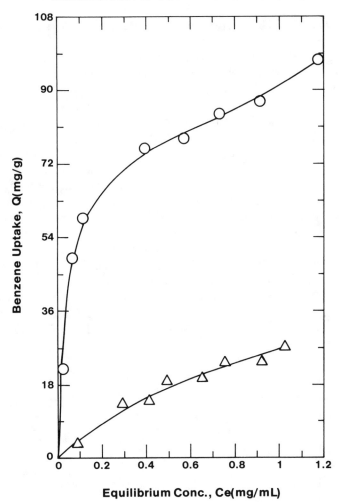

Figure 11.8. Adsorption of benzene from water by a high-charge (SAz) smectite (*triangles*) and a low-charge (SAC) smectite (*circles*) exchanged with tetramethylammonium cations. Reprinted from Lee et al.[6] Copyright 1990, Clay Minerals Society.

11.4, and 11.5, which show adsorption isotherms of benzene, ethylbenzene, and naphthalene by TMPA smectite (SAC) and TMA smectite (SAC). The strong affinity of TMPA smectite (SAC) for benzene, alkyl-substituted benzenes, and naphthalene is in sharp contrast to TMA smectite, which is an effective adsorbent for benzene but not for larger substituted benzenes or naphthalene. The different hydration properties of TMA and TMPA may account for the different selectivities of the corresponding clays. Tetramethylammonium would be expected to have a greater hydration energy (32 kcal/mole) than TMPA.[14] Greater hydration of TMA ions may explain the shape/

size selectivity of TMA smectite for aromatic hydrocarbon adsorption from water. No selectivity is evident in the adsorption of organic vapors by the dry TMA smectite.[8] Hence, the limited hydration of TMPA ions may account for the absence of shape/size selectivity in the adsorption of aromatic hydrocarbons from water by TMPA smectite.

The effects of charge density on the sorption of aromatic hydrocarbons by TMPA smectites was also evaluated by Jaynes and Boyd.[5] As with TMA smectite, the use of a high-charge smectite (SWa) decreased the effectiveness of the organo-clay as compared to a low-charge smectite (SAC) (Table 11.7).

The ion exchange of TMPA on a Mg^{2+} smectite was also quite favorable.[5] The added TMPA cations were shown to nearly stoichiometrically replace Mg^{2+} until about 90% of the cation exchange sites were occupied by TMPA. Thus, this clay would appear amenable to commercial preparation by the dry process (discussed earlier).

Three other related QUATs, benzyltrimethylammonium (BTMA), benzyltriethylammonium (BTEA), and benzyltributylammonium (BTBA), were used to prepare the corresponding organo-clays from smectite (SAC). The BTMA, BTEA, and BTBA smectites (SAC) were compared to TMPA smectite (SAC) as adsorbents of ethylbenzene from water. The adsorption isotherms presented in Figure 11.9 show TMPA smectite (SAC) to be the most effective

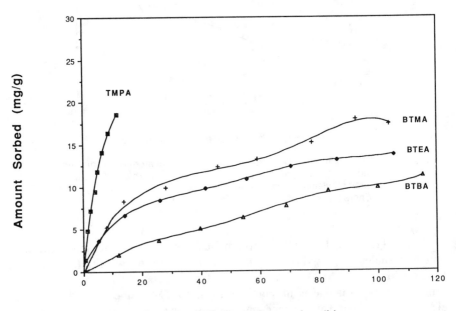

Figure 11.9. Adsorption isotherms of ethylbenzene on low-charge (SAC) smectite exchanged with trimethylphenylammonium (TMPA), benzyltrimethylammonium (BTMA), benzyltriethylammonium (BTEA), and benzyltributylammonium (BTBA).

adsorbent of ethylbenzene, followed in decreasing order by BTMA, BTEA, and BTBA smectite (SAC). As the size of trialkyl groups increases from trimethyl to triethyl to tributyl, the affinities of the corresponding organo-clays for ethylbenzene decreases and the isotherms appear to become increasingly linear. It may be that the benzyltributylammonium ions are sufficiently large to allow the formation of a partition phase analogous to that of HDTMA smectite.

Clearly, TMPA smectite (SAC) is the most effective organo-clay yet developed for the adsorptive removal of aromatic hydrocarbons from water. In comparing the effectiveness of HDTMA, TMPA, and TMA clays (see Figures 11.3, 11.4, and 11.5), TMA and TMPA smectites (SAC) are clearly more effective than HDTMA smectite (SAC), except at very high benzene concentrations. For alkyl-substituted benzenes, TMPA smectite (SAC) is very effective, and TMA smectite (SAC) is much less effective. As the water solubility of the solute decreases, the differences in TMPA smectite (SAC) and HDTMA smectite (SAz) diminish. Thus, naphthalene (water solubility, 31 ppm) sorption by TMPA smectite (SAC) and HDTMA smectite (SAz) is quite similar.

Applications

The results summarized here provide the basis for an effective, simple, and economically viable technology for the in situ treatment of contaminated soils and aquifer materials. This soil modification technology could be used as an immediate response to terrestrial chemical spills to limit the extent of contamination and to protect groundwater supplies. Organo-clays formed in situ from soil clays, or added as soil amendments, may also be used to stabilize contaminated soils at the surface or the near-surface of existing hazardous waste sites by assimilating toxic organic chemicals. The underground injection of organic cations may be used to increase the sorptive properties of subsurface materials and thus control the migration of organic contaminant plumes. For these applications, organophilic cations such as HDTMA are most suitable because of their excellent cation exchange properties, their effectiveness when associated with a variety of clay mineral types, and their sorptive capabilities for NOCs in general.

Organo-smectites, such as those derived from HDTMA or TMPA cations, may also be useful as components of clay containment barriers. Organo-clays used in conjunction with conventional Na smectites would provide a barrier with the ability to impede the movement of water *and* to sorb dissolved organic contaminants. This would represent a major advance in the performance of liners used in hazardous waste disposal facilities. These clays could also be used to create subsurface sorptive barriers in a fashion analogous to the installation of bentonite slurry walls. The design of organo-clays (e.g., TMPA smectite) with increased affinities for high water solubility "breakthrough" compounds further advances the potential effectiveness of this technology. The effectiveness of TMPA smectite as an adsorbent of water-soluble aromatic hydrocarbon petroleum constituents suggests applications as liner components in petroleum tank farms, underground storage tanks, and in the treatment of petroleum-contaminated water.

ACKNOWLEDGMENT

This research was supported by in part by grant 1 P42 ES04911 01 from the National Institute of Environmental Health Sciences, and the Michigan Agricultural Experiment Station.

REFERENCES

1. Chiou, C. T., L. J. Peters, and V. H. Freed. "A Physical Concept of Soil-Water Equilibria for Nonionic Organic Compounds," *Science* 206:831–832 (1979).
2. Karickhoff, S. W., D. S. Brown, and T. A. Scott. "Sorption of Hydrophobic Pollutants on Natural Sediments," *Water Res.* 13:241–248 (1979).
3. Chiou, C. T. "Roles of Organic Matter, Minerals and Moisture in Sorption of Nonionic Organic Compounds and Pesticides by Soil," in *Humic Substances in Soil and Crop Sciences,* P. MacCarthy, C. E. Clapp, R. L. Malcolm, and P. Bloom, Eds. (Madison, WI: American Society of Agronomy, 1990), pp. 111–160.
4. Jaynes, W. F., and S. A. Boyd. "Clay Mineral Type and Organic Compound Sorption by Hexadecyltrimethylammonium-Exchanged Clays," *Soil Sci. Soc. Am. J.* 55:43–48 (1991).
5. Jaynes, W. F., and S. A. Boyd. "Trimethylphenylammonium-Smectite as an Effective Adsorbent of Water Soluble Aromatic Hydrocarbons," *J. Air and Waste Management Assoc.* 40:1649–1653 (1990).
6. Lee, J. F., M. M. Mortland, C. T. Chiou, D. E. Kile, and S. A. Boyd. "Adsorption of Benzene, Toluene and Xylene by Two Tetramethylammonium-Smectites Having Different Charge Densities," *Clays and Clay Minerals* 38:113–120 (1990).
7. Lee, J. F., J. R. Crum, and S. A. Boyd. "Enhanced Retention of Organic Contaminants by Soils Exchanged with Organic Cations," *Environ. Sci. Technol.* 23:1365–1372 (1989).
8. Lee, J. F., M. M. Mortland, C. T. Chiou, and S. A. Boyd. "Shape Selective Adsorption of Aromatic Molecules from Water by Tetramethylammonium-Smectite," *J. Chem. Soc. Faraday Trans. I.* 85:2953–2962 (1989).
9. Boyd, S. A., M. M. Mortland, and C. T. Chiou. "Sorption Characteristics of Organic Compounds on Hexadecyltrimethylammonium-Smectite," *Soil Sci. Soc. Am. J.* 52:652–657 (1988).
10. Boyd, S. A., J. F. Lee, and M. M. Mortland. "Attenuating Organic Contaminant Mobility by Soil Modification," *Nature* 333:345–347 (1988).
11. Boyd, S. A., S. Sun, J. F. Lee, and M. M. Mortland. "Pentachlorophenol sorption by organo-clays," *Clays and Clay Minerals* 36:125–130 (1988).
12. Mortland, M. M., S. Sun, and S. A. Boyd. "Clay-Organic Complexes as Adsorbents for Phenol and Chlorophenols," *Clays and Clay Minerals* 34:581–586 (1986).
13. Lagaly, G., and A. Weiss. "The Layer Charge of Smectitic Layer Silicates," in *Proc. Int. Clay Conf. Mexico City, 1975,* S. W. Bailey, Ed. (Wilmette, IL: Applied Publishing, 1976), pp. 157–172.
14. Cotton, F. A., and G. Wilkinson. *Advanced Inorganic Chemistry,* 2nd ed. (New York: John Wiley and Sons, 1966), p. 43.

CHAPTER 12

Effects of Surfactants on the Mobility of Nonpolar Organic Contaminants in Porous Media

James A. Smith, David M. Tuck, Peter R. Jaffé, and Robert T. Mueller

INTRODUCTION

Surface-active agents (surfactants) are a class of organic compounds that have a two-component molecular structure. One component of the molecule is water soluble (hydrophilic); the other is relatively water insoluble (lipophilic). Surfactants are commonly classified by ionic type. Table 12.1 shows the molecular structures of several anionic, cationic, and nonionic surfactants. Amphoteric surfactants, although not shown in Table 12.1, possess both anionic and cationic functional groups. Commercial surfactant formulations are used in a number of broad applications: emulsifiers and demulsifiers, wetting and rewetting agents, foaming and defoaming agents, dispersants, detergents, and solubility enhancers.[1]

The two-component molecular structure of surfactants uniquely influences their behavior in aqueous solution. At dilute aqueous concentrations, surfactants position themselves at water-air and water-solid interfaces. The surfactant molecule is oriented at the interface so that the hydrophobic component faces the air or solid side of the interface and the hydrophilic component faces the aqueous side of the interface. Increasing the aqueous surfactant concentration causes a reduction in the surface tension of the water until the critical micelle concentration (CMC) of the surfactant is reached.[2] At concentrations less than CMC, the surfactant molecules exist as monomers in solution. At concentrations greater than CMC, the surfactant molecules group together in "bundles" (micelles), with the hydrophobic component of each molecule oriented toward the center of the micelle and the hydrophilic component oriented toward the bulk aqueous solution.[2-4] At concentrations greater than CMC, surface tension is independent of surfactant concentration.

The objective of this chapter is to discuss the effect of surfactants on the mobility of dissolved and non-aqueous-phase organic contaminants in

Table 12.1. Structural Diagrams of Six Surfactants

Surfactant	Ionic Type	Molecular Structure
Sodium dodecylsulfate	Anionic	$CH_3(CH_2)_{11}OSO_3Na$
Sodium dodecyl-benzenesulfonate	Anionic	$C_{12}H_{25}$–⬡–SO_3Na
Triton X-100	Nonionic	CH_3-$\overset{\underset{\mid}{CH_3}}{C}$-$CH_2$-$\overset{\underset{\mid}{CH_3}}{C}$-⬡-$(OCH_2CH_2)_xOH$
Tetramethyl-ammonium bromide	Cationic	$\left[CH_3-\overset{\underset{\mid}{CH_3}}{N}-CH_3\right]^+ Br^-$
Benzyltrimethyl-ammonium bromide	Cationic	$\left[CH_3-\overset{\underset{\mid}{CH_3}}{N}-CH_2-⬡\right]^+ Br^-$
Hexadecyltrimethyl-ammonium bromide	Cationic	$\left[CH_3-\overset{\underset{\mid}{CH_3}}{N}-CH_3(CH_2)_{15}\right]^+ Br^-$

Note: For Triton X-100, the average value of x is 10.

porous media. The first part of the chapter shows that certain anionic and nonionic surfactants can *increase* the mobility of nonionic organic contaminants in soil-water systems by increasing the contaminants' aqueous solubility, and thereby decreasing sorption to the soil. Similarly, surfactant solutions can mobilize non-aqueous-phase organic compounds at residual saturation in porous media by lowering the interfacial tension between water and the nonpolar organic liquid. These engineered processes can have great practical utility for the remediation of contaminated soil and groundwater. For example, surfactant solutions applied to the subsurface system can increase the removal rate of organic contaminants during a pump-and-treat groundwater remediation.

The second part of this chapter discusses how certain cationic surfactants can *decrease* the mobility of nonionic organic compounds in clay-water systems. Cationic surfactants are strongly retained on the mineral surfaces of natural soil and clay by an ion exchange process, and the modified mineral surfaces exhibit an increased sorption capacity for nonionic organic compounds in water relative to the natural mineral surfaces. Again, this process can have great practical significance: the application of cationic surfactants to clay used in liners for waste-disposal facilities and in slurry walls at groundwater contamination sites can increase its effectiveness in preventing contaminant migration.

SURFACTANT EFFECTS THAT INCREASE CONTAMINANT MOBILITY IN POROUS MEDIA

Because of their unique chemical properties, certain surfactants can increase the mobility of dissolved and non-aqueous-phase organic contaminants through porous media, which may be desirable during the remediation of contaminated soil and groundwater. For aqueous-phase contaminants, anionic and nonionic surfactants can enhance the contaminant's solubility and thereby decrease its sorption to the aquifer material. For non-aqueous-phase contamination, anionic and nonionic surfactants can change the interfacial tensions between the fluid-fluid and fluid-solid interfaces and thereby mobilize nonpolar organic liquids present in porous media at residual saturation. These two approaches are discussed in the following sections.

Solubility Enhancement and Reduced Sorption

Large organic macromolecules (e.g., humic and fulvic acids and commercial surfactants) are capable of increasing the apparent water solubility of nonionic organic compounds.[5-18] The accumulation of these macromolecules into micelles or an emulsified phase can create a nonpolar environment (relative to water) into which a contaminant can partition by a solubilization reaction. Nonionic organic compounds partition into this micellar or emulsified phase in inverse proportion to their aqueous (pure-water) solubilities. The modified bulk solution, which includes both aqueous and micellar or emulsified phases, can accommodate a larger mass of the solute per unit volume than the solute saturation in pure water. Therefore, the presence of dissolved humic and fulvic acids or commercial surfactants can increase the apparent water solubility of the solute.

The magnitude of this solubility-enhancing effect can be substantial and is a function of the molecular structure of the humic or fulvic acid or surfactant. Certain petroleum sulfonate surfactants have been shown to form stable emulsions in solution that are similar to a bulk organic phase (e.g., a pure organic liquid such as n-octanol).[18] This bulk organic phase can significantly enhance solute solubility over a wide range of surfactant concentrations. Natural humic and fulvic acids also can increase solute solubility over a wide range of dissolved organic matter concentrations.[14] Other commercial surfactants (e.g., Triton X-100, sodium dodecyl sulfate, hexadecyltrimethylammonium bromide) exhibit significant solubility enhancements of organic solutes only at concentrations above the surfactant CMC.[17] Table 12.2 gives the logarithms of experimentally determined distribution coefficients for p,p'-DDT and 1,2,3-trichlorobenzene distributed between water and a variety of commercial surfactants and humic substances. The distribution coefficients for dissolved organic matter and water, micelles and water, and emulsions and water are denoted by K_{dom}, K_{mc}, and K_{em}, respectively. For comparison, the logarithms of the octanol-water partition coefficients of p,p'-DDT and 1,2,3,-

Table 12.2. Comparison of Partition Coefficients for *p,p'*-DDT and 1,2,3-Trichloro-
benzene (TCB)

Surfactant	TCB	*p,p'*-DDT	Reference
	Log K_{em}		
Petronate L	3.94	6.32	18
Petronate HL	3.94	6.32	18
Pyronate 40	3.77	6.14	18
	Log K_{mc}		
Hexadecyltrimethylammonium bromide	3.80	5.88	17
Sodium dodecyl sulfate	3.54	5.38	17
Triton X-100	3.82	6.15	17
Triton X-114	3.95	6.18	17
	Log K_{dom}		
Fulvic acid	2.0	4.27	14
Humic acid	2.8	4.82	14

Note: K_{em}, K_{mc}, and K_{dom} are partition coefficients for emulsion-water, micelle-water, and dissolved organic matter–water systems, respectively.

trichlorobenzene are 6.73 and 4.20, respectively. The data indicate that the solubility-enhancing power of certain commercial surfactants (e.g., Petronate L, Petronate HL, and Pyronate 40) approaches that of octanol.[18]

Table 12.2 demonstrates the significant effect of dissolved organic matter and commercial surfactants on the aqueous solubility of nonionic organic compounds. This observation is relevant to environmental systems because, among the many properties of solutes and soils, the solubility of the solute in water repeatedly has been shown to be the primary factor controlling the sorption of nonionic organic solutes by natural soil from water.[19-23] Empirical sorption data have shown an inverse, linear relation between log (logarithm$_{10}$) aqueous solubility and log K_{oc} (log organic carbon normalized distribution coefficient) for a wide variety of solutes.[19,20] Therefore, the ability of dissolved organic matter and commercial surfactants to increase the aqueous solubility of a nonionic solute can also cause a reduction in solute sorption if the micellar or emulsified surfactant phase remains predominantly in the bulk solution in a soil-water mixture.

Table 12.2 shows that the solubility-enhancing effect of natural organic matter is small relative to commercial surfactants. The effect of anionic and nonionic commercial surfactants on the distribution of nonionic contaminants between soil and water is likely to be greater than the effect of natural organic matter, particularly because these surfactants are extremely water soluble and are not strongly sorbed by soil. This observation is supported by the tetrachloromethane sorption data shown in Figure 12.1.

The data of Figure 12.1 were generated at room temperature by a batch equilibration method similar to that described by Smith et al.[24] The sorbent is a natural soil collected from 1 m below land surface at Picatinny Arsenal in Morris County, New Jersey. The soil organic carbon content is 1.4%. In Figure 12.1 for the upper isotherm, 8 g of soil, 12 mL of water, and varying

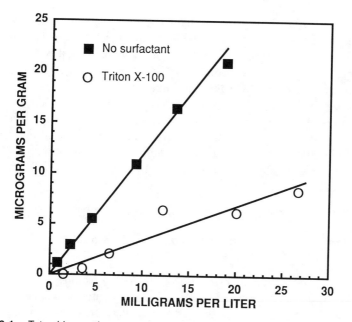

Figure 12.1. Tetrachloromethane sorption to natural soil from water, and from water with Triton X-100.

amounts of ^{14}C-tetrachloromethane (specific activity equal to 4.3 mCi/mmol) were combined in 15-mL (nominal volume) glass centrifuge tubes with Teflon-lined septum caps. The centrifuge tubes were equilibrated in the dark. During equilibration, the tubes were rotated continuously to promote mixing. The equilibration period was 48 hr. Following equilibration, the tubes were centrifuged at 2000 G for 60 min. One mL of the supernatant from each tube was transferred by pipette into 10 mL of scintillation cocktail, and the radioactivity of the sample was measured with a Packard Tri-Carb 1900CA liquid scintillation analyzer. The equilibrium aqueous tetrachloromethane concentrations (C_e) were determined from a standard curve relating concentration to the number of disintegrations per minute. The corresponding sorbed tetrachloromethane concentrations (C_s) were then determined by difference. For the lower isotherm in Figure 12.1, an identical procedure was used with one exception: 100 mg of Triton X-100 also was added to each centrifuge tube prior to the beginning of the equilibration period.

The results indicate that the addition of Triton X-100 to the soil-water-tetrachloromethane mixture decreases tetrachloromethane sorption to soil. Although the isotherm data are unique to the specific dose of the surfactant and its concentration on the soil (i.e., the surfactant sorption to the soil has not been quantified), the results are significant given the relatively high aqueous solubility of tetrachloromethane (800 mg/L). The reduction in sorption for more hydrophobic solutes, such as polychlorinated biphenyls, chlorinated insecticides, or polycyclic aromatic hydrocarbons, is likely to be larger. Moreover, the K_{oc} value for the upper isotherm in Figure 12.1 is 84.4 and agrees closely with empirical predictions based on aqueous solubility.[20,23]

Although dilute surfactant concentrations that are less than CMC are likely to affect only extremely hydrophobic solutes such as p,p'-DDT, concentrated surfactant solutions can have a pronounced effect on the transport of relatively water-soluble solutes. Such conditions exist during tertiary oil-recovery operations, and in runoff waters from farmlands following pesticide application or in leachate from waste-disposal facilities.[17,18] In addition, concentrated anionic or nonionic surfactant solutions can be injected into polluted aquifers to increase the contaminant removal rate during a pump-and-treat remediation.

Enhancement of Non-Aqueous-Phase Liquid Mobility

Removal of non-aqueous-phase liquids (NAPLs) from the subsurface is an important aspect of the remediation of many contaminated sites. Soil-gas venting is a technique that is being applied successfully for the removal of volatile residual organic compounds in the unsaturated zone. Different techniques are required for the recovery of residual NAPLs that are nonvolatile or that are present in the saturated zone. The oil industry has conducted a great deal of research to understand processes through which residual oil is mobilized in porous media. Since the 1950s and 1960s, much of this work has focused on tertiary oil recovery (i.e., those processes involved in trapping and establishing a residual saturation of hydrocarbon during secondary water-flooding operations and the later mobilization of that residual in tertiary recovery operations). Tertiary oil-recovery techniques are also known as enhanced oil recovery (EOR). This technology can also be applied to environmental remediation operations to remove nonvolatile NAPLs.

A residual (nonmobile) organic phase can remain in a porous medium when capillary forces are balanced against gravity forces and—in the presence of an additional moving fluid phase such as water—viscous forces. Several authors have examined the interfacial-tension relation between fluid phases and the forces required to mobilize residual hydrocarbons.[25-28] Two dimensionless numbers are used to characterize this relation: the Bond number (N_B) and the capillary number (N_{Ca}). These numbers represent, respectively, the ratio of gravity forces and viscous forces to capillary forces. These authors have demonstrated that the efficiency of flooding operations increases with the magnitude of these dimensionless ratios. Morrow and Songkran give the following definitions for the Bond and capillary numbers:[27]

$$N_B = \frac{\Delta\rho g R^2}{\gamma} \qquad (12.1)$$

$$N_{Ca} = \frac{v\mu}{\gamma} \qquad (12.2)$$

where: $\triangle\rho$ = fluid-density difference (g/cm^3)
        g = acceleration due to gravity (cm/sec^2)
        R = radius of the porous-medium particles (cm)
        v = displacing-fluid velocity (cm/sec)
        μ = displacing-fluid viscosity (poise or g/cm-sec)
        γ = interfacial tension (dynes/cm)

Morrow and Songkran found a direct correlation between the Bond and capillary numbers and residual saturations in their studies of vertical displacement of a nonwetting fluid by a wetting fluid in glass-bead media. Their results show that the residual saturations varied from normal values (about 14%) when the Bond number was less than 0.00667 and the capillary number was less than 3 × 10^{-6}, to near zero when either the Bond number exceeded 0.35 or the capillary number exceeded about 7 × 10^{-4}.

The Bond number, as defined above, is not commonly used to interpret flushing experiments because it remains constant as the residual saturation decreases, while the ratio of gravity to capillary forces decreases. The gravitational force acting on a given blob of oil at residual saturation is proportional to its volume because the mass of the blob is proportional to its volume. Chatzis et al.[29] demonstrated that blobs of residual oil commonly occupy up to tens of connected pores.

The majority of studies of NAPL mobility have focused on the capillary number. The capillary number can be increased, as is clear from Equation 12.2, by increasing the velocity or viscosity of the displacing fluid or by reducing the interfacial tension between the fluid phases. As discussed above, surfactants can reduce significantly the interfacial tension between fluid phases, thereby increasing the capillary number. Taber indicated that the interfacial tension between an oil and an aqueous phase needs to be reduced to the order of 10^{-2} to 10^{-3} dynes/cm in order to modify the capillary number sufficiently to produce significant tertiary oil recovery without increasing the pressure gradient. These values of interfacial tension are called "ultralow interfacial tensions."[25]

The oil industry has identified two surfactant-concentration ranges that yield the ultralow interfacial tensions necessary to increase the capillary number: a low concentration range (0.1 to 0.2% by weight) and a high concentration range (4 to 10% by weight).[30] Surfactants also can significantly increase the viscosity of the flushing solution if used at high concentrations, which can pose a problem for the engineered removal of nonvolatile NAPLs by surfactants from groundwater systems. Increasing the kinematic viscosity (μ/ρ) leads to a decrease in the hydraulic conductivity.[31] In order to maintain high flow rates, the hydraulic gradient must be increased comparably. This increase in kinematic viscosity places a limit on the concentration of surfactant that could be used in practice for the remediation of groundwater and soils. The low surfactant-concentration range, which does not increase the flushing-solution viscosity greatly, is therefore likely to be more appropriate for groundwater remedial applications than a high surfactant-concentration range.

The system under study consists of aqueous solutions of surfactant (Emphos CS-136) of varying concentration, tetrachloroethene (PCE), air, and glass. The surfactant was selected because of its ability to reverse the wetting relation among PCE, the water, and glass.[32,33]

Emphos CS-136 is a mixture of nonyl phenyl phosphate esters. Figure 12.2 illustrates the wetting reversal, as well as the interfacial-tension reduction between the liquid phases, as a function of surfactant concentration.

The interfacial tension measurements were made by the drop weight method.[34] The contact angle was measured under a binocular microscope with a precision rotating stage. A 1-mL volume of PCE was added to a glass container (88-mm diameter) filled to 1-mm depth with water. A glass slide was then inserted perpendicular to the bottom of the container to divide the PCE blob into two parts. The container was placed under the microscope, and the contact angle was measured as the angle between the water and PCE phases on the glass slide, which was oriented perpendicular to the field of vision of the microscope.

The potential utility of surfactant solutions to mobilize residual saturations of NAPLs from a porous medium was examined in laboratory column experiments. The model porous medium consisted of a 20-cm-long bed of 1-mm

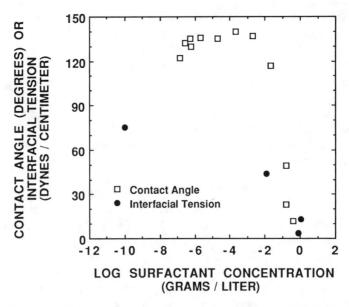

Figure   12.2. Equilibrium   contact   angle   and   aqueous-surfactant-solution–PCE (tetrachloroethylene) interfacial tension measurements as a function of surfactant concentration.

glass beads in a 1.27-cm-radius column. The initially dry medium was wetted from below with water. The water was allowed to drain, and a sequence of vacuums (approximately 2 cm of water) was applied until the capillary fringe at the base of the column was breached. PCE was then introduced to the top of the column by pipet. Output from the column, for both the initial PCE introduction and all subsequent samples, was collected into a separatory funnel, and the PCE was concentrated by liquid-liquid extraction with chloroform. The columns were then flushed five times with two pore volumes of distilled water to establish an initial residual saturation for each experiment. The vacuum procedure described above was applied after each introduction of flushing solution in order to maintain unsaturated conditions in the column. Baseline experiments were conducted by continuing to flush with distilled water. Solutions of 0.05 g/L and 0.15 g/L of the surfactant were used to evaluate the surfactant influence on the flushing efficiency. All chloroform extracts were analyzed with a Perkin-Elmer Model 3920B gas chromatograph with a flame-ionization detector. The residual PCE in the column was determined by mass balance.

The results of several experiments, comparing flushing with distilled water to flushing with surfactant solutions, are presented in Figure 12.3. The initial residual saturations varied from 4.2 to 6.8%. The results of all the experiments have been normalized to the residual saturation in each experiment after the

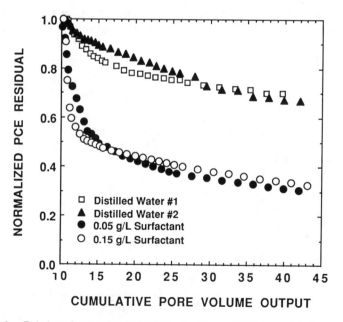

Figure 12.3. Relation of normalized residual saturations of PCE to cumulative pore volume output for column flushing experiments. Results are normalized to the mass of PCE remaining in the columns after approximately ten pore volumes of distilled water were passed through them.

initial 10-pore-volume flush with distilled water in order to facilitate comparison between experiments. The residual PCE remaining in the column decreased rapidly when the column was flushed with the surfactant solutions, compared with the reduction achieved when the column was flushed only with water. Most of the difference was observed within approximately three pore volumes.

This improved performance can be understood in terms of the Bond and capillary numbers for the different experiments. The interfacial tensions between each of the two surfactant flushing solutions and PCE were estimated from the results shown in Figure 12.2. They both were approximately an order of magnitude less than the measured interfacial tension between distilled water and PCE. Hence, the Bond and capillary numbers should both be approximately an order of magnitude higher for the surfactant flushing experiments than for the water flushing experiments. Melrose and Brandner reported that in order to reduce the residual oil saturation by a factor of 2, it is necessary to increase the capillary number by a factor of between 10 and 100.[26] The difference in the normalized residuals between the surfactant flushing results and the distilled water flushing results remains fairly constant at about 0.37 after the first three pore volumes. Therefore, these results are consistent with the order of magnitude range reported by Melrose and Brandner.[26]

Another aspect of the flushing results is illustrated in Figure 12.4, which shows for one of the distilled water flushing experiments a plot of the amount of PCE flushed divided by the pore-volume aliquot of flushing solution used to produce that sample against the cumulative pore volumes of flushing solu-

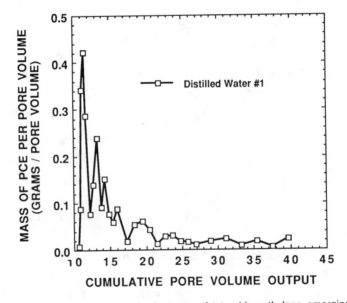

Figure 12.4.   "Cyclic" fluctuations in the mass of tetrachloroethylene emerging from the column per pore volume of flushing solution used to produce the sample.

tion passed through the column. Two phenomena are illustrated by this figure: first, the relative recovery of PCE per pore volume of flushing solution decreases as the flushing operation progresses; second, a cyclic (sinusoidal) fluctuation in the relative recovery is observed. These phenomena can both be explained in terms of the ratio of the gravitational to capillary forces.

The capillary number remains constant during a flushing operation as long as neither the flushing rate changes (i.e., no change occurs in the displacing-fluid velocity, v) nor the flushing solution changes (i.e., no change occurs in the displacing-fluid viscosity, μ, or in the interfacial tension between the fluids, γ). Continuous flushing at a uniform rate with distilled water, as was done in the baseline experiments, fulfills these conditions. The ratio of gravitational to capillary forces, on the other hand, changes as the size of the residual droplets of PCE in the column decreases. Smaller droplets, present during later stages of flushing, are less mobile than larger droplets, present during early stages of flushing, because of the smaller gravitational force acting on them. Hence, as flushing progresses, the column-averaged ratio of gravitational to capillary forces decreases, and, therefore, the amount of PCE flushed per pore volume of flushing solution decreases.

The cyclic fluctuations can be understood as a cascade effect. As PCE is mobilized from upper portions of the column it coalesces with the residual in the lower portion of the column. This behavior results in an increase in the size of the residual droplets in that part of the column, with a corresponding increase in the gravitational potential for the coalesced droplets. This behavior is the equivalent of forming an oil bank in enhanced oil-recovery techniques. The cyclic character probably reflects a progressive coalescing of droplets of residual PCE migrating downward through the column. As the droplets migrate, they break up through the processes of "snap-off" and "by-passing" discussed by Chatzis et al.[29] and Wilson and Conrad,[28] leaving a secondary residual. This secondary residual is then mobilized so that the cycle is repeated continuously. The Bond-number definition of Morrow and Songkran[27] cannot account for this cyclic behavior because of its restricted focus on singlet droplets, as was discussed previously.

This cyclic mobilization is avoided during enhanced oil-recovery operations by industry through the use of a high-viscosity drive solution. In addition, oil-recovery operations do not necessarily require that the direction of oil movement parallel the direction of the gravity force vector. For these cases, the ratio of gravitational to capillary forces is less important.

SURFACTANT EFFECTS THAT DECREASE CONTAMINANT MOBILITY IN POROUS MEDIA

Because of its low cost, land disposal continues to be the preferred method for the management of municipal and hazardous waste. To protect the underlying groundwater at waste-disposal sites, a compacted clay liner, or in some cases, a compacted clay liner in combination with a synthetic liner, are constructed beneath the waste-disposal area. Clay also has been used as a component of slurry walls to control the migration of contaminated groundwater. The utility of compacted clay liners and slurry walls results from their low hydraulic conductivities, which typically are less than 10^{-6} cm/s.[35,36] Because of its polar mineral surface, however, natural clay is not an effective sorbent for nonionic contaminants. Water molecules are adsorbed preferentially by the negatively charged clay-mineral surface and thereby prevent significant adsorption of nonionic organic compounds. Sorption is limited to nonionic solute partition into the natural organic matter of the clay; this uptake is small for solutes with high solubilities and for clays with low organic carbon contents.

The purpose of this section is to discuss how certain cationic surfactants can be applied to montmorillonite-water mixtures to increase the sorptive capacity of the montmorillonite for nonionic organic solutes. Unlike other clay minerals, such as kaolinite, montmorillonite is a layered silicate that expands (swells) when wet with water. Larger, cationic surfactants readily replace inorganic cations such as Na^+ or Ca^{2+} both on the outer surfaces of the clay particle and between the silicate layers of the montmorillonite (the interlamellar space). By replacing inorganic cations in the interlamellar space of the clay mineral by an ion exchange process, the quaternary ammonium cations are said to be "intercalated" by the clay. The resulting organo-clay complex functions as a two-component sorbent with regard to the sorption of nonionic organic compounds.

The first component of the organo-clay sorbent consists of both the interlamellar and external mineral surfaces. These mineral surfaces provide sites for sorbate adsorption. In the presence of water, however, nonionic solutes are not strongly adsorbed to these sites because of the competitive adsorption of water molecules onto the polar mineral surface. The second component of the organo-clay sorbent is created by the quaternary ammonium cations. Its role in the sorption process is a function of the molecular structure of the surfactant cation. Substitution of quaternary ammonium cations having one or more alkyl chains with greater than 12 carbon atoms onto the clay creates an organic medium capable of solute uptake by a partition interaction. By contrast, substitution of smaller quaternary ammonium cations, such as tetramethylammonium (TMA), onto the clay creates a hydrophobic surface capable of solute uptake by adsorption. The extent of intercalation and subsequent adsorption of solutes from water by this latter mechanism can in some instances be highly sensitive to the size and shape of the solute molecule. The discussion of the above phenomenon will be limited largely to montmorillonite clay, but additional discussion of the use of other clay minerals and natural soil will be provided.

Surfactant-Clay Interaction

One of the most common clays used as a component of liners for waste disposal and in slurry walls is Wyoming bentonite. Bentonite is composed primarily of Na montmorillonite that is formed by the in situ alteration of volcanic ash.[37] Montmorillonite is a layered silicate that can swell to many times its original volume when exposed to water. Each layer is composed of an octahedral sheet of the general form $M_{2-3}(OH)_6$ (where M is typically Al^{3+}) sandwiched between two SiO_4 tetrahedral sheets.[38] The substitution of Al^{3+} for Si^{4+} in the tetrahedral layer and Mg^{2+} or Zn^{2+} for Al^{3+} in the octahedral layer results in a net negative surface charge on the clay.[39] To maintain electroneutrality, this charge imbalance typically is offset by exchangeable cations such as H^+, Na^+, or Ca^{2+} at the clay surface.[40] Larger organic cations can also be exchanged to neutralize the mineral-surface charge. A structural diagram of a montmorillonite layer is given in Figure 12.5.

Before discussing the sorption of nonionic organic contaminants to montmorillonite modified by organic cations, an understanding of the interaction of the clay mineral surface with the cationic surfactant in water is necessary. The layered structure of montmorillonite allows for the intercalation of water and subsequent expansion (swelling) after wetting, which in turn exposes additional mineral surface capable of cation retention. As a result, montmorillonite has a high cation-exchange capacity relative to other natural soils and clays. The retention of surfactant cations by montmorillonite is very strong when the amount of added surfactant is less than the cation-exchange capacity

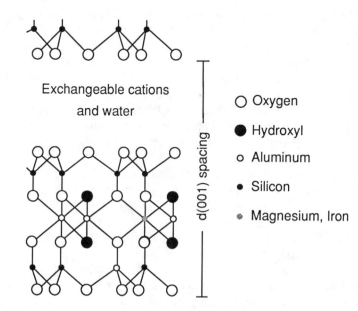

Figure 12.5. Typical structure of a montmorillonite layer viewed along the a-axis.[38]

of the clay.[24,41–44] For example, Smith et al. showed that greater than 99% of the tetraethylammonium cation added to a water–Wyoming bentonite mixture was removed from solution and retained by the clay when the added surfactant mass was less than 90% of cation-exchange capacity.[24] These researchers also noted a similarly strong uptake for nine other quaternary ammonium cations. Attempts to displace surfactant cations from montmorillonite by washing with a hydrochloric acid solution have been shown ineffective, and only organic cations of similar size are capable of exchange with the surfactant cation.[41]

The d(001) (basal) spacing of montmorillonite is a function of both the exchangeable cations present and the relative humidity. In the absence of water, Na-montmorillonite has a basal spacing of approximately 0.95 nm, and Cs-montmorillonite has been reported to have a basal spacing of 1.2 nm.[38] In dilute aqueous solution, these spacings may increase to greater than 4 nm.[38]

Inorganic cations such as Na^+ and Ca^{2+} on montmorillonite can be exchanged with surfactant cations to further increase the d(001) spacing. For example, Gieseking reported basal spacings as large as 3.0 nm for air-dried Wyoming bentonite saturated with substituted ammonium cations.[41] Table 12.3 presents the d(001) spacings for montmorillonite substituted with different amounts and types of quaternary ammonium and diammonium cations.

Table 12.3 illustrates several important points. First, comparison of the d(001)-spacing data in Table 12.3 to the d(001) spacing of Na^+ montmorillonite (approximately 0.95 nm) indicates that the organic cations are intercalated by the clay, and that increasing the size of the substituted organic cation generally increases the d(001) spacings. Neither of these effects has been observed for surfactant anions.[45] The data of Boyd et al. in Table 12.3 also show that the application of increasing amounts (up to cation-exchange capacity) of the hexadecyltrimethylammonium (HTMA) cation to montmorillonite increases the d(001) spacing.[44] This latter observation is supported by the data of Barrer and Brummer for the monomethylammonium cation.[46]

Second, the data of Theng et al.[43] and Gieseking[41] in Table 12.3 suggest that after quaternary ammonium cations are intercalated by montmorillonite, water no longer significantly influences the interlamellar spacings. Presumably, the nonpolar hydrocarbon component of the quaternary ammonium surfactant is not strongly hydrated. By contrast, inorganic cations such as Na^+, Ca^{2+}, or NH_4^+ are strongly hydrated because of their small size, the absence of a nonpolar moiety, and the dipole induced in the cation by the electric field created by the negative charge at the clay surface.

Third, despite their larger molecular size, pentyl-, octyl-, and dodecyldiammonium cations cause smaller d(001) spacings than the respective monoammonium cations. This effect is likely caused by two factors. First, a single diammonium cation satisfies two equivalents of charge. Therefore, fewer organic cations are required to satisfy 100% of the cation-exchange capacity. Second, each end of the alkyl chain of the diammonium cation is anchored to the clay-mineral surface by the positively charged ammonium ions. This can cause the alkyl chain to lie in a relatively flat position on the mineral surface or to form a

Table 12.3. The d(001) Spacings of Montmorillonite Substituted with Quaternary Ammonium Cations

Organic Cation	Percent of Cation Exchange Capacity Satisfied by Cation	d(001) Spacing, nm	Reference
Monomethylammonium	50	1.14	46
Monomethylammonium	100	1.24	46
Monomethylammonium[a]	100	1.25	43
Monomethylammonium[b]	100	1.27	43
Trimethylammonium[a]	100	1.33	43
Trimethylammonium[b]	100	1.37	43
Tetramethylammonium	100	1.35	47
Tetramethylammonium	100	1.38	43
Tetramethylammonium	100	1.38	60
Monoethylammonium[a]	100	1.27	43
Monoethylammonium[b]	100	1.28	43
Triethylammonium	100	1.32	43
Tetraethylammonium	100	1.39	47
Tetraethylammonium	100	1.40	43
Propylammonium	100	1.31	61
Pentylammonium	100	1.34	54
Hexylammonium	100	1.36	62
Octylammonium	100	1.86	62
Decylammonium	100	1.86	62
Dodecylammonium	100	1.46	54
Dodecylammonium	100	1.51	61
Dodecylammonium	100	2.0	62
Pentyldiammonium	100	1.33	54
Octyldiammonium	100	1.33	54
Dodecyldiammonium	100	1.34	54
Dodecyldiammonium	100	1.32	61
Tetradecyltrimethylammonium	100	1.83	63
Hexadecyltrimethylammonium	35	1.4	44
Hexadecyltrimethylammonium	70	1.6	44
Hexadecyltrimethylammonium	100	1.73	44
Tributylheptylammonium[c]	100	1.73	41
Tributylheptylammonium[d]	100	1.80	41
Tributylheptylammonium[e]	100	1.72	41
Tributylheptylammonium[f]	100	1.82	41

[a]Oven-dry organo-clay.
[b]Moist organo-clay.
[c]Dried and exposed over $Mg(ClO_4)_2$.
[d]Dried at 105° C.
[e]Air-dried.
[f]Under water.

bridge across the interlamellar space. In either case, the divalent cation can produce a more compact arrangement of the nonpolar component of the molecule in the interlamellar space relative to a monovalent surfactant cation with its alkyl chain anchored at only one end.

As indicated by the above discussion, quaternary ammonium cations are intercalated by the expandable clay, resulting in an increase of the d(001) spacing. The magnitude of the increase is a function of the molecular structure of the cation. Once the substitution reaction is complete, water has little subsequent effect on the d(001) spacing. Given these observations, it is not surprising to discover that the substitution of large quaternary ammonium cations for inorganic ions such as Na^+ or Ca^{2+} may have a profound effect on the sorption of nonionic organic compounds by montmorillonite.

Nonionic-Compound Sorption by Organo-Clays

The sorption of nonionic organic compounds by montmorillonite substituted with quaternary ammonium compounds was first studied from the vapor phase in the 1950s and 1960s.[46-55] Organic cations such as tetramethyl- and tetraethylammonium were found to act as props which held apart the layers of montmorillonite, thereby allowing the intercalation and sorption of large amounts of sorbate relative to Na-montmorillonite.[47] In the absence of these quaternary ammonium cations, sorption of nonionic organic vapors such as dibromoethene is limited to the external surfaces of the clay mineral.[56] Subsequent studies showed that monomethyl-, dimethyl-, trimethyl-, monoethyl-, diethyl-, and triethylammonium cations also could selectively increase the intercalation of some organic vapors.[49,50] The selective uptake was found to be influenced by the size and shape of the organic sorbate,[48,52] and gas chromatographic applications of these types of sorbents have been investigated.[55]

Figure 12.6 demonstrates more clearly the effect of quaternary ammonium cations on the sorption of an organic vapor by montmorillonite.[44] Specifically, the data quantify the sorption of benzene vapor to Ca-montmorillonite substituted with different amounts of hexadecyltrimethylammonium (HTMA) cations. The organic carbon contents of 17.3, 13.0, and 7.1% correspond to 100, 70, and 35% of the cation-exchange capacity that is satisfied by the HTMA cations. The data for Ca-montmorillonite were generated without the substitution of quaternary ammonium cations onto the clay. The ordinate in the figure is the milligrams of benzene sorbed per gram of organo-clay and the abscissa is the equilibrium vapor pressure of benzene (P) normalized by its saturation vapor pressure (P°).

The isotherms in Figure 12.6 indicate that benzene uptake by the Ca-montmorillonite is low relative to uptake by the organo-montmorillonites, and this weak uptake is caused by the limitation of sorption to the external surfaces of the clay.[44] Replacement of Ca^{2+} ions by HTMA ions causes expansion of the layered silicate and thereby allows intercalation of the benzene molecules.

Examination of the 17.3, 13.0, and 7.1% organic carbon isotherms in Figure 12.6 also suggests that two mechanisms of sorption occur simultaneously. For the sorbent with 7.1% organic carbon, the isotherm exhibits strong curvature at low relative pressures, and the isotherm shape is typical of a Type II adsorption isotherm from Brunauer's classification scheme.[34] For this case, the HTMA ions act as props that hold apart the silicate layers to allow the benzene molecules access to additional mineral-adsorption sites in the interlamellar space. Therefore, the shape of the isotherm is similar to the shape observed for benzene sorption to the Ca montmorillonite, but the magnitude of the uptake is much higher. For the sorbents with 13.0 and 17.3% organic carbon, the HTMA cations again hold apart the silicate layers to allow access to the

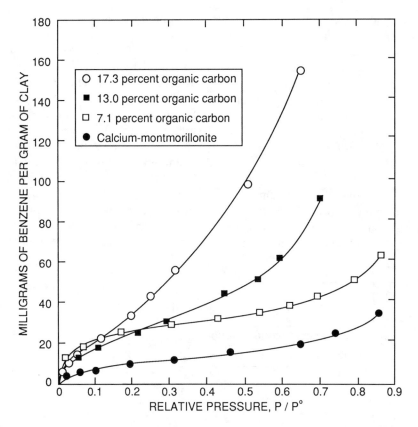

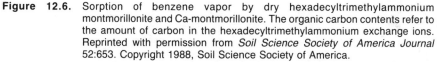

Figure 12.6. Sorption of benzene vapor by dry hexadecyltrimethylammonium montmorillonite and Ca-montmorillonite. The organic carbon contents refer to the amount of carbon in the hexadecyltrimethylammonium exchange ions. Reprinted with permission from *Soil Science Society of America Journal* 52:653. Copyright 1988, Soil Science Society of America.

interlamellar space by benzene, but more of the mineral-adsorption sites are occupied by HTMA cations and are unavailable for benzene adsorption. Therefore, adsorption is weakened relative to the sorbent with 7.1% organic carbon, and the isotherm curvature at low relative pressure is less pronounced.

At the same time, the conglomeration of the flexible alkyl chains of the HTMA ions in the interlamellar space serves as a partition medium for the uptake of benzene.[44] The increased substitution of the HTMA ions for Ca^{2+} ions, as reflected by the increase in organic carbon contents from 7.1 to 17.3%, causes a larger (and perhaps denser) partition medium for benzene uptake, as seen by the increasing linearity of the isotherms at low relative pressures with increasing organic carbon content. Although an increase in organic carbon content from 7.1 to 17.3% decreases sorbent adsorption capacity, the partition uptake of benzene from the vapor phase is increased.

The above observations suggest that montmorillonite substituted with HTMA cations can be regarded as a dual sorbent for nonionic organic compounds. The conglomeration of the alkyl chains of the HTMA ions, along with natural organic matter, forms a partition medium, and the mineral surface of the montmorillonite serves as an adsorption surface. The isotherm data of Figure 12.7, which shows a comparison of benzene sorption to HTMA-clay with 17.3% organic carbon and with 7.1% organic carbon from the gaseous and aqueous phases, lend additional support to this hypothesis. Isotherm comparison is facilitated by normalizing the equilibrium vapor pressure (P) by the saturation vapor pressure (P°), and the equilibrium aqueous concentration (C_e) by the aqueous solubility (S). In this manner, water and vapor isotherms can be compared on the same graph.

The data show that, for each sorbent, water reduces benzene sorption. The polar water molecules are strongly adsorbed by the hydrophilic clay-mineral surfaces and thereby prevent adsorption of the nonionic benzene molecules.

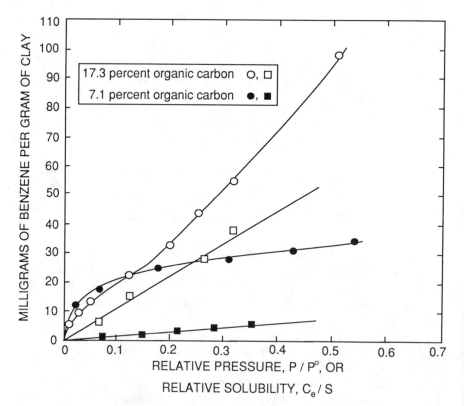

Figure 12.7.   Comparative sorption of benzene as a vapor (circles) and as a solute from water (squares) by hexadecyltrimethylammonium montmorillonite. The organic carbon contents refer to the amount of carbon in the hexadecyltrimethylammonium exchange ions. Reprinted with permission from *Soil Science Society of America Journal* 52:655. Copyright 1988, Soil Science Society of America.

The observed reduction in sorption is greatest for the sorbent with only 7.1% organic carbon because a larger percentage of its mineral surface is available for adsorption than for the sorbent with 17.3% organic carbon. Therefore, in the presence of water, benzene sorption is limited to partition into the organic medium formed by the alkyl chains of the HTMA cations.[44] Similarly, Smith et al. have shown that tetrachloromethane sorption from water by dodecyltrimethyl-, tetradecyltrimethyl-, hexadecyltrimethyl-, benzyldimethyl-hexadecyl-, and dodecyldimethyl(2-phenoxyethyl) ammonium clays is characterized by relatively weak solute uptake, isotherm linearity, and noncompetitive sorption.[24] The measured organic-carbon-normalized sorption coefficients (K_{oc}) are comparable to K_{oc} values that have been measured for tetrachloromethane sorption to natural soil from water. In addition, the K_{oc} values were shown to increase with the size of the nonpolar hydrocarbon functional group(s) of the quaternary ammonium cation used to treat the clay.[24] These results strongly suggest that nonionic solute uptake by clays treated with quaternary ammonium cations with large alkyl functional groups (such as HTMA) is caused by a partition interaction of the sorbate between water and the organic medium formed by the surfactant cations. Nonionic solute uptake from water by Na- and Ca-montmorillonite is negligible.[24,44]

To this point, the discussion of the mechanistic effect of quaternary ammonium cations on nonionic compound sorption has been limited to cationic surfactants with at least one long-chain alkyl functional group (i.e., 12 or more carbon atoms). Figure 12.8 shows, however, that the modification of montmorillonite with quaternary ammonium cations having smaller functional groups can change both the magnitude and mechanism of nonionic solute sorption.[24] Graph A quantifies tetrachloromethane sorption to benzyltrimethylammonium clay (BTMA clay) in both the presence and absence of a second solute, trichloroethene. Tetrachloromethane sorption by BTMA clay is characterized by relatively strong solute uptake, isotherm nonlinearity, and competitive sorption. Similar results were observed for clay modified with tetramethylammonium (TMA), tetraethylammonium, and benzyltriethylammonium. This behavior is typical of an adsorption process. By contrast, Graph B quantifies tetrachloromethane sorption to HTMA clay, which is characterized by relatively weak solute uptake, isotherm linearity, and noncompetitive sorption. Results similar to those shown in Graph B were also observed for clay modified with dodecyltrimethyl-, tetradecyltrimethyl-, hexadecyltrimethyl-, benzyldimethylhexadecyl-, and dodecyldimethyl(2-phenoxyethyl)-ammonium. This behavior is typical of a partition process.

Therefore, the data of Smith et al.[24] suggest that the difference in the sorption mechanism for the two groups of organo-clays is caused by differences in the lengths of the alkyl chains: organic cations that caused solute uptake by an adsorption process all have alkyl chains with less than 3 carbon atoms; organic cations that caused solute uptake by a partition process all have one or more alkyl chains with more than 11 carbon atoms. The benzyl functional group, by contrast, appears to have no effect on the mechanism of sorption. This conclu-

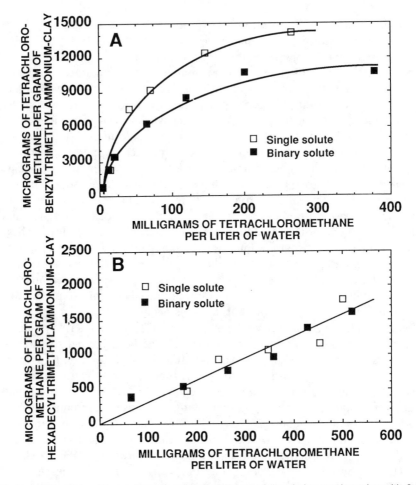

Figure 12.8.   Tetrachloromethane sorption to (*A*) benzyltrimethylammonium clay with 3.8% organic carbon and (*B*) hexadecyltrimethylammonium clay with 6.16% organic carbon. The binary solute is trichloroethene. Reprinted from *Environmental Science and Technology* 24:1167.

sion is confirmed by the observation that tetrachloromethane sorption to TMA, benzyltrimethylammonium, tetraethylammonium, and benzyltriethylammonium clay is characterized by strong solute uptake, isotherm nonlinearity, and competitive sorption.[24]

These observations are supported by data from other researchers.[44,50,51,57-60] Barrer and Kelsey measured strongly exothermic heats of sorption for the uptake of organic vapors by monomethylammonium montmorillonite,[50] but the uptake of organic vapors by dimethyldioctadecylammonium montmorillonite was characterized by comparatively low heats of sorption that were at times endothermic.[51] Strongly exothermic heats of sorption are requisite for adsorption processes to balance the loss in entropy resulting from solute adsorption. The heat of sorption must be at least as great as the solute's heat of condensation from water. A partition process is characterized by relatively low heats of sorption that equal the difference between the molar enthalpies of solution of the solute in the aqueous and organic phases. The heat of sorption typically is less exothermic than the solute's heat of condensation from water and can be endothermic. Therefore, the data of Barrer and Kelsey[51] and Barrer and Perry[52] indicate that the organic vapors are taken up by monomethylammonium clay by an adsorption-dominated process, and by dimethyldioctadecylammonium clay by a partition-dominated process. The two 18 carbon-chain functional groups on the latter sorbent apparently are capable of forming an organic partition medium for the uptake of nonionic organic vapors.

Lee et al. have observed that the magnitude of sorption from water for a variety of nonionic organic compounds to soil modified with nonyltrimethylammonium, dodecyltrimethylammonium, or HTMA cations varies inversely with solute solubility.[58] This solubility dependence is again characteristic of a partition process wherein the solute distributes itself between water and organic phases in proportion to its solubility in the two phases. Because the percent variation in aqueous solubility generally is much greater than the percent variation in solubility in a nonpolar, organic medium for a group of nonionic organic compounds, aqueous solubility alone generally can be used to accurately predict the partition of a solute between aqueous and organic phases. Lee et al. also observed isotherm linearity for all nonionic solute sorption data.[58] Linear isotherms for nonionic solute sorption from water to soil and clay modified with HTMA cations have been reported in other studies.[44,57] By contrast, nonlinear isotherms (characteristic of an adsorption process) have been reported for benzene sorption from water to montmorillonite modified by TMA cations.[59,60]

Figures 12.9 and 12.10 provide additional sorption data that are particularly relevant from an engineering perspective. Figure 12.9 presents previously unpublished data generated experimentally by the batch equilibration methods described by Smith et al.[24] Each data point on the graph is derived from a complete isotherm quantifying tetracloromethane sorption from water to an organo-clay. The organo-clay is Wyoming bentonite with varying percentages of its cation-exchange capacity satisfied by benzyltriethylammonium cations. As discussed earlier, tetrachloromethane sorption to this sorbent from water is characterized by nonlinear isotherms, competitive sorption, and relatively strong solute uptake.[45] The sorption data were fit to the linear form of the Langmuir equation given by

$$\frac{C_e}{C_s} = \frac{1}{ab} + \frac{C_e}{a} \qquad (12.3)$$

where C_e is the equilibrium aqueous solute concentration (mmol/L) and C_s is the equilibrium sorbed concentration of the contaminant (mmol/g). The

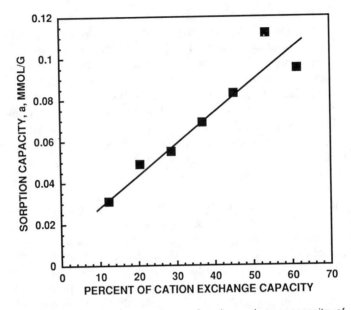

Figure 12.9. Relation between the percent of cation-exchange capacity of Wyoming bentonite satisfied by benzyltriethylammonium cations and the tetrachloromethane sorption capacity of the organo-clay.

parameters a and b are determined by a linear regression of C_e against C_e/C_s. The parameter a is the sorption capacity (mmol/g) of the organo-clay. Figure 12.9 shows the relation between the percentage of the cation-exchange capacity of the clay satisfied by benzyltriethylammonium cations and the sorption capacity, a, of the clay.

As evidenced by the data, the sorption capacity increases in an approximately linear fashion with increased substitution of the quaternary ammonium cation for inorganic cations. Therefore, doubling the mass of organic cations applied to a clay liner will likely double the sorption capacity of the clay for tetrachloromethane, provided that less than 100% of the cation-exchange capacity of the clay is satisfied by the quaternary ammonium cations. More fundamentally, the data reinforce the hypothesis that nonionic solute uptake by the inorganic-cation-treated mineral surface is negligible relative to uptake by the organic-cation-treated mineral surface.

Figure 12.10 illustrates another important consideration regarding the use of TMA cations to increase the magnitude of sorption of nonionic organic compounds from water by clay. The figure presents sorption data for eight aromatic solutes from water to TMA-montmorillonite. To allow a direct comparison of all the isotherms, the equilibrium aqueous concentrations, C_e, of each solute have been normalized by their respective aqueous solubilities, S. The data show that TMA-montmorillonite is a selective sorbent.[59] Uptake of benzene by the organo-clay from water is very strong relative to that by the larger

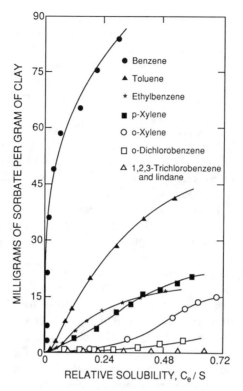

Figure 12.10.   Adsorption of eight organic compounds from water by tetramethylammonium-montmorillonite. Reprinted with permission from *Journal of the Chemical Society Faraday Transactions* 85:2953. Copyright 1989, Royal Society of Chemistry.

solutes. The addition of small functional groups to the benzene ring presumably prevents significant intercalation and adsorption of the solutes. No measurable uptake of the largest solutes (1,2,3-trichlorobenzene and lindane) was observed even though these solutes have the lowest aqueous solubilities of the group. The size- and shape-selective sorption of nonionic organic compounds from water by TMA-montmorillonite also has been noted by Smith et al.,[24] who reported strong uptake of tetrachloromethane by TMA-montmorillonite relative to trichloroethene uptake by the same sorbent.

Although it has been shown that small quaternary ammonium cations such as TMA and benzyltriethylammonium can modify the external and internal surfaces of montmorillonite to create a strong adsorbent for benzene and tetrachloromethane sorption from water, these sorbents may be unable to effectively remove larger solute molecules from solution. The work of Lee et al. clearly demonstrates this point for TMA-montmorillonite, suggesting that the treatment of clays for use in landfills by TMA cations is not suitable if the leachate contains a variety of nonionic organic solutes.[59] No shape or size selectivity has been reported for tetraethylammonium, benzyltrimethylammonium, or benzyltriethylammonium montmorillonite, however; in fact, these three sorbents have been shown to effectively remove both tetrachloromethane and trichloroethene from water.[24] Additional studies quantifying the sorption of different solutes to these three sorbents are required to determine their utility for use with clay as liners at waste-disposal facilities or as a component of slurry walls.

Application of larger quaternary ammonium cations with at least one alkyl functional group having 12 or more carbon atoms (e.g., HTMA) to montmorillonite also increases nonionic solute uptake from water relative to Na or Ca montmorillonite by creating an organic partition medium at the clay-mineral surface. No competitive sorption is observed, and the magnitude of solute uptake varies inversely with the aqueous solubility of the solute. Certainly, application of these larger quaternary ammonium cations to clay may be a viable technology to retard contaminant migration through earthen liners and slurry walls.

The benefits derived from the application of quaternary ammonium cations to clay are not limited to a specific mineral type. The advantage of using montmorillonite is that the expandable clay allows for the intercalation of both the organic cation and the solute. Therefore, exceptionally large sorption capacities are possible. For nonexpandable clay minerals such as illite, however, application of quaternary ammonium cations to the clay can also increase solute uptake, although sorption is limited to the external surfaces of the organo-clay. This observation is supported by data that show the uptake of a variety of solutes by TMA-illite is significantly reduced relative to uptake by TMA-montmorillonite and is increased relative to Ca-montmorillonite.[59] Silt- and sand-size particles also have a negative charge on their mineral surfaces and therefore are capable of retaining appreciable amounts of quaternary ammonium cations. Specifically, application of nonyl-, dodecyl-, and hexadecyltrimethylammonium cations to natural soils has been shown to increase the soils' sorptive capacity for a variety of nonionic organic compounds.[57,58] The resulting K_{oc} values indicated that the organic medium formed by the cationic surfactants is at least 10 times more effective than natural organic matter with respect to the sorption of nonionic organic compounds from water when approximately 100% of the cation-exchange capacity of the soil is satisfied by the organic cations.[57,58]

Other Transport Considerations: Diffusion, Dispersion, and Permeability

The primary physicochemical factors affecting the transport of nonionic organic contaminants through clay are the solute retardation caused by sorption, solute diffusion and dispersion, and the hydraulic conductivity (permeability) of the porous media. The above discussion shows that nonionic solute retardation by sorption can be increased significantly by the application of certain quaternary ammonium compounds to soil and clay. However, the effect of this chemical treatment on solute diffusion/dispersion and the hydraulic conductivity of the porous medium is unknown.

Because of their negative surface charge, the external surfaces of clay particles in an aqueous solution are believed to be surrounded by a diffuse double layer of charge, which consists of water molecules and cations. The cations (typically Na^+ and Ca^{2+}) are attracted to the surface of the clay by electrostatic forces in accord with Coulomb's Law. At the same time, however, this electrostatic attraction is partially counteracted by the diffusion of the cations into the bulk solution. The combined effect of electrostatic attraction and diffusion results in a decrease in cation concentration with distance from the surface of the clay. The layer of negative charge at the surface of the clay particle and the diffuse layer of cations surrounding the clay constitute the diffuse double layer.

Water molecules in the diffuse double layer are tightly bound to the clay particle, and therefore do not constitute part of the flow field around the clay. By increasing the thickness of the double layer, the hydraulic conductivity of a porous clay medium (and, consequently, the advection of contaminants) can be reduced. This hypothesis is supported by the observed greater hydraulic conductivity of Ca-clay relative to Na-clay. Because one Ca^{2+} ion satisfies two equivalents of charge, only half as many Ca^{2+} as Na^+ ions are needed for electroneutrality. The smaller number of Ca^{2+} ions can cause the thickness of the diffuse double layer to be reduced, and the hydraulic conductivity can subsequently be increased. Therefore, the substitution of large organic cations (e.g., quaternary ammonium surfactants) for Na^+ and Ca^{2+} ions at the surface of a montmorillonitic clay may increase the thickness of the double layer and consequently decrease the permeability of the porous medium.

Conversely, it has been shown that, following the substitution of quaternary ammonium cations for organic ions, the interlamellar expansion (swelling) of montmorillonite is increased by nonpolar organic solvents[61] and is no longer affected by water.[41] Similarly, Barrer and Millington observed that further expansion of dry organo-clays appears to be a function of the sorbate molecular structure.[54] Therefore, the adverse effect of quaternary ammonium cations on the expansion of montmorillonite in water can contribute to an increase in the hydraulic conductivity of the clay medium.

SUMMARY

Nonionic and anionic surfactants can increase the mobility of dissolved and non-aqueous-phase organic compounds in porous media. For dissolved constituents, surfactant concentrations greater than critical micelle concentration can increase solute solubility in water by creating a nonpolar, micellar phase in water favorable for contaminant uptake by partition. The enhanced solubility of the solute results in reduced sorption to natural soil and increased mobility in porous media. For non-aqueous-phase organic liquids at residual saturation in vadose-zone soil, anionic surfactant solutions can lower the interfacial fluid tensions so that gravity and viscous forces can transport the organic liquid downward to a collection system.

Conversely, cationic surfactants can decrease the mobility of nonionic organic compounds in clay-water and soil-water systems by increasing contaminant sorption. Cationic surfactants are exchanged easily with inorganic ions such as Ca^{2+} or Na^+ at the mineral-water interface and can increase significantly the sorption of nonionic organic compounds to the soil or clay from water. Sorption of nonionic organic solutes by clay modified with quaternary ammonium surfactants with small functional groups (e.g., methyl, ethyl, or benzyl groups) is characterized by isotherm nonlinearity, competitive sorption, and relatively strong solute uptake. For these cases, the sorption capacity appears to be a linear function of the percentage of the clay's cation-exchange capacity that is satisfied by the surfactant cations. Sorption of nonionic organic solutes by clay modified with quaternary ammonium surfactants with large functional groups (e.g., hydrocarbon chains with more than 11 carbon atoms) is characterized by isotherm linearity, noncompetitive sorption, and relatively weak solute uptake. This behavior is typical of a partition process wherein the solute is distributed between two phases in proportion to the compound's solubility in each phase.

DISCLAIMER

ACKNOWLEDGMENTS

The authors thank Drs. Patrick Witkowski, Julia Barringer, and David Crerar for helpful technical reviews of the manuscript. This research was cooperatively supported by the Division of Science and Research of the New Jersey Department of Environmental Protection and the U.S. Geological Survey.

REFERENCES

1. Karsa, D. R., Ed. *Industrial Applications of Surfactants* (London: The Royal Society of Chemistry, 1987).
2. Hall, D. G., and G. J. T. Tiddy. "Surfactant Solutions: Dilute and Concentrated," in *Anionic Surfactants: Physical Chemistry of Surfactant Action*, E. H. Lucassen-Reynders, Ed. (New York: Marcel Dekker, Inc., 1981), pp. 55–108.

3. Tanford, C. "The Hydrophobic Effect and the Organization of Living Matter," *Science* 200:1012–1018 (1978).

4. Tanford, C. *The Hydrophobic Effect: Formation of Micelles and Biological Membranes* (New York: John Wiley and Sons, 1980).

5. McBain, J. W., and P. H. Richards. "Solubilization of Insoluble Organic Liquids by Detergents," *Ind. Eng. Chem.* 38:642 (1946).

6. Elworthy, P. E., A. T. Florence, and C. B. McFarlane. *Solubilization by Surface Active Agents* (New York: Academic Press, 1968).

7. Wershaw, R. L., P. J. Burcar, and M. C. Golberg. "Interaction of Pesticides with Natural Organic Material," *Environ. Sci. Technol.* 3:271–273 (1969).

8. Matsuda, K., and M. Schnitzer. "Reactions between Fulvic Acid, a Soil Humic Material, and Dialkyl Phthalates," *Bull. Environ. Contam. Toxicol.* 6:200–204 (1971).

9. Boehm, P. D., and J. G. Quinn. "Solubilization of Hydrocarbons by the Dissolved Organic Matter in Sea Water," *Geochim. Cosmochim. Acta* 37:2459–2477 (1973).

10. Schwartz, A. M., and J. W. Perry. *Surface Active Agents—Their Chemistry and Technology* (Huntington, New York: Robert E. Krieger Publishing Co., 1978).

11. Hassett, J. P., and M. A. Anderson. "Association of Hydrophobic Organic Compounds with Dissolved Organic Matter in Aquatic Systems," *Environ. Sci. Technol.* 13:1526–1529 (1979).

12. Carter, C. W., and I. H. Suffet. "Binding of DDT to Dissolved Humic Materials," *Environ. Sci. Technol.* 16:735–740 (1982).

13. Caron, G., I. H. Suffet, and T. Belton. "Effect of Dissolved Organic Carbon on the Environmental Distribution of Nonpolar Organic Compounds," *Chemosphere* 14:993–1000 (1985).

14. Chiou, C. T., R. L. Malcolm, T. I. Brinton, and D. E. Kile. "Water Solubility Enhancement of Some Organic Pollutants and Pesticides by Dissolved Humic and Fulvic Acids," *Environ. Sci. Technol.* 20:502–508 (1986).

15. Smith, J. A., P. J. Witkowski, and T. V. Fusillo. "Manmade Organic Compounds in the Surface Waters of the United States—A Review of Current Understanding," *U.S. Geological Survey Circular 1007* (1988).

16. Smith, J. A., P. J. Witkowski, and C. T. Chiou. "Partition of Nonionic Organic Compounds in Aquatic Systems," *Rev. Environ. Contam. Toxicol.* 103:127–151 (1988).

17. Kile, D. E., and C. T. Chiou. "Water Solubility Enhancements of DDT and Trichlorobenzene by Some Surfactants Below and Above the Critical Micelle Concentration," *Environ. Sci. Technol.* 23(7):832–838 (1989).

18. Kile, D. E., C. T. Chiou, and R. S. Helburn. "Effect of Some Petroleum Sulfonate Surfactants on the Apparent Water Solubility of Organic Compounds," *Environ. Sci. Technol.* 24(2):205–208 (1990).

19. Karickhoff, S. W., D. S. Brown, and T. A. Scott. "Sorption of Hydrophobic Pollutants on Natural Sediments," *Water Res.* 13:241–248 (1979).

20. Chiou, C. T., L. J. Peters, and V. H. Freed. "A Physical Concept of Soil-Water Equilibria for Nonionic Organic Compounds," *Science* 206:831–832 (1979).

21. Senesi, N., and C. Testini. "Adsorption of Some Nitrogenated Herbicides by Soil Humic Acids," *Soil Sci.* 130:314–320 (1980).

22. Sharom, M. S., J. R. W. Milers, C. R. Harris, and F. L. McEwen. "Behaviour of 12 Pesticides in Soil and Aqueous Suspensions of Soil and Sediment," *Water Res.* 14:1095–1100 (1980).

23. Chiou, C. T., P. E. Porter, and D. W. Schmedding. "Partition Equilibria of Nonionic Organic Compounds between Soil Organic Matter and Water," *Environ. Sci. Technol.* 17:227–231 (1983).
24. Smith, J. A., P. R. Jaffé, and C. T. Chiou. "Effect of Ten Quaternary Ammonium Cations on Tetrachloromethane Sorption to Clay from Water," *Environ. Sci. Technol.* 24:1167–1172 (1990).
25. Taber, J. J. "Dynamic and Static Forces Required to Remove a Discontinuous Oil Phase from Porous Media Containing Both Oil and Water," *Soc. Pet. Eng. J.* 9(1):3–12 (1969).
26. Melrose, J. C., and C. F. Brandner. "Role of Capillary Forces in Determining Microscopic Displacement Efficiency for Oil Recovery by Waterflooding," *J. Can. Pet. Tech.* 13(4):54–62 (1974).
27. Morrow, N. R., and B. Songkran. "Effect of Viscous and Buoyancy Forces on Nonwetting Phase Trapping in Porous Media," in *Surface Phenomena in Enhanced Oil Recovery*, D. O. Shah, Ed. (New York: Plenum Press, 1981), pp. 387–411.
28. Wilson, J. L., and S. H. Conrad. "Is Physical Displacement of Residual Hydrocarbons a Realistic Possibility in Aquifer Restoration?" in *Proceedings of the NWWA/API Conference on Petroleum Hydrocarbons and Organic Chemicals in Ground Water—Prevention, Detection, and Restoration* (Dublin, OH: National Water Well Association, 1984), pp. 274–298.
29. Chatzis, I., N. R. Morrow, and H. T. Lim. "Magnitude and Detailed Structure of Residual Oil Saturation," *Soc. Pet. Eng. J.* 23(2):311–326 (1983).
30. Gogarty, W. B. "Oil Recovery with Surfactants: History and a Current Appraisal," in *Improved Oil Recovery by Surfactant and Polymer Flooding*, D. O. Shah and R. S. Schechter, Ed. (New York: Academic Press, 1977).
31. Lohmann, S. W. "Ground-Water Hydraulics," *U.S. Geological Survey Professional Paper 708*, U.S. Government Printing Office (1972).
32. Tuck, D. M., P. R. Jaffé, D. A. Crerar, and R. T. Mueller. "Enhancing Recovery of Immobile Residual Non-Wetting Hydrocarbons from the Unsaturated Zone Using Surfactant Solutions," in *Proceedings of the NWWA/API Conference on Petroleum Hydrocarbons and Organic Chemicals in Ground Water—Prevention, Detection, and Restoration* (Dublin, OH: National Water Well Association, 1988), pp. 457–478.
33. Jaffé, P. R., D. M. Tuck, J. D. Hohmann, D. A. Crerar, and M. P. Borcsik. "Investigation of Residual Solvent Recovery in Soils Using Surfactants," draft technical report prepared for the New Jersey Department of Environmental Protection, Trenton, NJ (1988).
34. Adamson, A. W. *Physical Chemistry of Surfaces*, 4th ed. (New York: John Wiley and Sons, 1982).
35. Daniel, D. E. "Predicting Hydraulic Conductivity of Clay Liners," *J. Geotech. Eng.* 110:285–300 (1984).
36. Boynton, S. S., and D. E. Daniel. "Hydraulic Conductivity Tests on Compacted Clay," *J. Geotech. Eng.* 111:465–478 (1985).
37. Grim, R. E. *Applied Clay Mineralogy* (New York: McGraw-Hill, 1962).
38. Theng, B. K. G. *The Chemistry of Clay-Organic Reactions* (New York: John Wiley and Sons, 1974).
39. Stumm, W., and J. J. Morgan. *Aquatic Chemistry. An Introduction Emphasizing Chemical Equilibria in Natural Waters* (New York: John Wiley and Sons, 1981).

40. Worrall, W. E. *Clays: Their Nature, Origin and General Properties* (New York: Transatlantic Arts, 1968).

41. Gieseking, J. E. "The Mechanism of Cation Exchange in the Montmorillonite-Beidellite-Nontronite Type of Clay Minerals," *Soil Sci.* 47:1–13 (1939).

42. Cowan, C. T., and D. White. "The Mechanism of Exchange Reactions Occurring between Sodium montmorillonite and Various *n*-Primary Aliphatic Amine Salts," *Trans. Faraday Soc.* 54:691–697 (1958).

43. Theng, B. K. G., D. J. Greenland, and J. P. Quirk. "Adsorption of Alkylammonium Cations by Montmorillonite," *Clay Miner.* 7:1–17 (1967).

44. Boyd, S. A., M. M. Mortland, and C. T. Chiou. "Sorption Characteristics of Organic Compounds on Hexadecyltrimethylammonium-Smectite," *Soil Sci. Soc. Amer. J.* 52:652–657 (1988).

45. Law, J. R., Jr., and G. W. Kunze. "Reactions of Surfactants with Montmorillonite: Adsorption Mechanisms," *Soil Sci. Soc. Am. Proc.* 30:321–327 (1966).

46. Barrer, R. M., and K. Brummer. "Relations between Partial Ion Exchange and Interlamellar Sorption in Alkylammonium Montmorillonites," *Trans. Faraday Soc.* 59:959–968 (1963).

47. Barrer, R. M., and D. M. MacLeod. "Activation of Montmorillonite by Ion Exchange and Sorption Complexes of Tetra-Alkyl Ammonium Montmorillonites," *Trans. Faraday Soc.* 51:1290–1300 (1955).

48. Barrer, R. M., and M. G. Hampton. "Gas Chromatography and Mixture Isotherms in Alkyl Ammonium Bentonites," *Trans. Faraday Soc.* 53:1462–1475 (1957).

49. Barrer, R. M., and J. S. S. Reay. "Sorption and Intercalation by Methylammonium Montmorillonites," *Trans. Faraday Soc.* 53:1253–1261 (1957).

50. Barrer, R. M., and K. E. Kelsey. "Thermodynamics of Interlamellar Complexes. Part 1. Hydrocarbons in Methylammonium Montmorillonites," *Trans. Faraday Soc.* 57:452–462 (1961).

51. Barrer, R. M., and K. E. Kelsey. "Thermodynamics of Interlamellar Complexes. Part 2. Sorption by Dimethyldioctadecylammonium Bentonite," *Trans. Faraday Soc.* 57:625–640 (1961).

52. Barrer, R. M., and G. S. Perry. "Sorption of Mixtures, and Selectivity in Alkylammonium Montmorillonites. Part I. Monomethylammonium Bentonite," *J. Chem. Soc.* (Part 1):842–849 (1961).

53. Barrer, R. M., and G. S. Perry. "Sorption of Mixtures, and Selectivity in Alkylammonium Montmorillonites. Part II. Tetramethylammonium Montmorillonite," *J. Chem. Soc.* (Part 1):850–858 (1961).

54. Barrer, R. M., and A. D. Millington. "Sorption and Intracrystalline Porosity in Organo-Clays," *J. Coll. Interface Sci.* 25:359–372 (1967).

55. White, D., and C. T. Cowan. "The Sorption Properties of Dimethyldioctadecylammonium Bentonite Using Gas Chromatography," *Trans. Faraday Soc.* 54:557–561 (1958).

56. Jurinak, J. J. "The Effect of Clay Minerals and Exchangeable Cations on the Adsorption of Ethylene Dibromide Vapor," *Soil Sci. Soc. Amer. Proc.* 21:599–602 (1957).

57. Boyd, S. A., J.-F. Lee, and M. M. Mortland. "Attenuating Organic Contaminant Mobility by Soil Modification," *Nature* 333:345–347 (1988).

58. Lee, J.-F., J. R. Crum, and S. A. Boyd. "Enhanced Retention of Organic Contam-

inants by Soils Exchanged with Organic Cations," *Environ. Sci. Technol.* 23:1365–1372 (1989).

59. Lee, J.-F., M. M. Mortland, S. A. Boyd, and C. T. Chiou. "Shape-Selective Adsorption of Aromatic Molecules from Water by Tetramethylammonium-Smectite," *J. Chem. Soc. Faraday Trans. I* 85:2953–2962 (1989).

60. Lee, J.-F., M. M. Mortland, C. T. Chiou, D. E. Kile, and S. A. Boyd. "Adsorption of Benzene, Toluene and Xylene by Two Tetramethylammonium-Smectites Having Different Charge Densities," *Clays Clay Miner.* 38:113–120 (1990).

61. Wolfe, T. A., T. Demirel, and E. R. Baumann. "Interaction of Aliphatic Amines with Montmorillonite to Enhance Adsorption of Organic Pollutants," *Clays Clay Miner.* 33(4):301–311 (1985).

62. Stul, M. S., A. Maes, and J. B. Uytterhoeven. "The Adsorption of *n*-Aliphatic Alcohols from Dilute Aqueous Solutions on RNH$_3$-Montmorillonites. Part I. Distribution at Infinite Dilution," *Clays Clay Miner.* 26(5):309–317 (1978).

63. Harper, M., and C. J. Purnell. "Alkylammonium Montmorillonites As Adsorbents for Organic Vapors from Air," *Environ. Sci. Technol.* 24(1):55–61 (1990).

CHAPTER 13

The Effects of Pore-Water Colloids on the Transport of Hydrophobic Organic Compounds from Bed Sediments

G. J. Thoma, A. C. Koulermos, K. T. Valsaraj, D. D. Reible, and L. J. Thibodeaux

INTRODUCTION

Anthropogenic pollutants are ubiquitous in the environment. Our ability to protect ourselves and other life from exposure to these chemicals depends partly on our ability to predict their fate and transport within and between environmental compartments—air, water, and soil. These predictions, in turn, depend on accurate descriptions of the transport mechanisms involved and the pollutant properties.

Aquatic systems often carry a heavy burden of human waste. Hydrophobic compounds, in particular, tend to adsorb to the organic fraction of the sedimentary material found in aquatic systems. Thus, the sediment is a likely sink for these pollutants. If the source of pollution (industrial or other) is eliminated, the concentration gradient between the water column and the sediment will reverse, and the sediment will become a pollutant source. Thus, the influence of these chemicals may outlast the time required to dilute the polluted water body.

Many transport processes influence the fate of organic compounds in aquatic systems. Due to the tendency of these compounds to sorb to the sediment phase, the transport mechanisms between water column and sediment bed are generally the most important. The primary transport phenomena within bed sediments are adsorption-retarded diffusion, advection, erosion or deposition, and bioturbation (benthic organisms circulating sediment particles and pore water). Figure 13.1 presents a schematic representation of several of these and related processes.

Situations will exist where the transport is dominated by molecular diffusion retarded by sorption/desorption with the immobile sediment. Under these conditions the presence of organic colloids in the pore water may enhance the

231

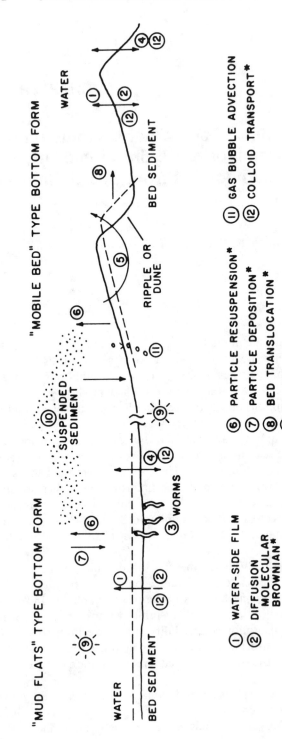

Figure 13.1.   Contaminant transport mechanisms between sediment and water.

flux of hydrophobic organic compounds (HOCs) over that of simple retarded diffusion.[1,2] Organic colloids have a large capacity to bind HOCs.[3,4] Adsorbed HOCs will be carried across the sediment-water interface by the Brownian movement of the colloidal particles, in addition to molecular diffusion of the solute out of the pore water.

Dissolved organic compounds (DOCs), originating from decaying plant and animal matter, are the primary source of colloidal material in sediments. Colloidal particles are primarily comprised of aggregates of humic acids that are stable in low-salinity environments.[5] DOC concentrations from 10 to 390 mg/L for anaerobic interstitial sediment waters have been reported.[5] Organic carbon in natural waters is divided into two size classes. Dissolved organic carbon is arbitrarily defined as aquatic carbon that passes a 0.45-μm filter.[5,6]

This chapter quantifies some transport properties of naturally occurring organic colloids and predicts their effect on the leaching of hydrophobic contaminants from bed sediments. We investigate the transport of natural colloids from bed sediments in terms of sorption/desorption kinetics and equilibrium. The effect of sediment concentration on DOC concentration and on the partitioning of naphthalene is also studied. Finally, a simple mathematical model is proposed that quantifies the transport of HOCs associated with sedimentary DOC.

EXPERIMENTAL

In this study, DOC concentration was considered a surrogate measure of the HOC-binding colloidal matter concentration. A series of laboratory experiments was designed to evaluate some of the characteristics of DOC important in defining their role in the transport and fate of organic pollutants in aquatic systems. Emphasis was placed on processes within the sediment bed, since transport between sediment and water is generally sediment-side controlled.[7]

Sediment Physical Properties and Pore-Water DOC Properties

Air-dried sediment samples obtained from the U.S. EPA Research Laboratory in Athens, Georgia, the Institute of Environmental Studies at Louisiana State University (LSU), and a bottom sediment from the University Lake at Baton Rouge were used. The bed porosity and bulk density of each sediment under laboratory conditions were estimated by adding water to known masses of dry sediment and measuring the bed volume after 2 days. The sediment characteristics are presented in Table 13.1.

The lake sediment was prepared by wet sieving (6-mm mesh size) to remove larger particles and all large organic material such as leaves and twigs. The LSU Soil Testing Laboratory measured the easily oxidizable organic matter (EOOM). EOOM is approximately 71% of the total organic matter, which is in turn approximately 1.7 times the total organic carbon;[8] thus, TOC $\cong$ EOOM/

Table 13.1. Sediment Characteristics

Sediment	pH	Organic Carbon %	Sand %	Clay %	Silt %	Bed Porosity	Bulk Density g/cm^3
EPA5	7.4	1.55[a]	33.6	31.0	35.4	0.47	0.946
Choptank River	5.7	1.36	57.0	7.0	36.0	0.56	0.88
Drum Point	3.7	0.7	21.0	11.0	68.0	0.48	0.99
Turkey Point	6.4	2.00	25.0	7.0	68.0	0.55	0.85
Univ. Lake	5.7	2.48	—	—	—	0.58	0.76

Source: Data for pH and organic carbon, except University Lake, and for sand, clay, and silt content from Means et al.[25]
[a]Means et al. report 2.28;[25] organic carbon may have oxidized in dry storage.

(1.724 × 0.71). The bulk of the University Lake sediment was air-dried several days, then ground into a fine powder by mortar and pestle. After sieving (0.33-mm pore size) the dry sediment, it was stored in an airtight plastic bucket.

DOC Size Fractionation

A 10-mL Amicon ultrafiltration cell with a 0.4-μm pore-size Nuclepore membrane filter was used to remove the particulate fraction from the water sample.[9,10] It was operated at a pressure of 30 psi of nitrogen. Filter membranes were prewashed by filtering at least 50 mL of deionized water through them to minimize organic carbon contamination of the filtrate from the filter.[10,11] The particle concentration above the filter membrane was kept below recommended levels to prevent breakthrough of larger particles,[12,13] and the solution was stirred by a magnetic stirring bar during filtration to reduce clogging of the membrane pores.[6] DOC size fractions were obtained by the same method, using 0.15-μm and 0.05-μm Nuclepore filters, and UM10 (0.003-μm pore size) and UM2 (0.0013-μm pore size) Amicon filters with N_2 pressures up to 55 psi for the smallest pore-size filters.

DOC Concentration Measurements

Organic carbon concentration measurements were made using a Technicon AutoAnalyzer II, which measures the total organic carbon of an aqueous sample by the UV-promoted persulfate oxidation method.[14] The 20-mL samples were frozen prior to analysis to prevent contamination or loss of carbon.

Sorption/Desorption of DOC to Suspended Sediment

Batch experiments were performed to elucidate some of the fundamental processes in the interaction between DOC and suspended sediment. Both the equilibrium behavior and the kinetics were investigated.

Equilibrium Partitioning

The equilibrium partitioning of DOC between suspended sediment and water was studied by both desorption and sorption experiments. A desorption experiment was performed with each of the five sediments characterized in Table 13.1. Dry sediment was suspended in distilled water in a 150-mL bottle sealed with a Teflon-lined rubber septum and aluminum crimp cap. The mixture was placed in a 25°C constant-temperature water bath for 30 days. Aqueous samples were taken and analyzed for DOC as before. A range of sediment concentrations was used to evaluate the DOC partitioning isotherm. In addition, low-shear replicate samples from the DOC desorption kinetics study were kept in a 25°C water bath for 140 days, then sampled to determine the equilibrium DOC concentration.

Equilibrium sorption experiments were only performed with EPA5 sediment. Different volumes of a stock DOC solution were added, in replicate, to a fixed mass of dry sediment and brought to a volume of 40 mL, providing different points on the sorption isotherm. After addition of the DOC to the sediment solution, the bottles were sealed and gently inverted to suspend the sediment, then kept in a 25°C temperature water bath for 3 days prior to sampling. Samples were size-fractionated as previously described.

Desorption Kinetics

DOC desorption kinetics was studied by suddenly wetting dry sediment in a 150-mL glass bottle and following the release of DOC with time. The aqueous solution was sampled from replicate bottles, then filtered to obtain the DOC size fraction and frozen until analysis. Two sediment concentrations and two shear rates were studied. A low-shear environment was simulated by gently inverting the sample bottle every 8 min; high shear was generated by a mechanical shaker operated at 350 rpm.

Sorption Kinetics

The sorption kinetics of DOC were studied by first washing EPA5 sediment to remove weakly bound DOC. Distilled wash water was added to suspend 2 g of dry sediment in a 150-mL glass bottle, and the supernatant was decanted after 1-day settling time. This procedure was repeated each day for 15 days. The DOC concentration in the supernatant water dropped from 4 mg/L for the first wash to below 0.1 mg/L after the fifteenth. The overlying water was then removed by pipette. A concentrated DOC solution was prepared from EPA5 sediment by filtering through a YM2 Amicon filter. This DOC solution was then added to the washed sediment. Replicate bottles were sealed and the sediment suspended by gentle hand shaking. Samples were taken periodically and filtered to obtain the DOC size fraction in order to follow the decrease of aqueous DOC with time.

DOC Transport in Bed Sediment

Two experiments were performed to study DOC transport in bed sediment. These experiments had a twofold purpose: (1) to show that DOC are mobile in interstitial sediment waters and (2) to provide estimates of the DOC diffusivity in water.

DOC Concentration Profile

One liter of distilled water was carefully added (to prevent suspension) to 600 g dry EPA5 sediment in a beaker and kept in a 25°C water bath. The sediment bed depth was 7 cm. Every 2 days, beginning on the third day, for 97 days, the water was completely removed by pipette and replaced with fresh distilled water to keep the overlying water DOC concentration near zero. After 97 days, the water was completely removed, and the sediment cored using a wax-coated paper cylinder. The sediment core was sliced into 1-cm sections and centrifuged at 6000 g for 2 min to obtain a pore-water sample,[15] which was then filtered through a 0.4-μm filter to obtain the DOC fraction. The DOC concentration in this fraction was analyzed as described previously.

DOC Accumulation in Overlying Water

EPA5 and University Lake sediment were placed in beakers as described above. The beakers were then covered with plastic film and kept in a 25°C water bath. The overlying water was periodically sampled, and the 20-mL sample was replaced with fresh deionized water. Prior to sampling, the water was gently mixed to give a uniform DOC concentration. The DOC concentration was measured as described previously.

Naphthalene Partitioning

We performed experiments to measure the partition coefficients of a model hydrophobic compound, naphthalene, to suspended sediment. Naphthalene concentration was measured by direct aqueous injection on a Hewlett Packard 5890 gas chromatograph equipped with an FID detector and calibrated using external standards. Gas chromatographic conditions have been previously described.[16] Sediment and distilled water were added to 150-mL shaker bottles, which were then sealed and crimped. A stock solution of naphthalene in acetone (~ 25,000 mg/L) was added to the bottles by microliter syringe. The bottles were shaken thoroughly by hand and then left for 10 days in a 25°C water bath. Supernatant samples were analyzed by direct GC injection. The overlying water was then removed by pipette. Two successive hexane extractions (5 mL/g sediment; 24-hr shaking) were used to remove sediment-bound naphthalene. The distribution coefficients were calculated by mass balance.

RESULTS AND DISCUSSION

DOC Size Distribution

The size distribution of DOC from EPA5 sediment determined by ultrafiltration is shown in Table 13.2. The total concentration obtained by summing

Table 13.2. DOC Size Distribution (EPA5 Sediment) and Overall Diffusion Coefficient

Size Fraction (μm)	DOC Conc. (mg/L)	D_i (cm²/sec)
0.05–0.4	14.0	2.84×10^{-8}
0.015–0.05	4.3	1.97×10^{-7}
0.003–0.015	18.3	7.11×10^{-7}
0.0013–0.003	12.3	2.98×10^{-6}
0–0.0013	5.0	9.85×10^{-6}
	Sum 53.9	

$$\overline{D_c} = \sum_{i=1}^{n} D_{Ci} C_i \, / \sum_{i=1}^{n} C_i$$

the individual size fractions, 53.9 mg/L, agrees well with the measured total of 50.1 mg/L, indicating that filter contamination was not significant.

The size distribution can be used to estimate the diffusivity of the colloids in water, D_c^w, from the Stokes-Einstein equation.[17] Defining an average radius for each size fraction allows calculation of an overall concentration-averaged diffusion coefficient (Table 13.2). The smaller size fractions tend to dominate the overall estimate, 1.86×10^{-6} cm²/sec, which corresponds to an effective particle diameter of 0.0034 μm.

This estimate of the diffusion coefficient of DOC in water is at a sediment concentration near that of the bed sediment. However, the possibility of an effect of total DOC concentration should be recognized. In a system with a low DOC concentration, fewer particle-particle collisions are expected (less agglomeration), and therefore a greater proportion of smaller particles might be expected than for a high-concentration system. This simplified view ignores other factors that may affect the size distribution, such as electrical repulsion and attraction; however, it illustrates the expected trend. The predicted DOC diffusion coefficient should decrease with increasing DOC concentration.

Two methods, described below, were also used to estimate the DOC diffusion coefficient in water, D_c^w. The first is based on the measured concentration profile of the DOC in a layer of sediment exposed for 97 days to DOC-free water, and the second is based on the accumulation of DOC in initially DOC-free water overlying a layer of sediment.

DOC Concentration Profile

The sediment concentration profile (see Figure 13.2) was modeled with a semiinfinite slab diffusion model:[18]

$$\frac{C_c}{C_{C0}} = \text{erf}\left(\frac{x}{\sqrt{4D_{C,eff}t}}\right) \tag{13.1}$$

where C_C = the pore-water DOC concentration (mg/cm³ water)
    C_{C0} = the initial pore-water DOC concentration (mg/cm³ water)

$D_{C,eff}$ = the DOC effective diffusivity in the sediment (cm²/sec)
x = the depth into the sediment (cm)
t = time (sec)

This model is appropriate for analyzing these results because the overlying water was removed regularly, satisfying the boundary condition of zero DOC in the overlying water, and the profile shows no depletion of DOC at depths below 5 cm, satisfying the semiinfinite condition.

The parameter $D_{C,eff}$ was adjusted to fit the model to the experimental concentration profile by minimizing

$$\sum_{n=0}^{6} \left(C_{C0} \int_{n}^{n+1} \mathrm{erf} \left(\frac{x}{\sqrt{4D_{C,eff}t}} \right) dx - \text{observed value} \right)^2 \qquad (13.2)$$

The estimated effective diffusivity was 6.7×10^{-7} cm²/sec. The agreement between the fitted and observed concentration profiles, shown in Figure 13.2, suggests that a simple Fickian diffusion model is a reasonable approach to analyzing the transport of DOC from a sediment bed. If DOCs exhibit linear and reversible sorption to the sediment phase (discussed below), DOC diffusion through bed sediment may be considered identical to diffusion of HOCs

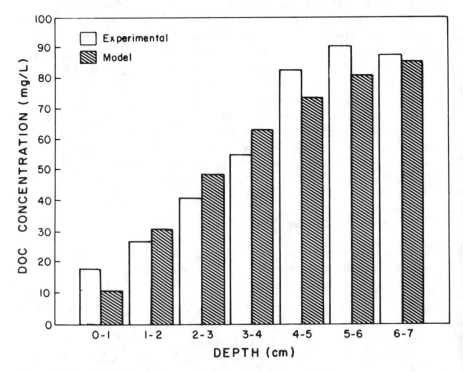

Figure 13.2. DOC concentration profile in a sediment layer.

through bed sediment. We may then estimate an adsorption-retarded effective diffusion coefficient from the diffusion coefficient of DOC in water, the sorption isotherms, and sediment bed characteristics using Equation 13.3, which is the conventional form used in retarded diffusion models:[19-21]

$$D_{C,eff} = \frac{D_C^w \, \epsilon^{4/3}}{\epsilon + \rho_b K_r} \qquad (13.3)$$

where ρ_b = the bulk density of the bed (g sediment/cm^3 total)
         ϵ = the sediment bed porosity (cm^3 water/cm^3 total)
         K_r = the partition coefficient for the weakly bound organic
              carbon between the sediment and the interstitial water
              (L water/kg sediment)

Conversely, the diffusion coefficient in water may be estimated from an experimental value of the effective diffusivity in the bed sediment through Equation 13.3. Thus, a second estimate of the DOC diffusion coefficient in water was made: $D_c^w = 5.7 \times 10^{-6}$ cm^2/sec, which agrees reasonably well with the value obtained from size fractionation.

DOC Transport to Overlying Water

A model was developed to predict the accumulation of DOC in initially DOC-free water overlying a sediment. Analysis of water-side mass-transfer resistances indicated that diffusion in the sediment layer controlled DOC transport. Diffusion from a finite layer with a no-flux condition at the sediment bottom and a finite well-mixed water layer at the sediment surface was modeled.[22] The DOC diffusivity in water was fit to the experimental data by a least-squares analysis. The estimated water diffusivity was 1.78×10^{-5} cm^2/sec for EPA5 sediment and 2.43×10^{-6} cm^2/sec for University Lake sediment. The fitted model and experimental data are presented in Figure 13.3 for both sediments.

Table 13.3 summarizes the DOC diffusivities. The variability in the estimated DOC diffusivity (particularly for EPA5) prevents a clear interpretation of whether the DOC behaves similarly to HOC by exhibiting retarded diffusion. Further experimentation is necessary to clarify this behavior. However, these experiments clearly show that the DOC is mobile in sedimentary systems, and therefore the possibility of enhanced HOC transport should be accounted for in modeling efforts.

Sorption/Desorption of DOC to Suspended Sediment

A conventional isotherm for DOC sorption/desorption from EPA5 sediment is plotted in Figure 13.4. The equilibrium sediment organic carbon concentration, S_{oc}, was calculated by mass balance using the measured DOC

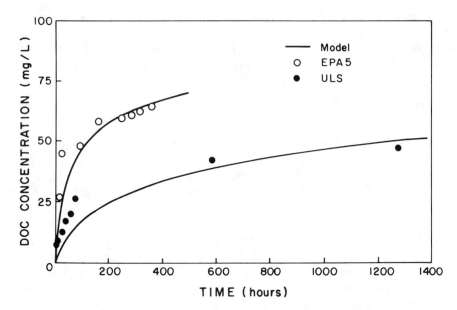

Figure 13.3. DOC accumulation over a sediment layer.

concentration and the organic carbon content of the dry sediment. The isotherm appears to be reversible since both the sorption and desorption data are collinear. Linear desorption isotherms were also observed for the other sediments studied.

To explain the nonzero intercept, we invoke the concept of strongly and weakly bound organic carbon. The strongly bound fraction of the organic carbon, f_{oc}^{sb}, of the dry sediment is defined by the zero DOC intercept of the SOC/DOC relationship. Repeated washing will not remove this organic carbon from the sediment.[14] The isotherm, then, describes only the behavior of the weakly bound fraction of organic carbon. The sum of the weakly and strongly bound fractions must equal the total organic fraction on the dry sediment.

This suggests a simple model

$$S_{oc} = K_r C_C + f_{oc}^{sb} \qquad (13.4)$$

Table 13.3. DOC Diffusivity

Method	$D_{C,eff}$ (cm²/sec)	D_C^w (cm²/sec)	Sediment Source
Size fractionation	2.24×10^{-7}	1.86×10^{-6}	EPA5
Sediment profile	6.7×10^{-7}	5.7×10^{-6}	EPA5
Accumulation in	2.14×10^{-6}	1.78×10^{-5}	EPA5
overlying water	4.74×10^{-7}	2.43×10^{-6}	ULS

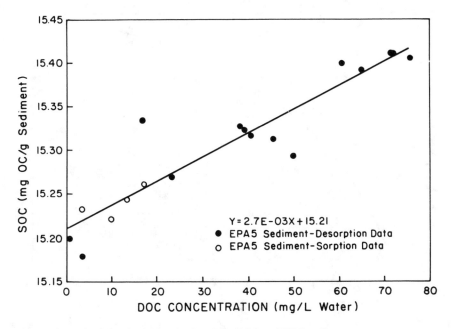

Figure 13.4. Sorption/desorption isotherm for DOC on EPA5 sediment.

where S_{oc} is the equilibrium concentration of total organic carbon on the sediment (kg OC/kg sediment) and C_C is the equilibrium concentration of organic carbon in the aqueous phase (kg OC/L water).

This can be used to model the effect of sediment concentration, S, on DOC concentration. Combining the above relationship with a mass balance on the organic carbon in a sediment-water system at equilibrium gives

$$C_C = \frac{f_{oc}^{wb} S}{1 + K_r S} \qquad (13.5)$$

where S is the sediment concentration (kg sediment/L water).

The constants K_r and f_{oc}^{wb}, determined by linear regression, are presented in

Table 13.4. Organic Carbon Sorption to Sediment-Model Parameter Estimates

Sediment	f_{oc}^{wb} (kg OC/kg sed)	K_r (L water/kg sed)
EPA5	2.9×10^{-4}	2.7
Drum Point	1.09×10^{-4}	1.5
Turkey Point	9.0×10^{-5}	0.61
Choptank River	3.59×10^{-4}	0.35
ULS	2.13×10^{-4}	2.5

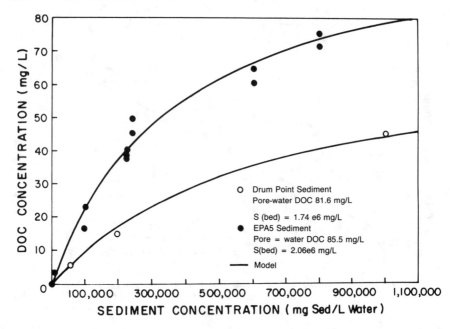

Figure 13.5.   The effect of sediment concentration on DOC concentration.

Table 13.4, and the results are shown in Figure 13.5. The agreement between the model and the data suggests that this form adequately describes the system behavior. f_{oc}^{wb} varies by a factor of 4, and K_r by a factor of 8, for the sediments studied.

We have shown that the DOC partitions between the sediment and aqueous phases. It is reasonable to ask whether the partitioning is equilibrium or kinetically controlled. Figure 13.6 shows the time variation of DOC concentration after wetting the sediment. An equilibrium concentration was reached within 2 min (the time required to filter a sample) in each case, indicating that the kinetics of sorption/desorption are fast. Similar results (not presented) were observed for the sorption of DOC to sediment from which natural DOC had been removed by repeated washing—the aqueous concentration dropped within the first minutes of the experiment and then remained constant. These kinetic studies support a modeling assumption of equilibrium sorption for DOC onto sediment.

We did not observe significant differences in the steady-state DOC concentration between the two shear rates for either sediment concentration. This indicates that neither liquid shear nor increased particle-particle collisions (at the higher sediment concentration) significantly affect the desorption equilibrium of organic carbon from sediment.

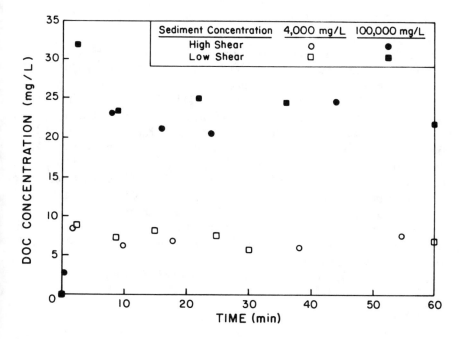

Figure 13.6. Release of DOC from a suddenly wetted sediment.

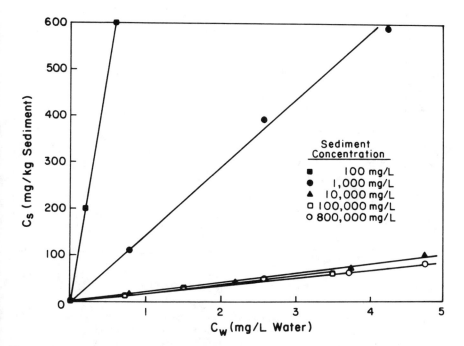

Figure 13.7. Naphthalene sorption isotherms on EPA5 sediment.

Naphthalene Partitioning

Figure 13.7 shows a linear distribution of naphthalene between the sediment and aqueous phase which contains DOCs. The sediment-water partition coefficient can be calculated from the measured distribution coefficient, K_d, following the method of Gschwend and Wu.[14] This calculation requires knowledge of the DOC-water partition coefficient, K_{DOC}, for the pollutant. Due to uncertainties associated with our measurements of K_{DOC}, we estimated K_{DOC} from a correlation given by Sigleo and Means.[23] As shown in Figure 13.7, K_d decreased with increasing sediment concentration up to ~ 10,000 mg/L, becoming constant above that. Figure 13.8 shows K_d and K_p normalized to the organic carbon content of the sediment. In this case, correction for the presence of DOC does not significantly affect the value of K_{oc}. These results suggest that the sediment-water partition coefficient is dependent on the sediment concentration below 10,000 mg/L. For the purposes of this study, only the range above 10,000 mg/L, where the partition coefficient is constant, is of interest because the modeling effort is directed at bed sediment conditions.

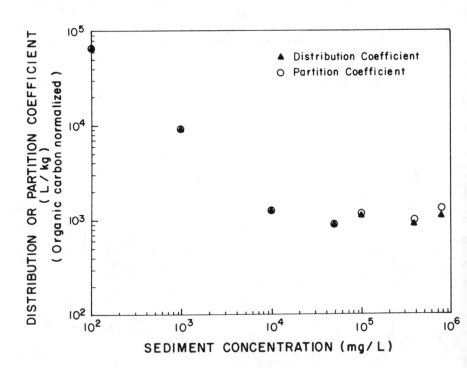

Figure 13.8. Organic carbon normalized naphthalene distribution/partition coefficient to sediment. The partition coefficient is obtained by correcting the distribution coefficient for the presence of DOC in the aqueous phase after Gschwend and Wu.[14]

MODELING THE EFFECT OF DOC ON POLLUTANT TRANSPORT

We have examined the rate of transport of DOCs from bed sediments to aquatic environments. We now develop a model for DOC-facilitated transport of hydrophobic organic pollutants which may be sorbed to the DOC.

The equation of continuity for component A in the system is

$$\frac{\partial C_{At}}{\partial t} + \nabla \cdot n_A = R_A \tag{13.6}$$

where n_A = the mass flux of A across the system boundaries (g A/ cm^2 hr)

C_{At} = the total quantity of chemical in the system (g A/cm^3 total)

R_A = the rate of appearance or disappearance of A (g A/cm^3 hr)

In a sediment bed, which is a three-phase system, the total concentration of component A is

$$C_{At} = \epsilon C_A + W_{As}\rho_b + \epsilon W_{Ac}C_C \tag{13.7}$$

where C_A = the contaminant pore-water concentration (g A/cm^3 water)

W_{As} = amount of A sorbed to the sediment (g A/g sediment)

W_{Ac} = the amount of A sorbed to DOC (g A/g DOC)

If the partitioning between phases is fast, the local equilibrium assumption for HOC partitioning is justified.[24] Thus, we may write the following expressions for W_{As} and W_{Ac}:

$$W_{As} = K_p C_A \tag{13.8}$$

$$W_{Ac} = K_C C_A$$

where K_C is the partition constant for A between DOC and pore water.

For a conserved species the rate of formation or disappearance, R_A, is zero. The mass flux of chemical A in an abiotic sediment is accounted for by four transport mechanisms:

1. molecular diffusion of A in the pore water
2. Brownian diffusion of A associated with colloids
3. advection of A with the aqueous-phase apparent velocity V_w
4. advection of A associated with colloids due to the apparent colloid velocity V_c

The solid sediment phase is assumed to be stationary, and both C_A and W_{Ac} are small. The mass flux can then be expressed as

$$n_A = -D_A \nabla \cdot C_A - D_C \nabla \cdot (K_r C_A C_C) + V_w C_A + V_C(K_C C_A C_C) \quad (13.9)$$

with the following correction for porosity and tortuosity:

$$D_A = D_A^w \epsilon^{4/3}$$

$$D_C = D_C^w \epsilon^{4/3} \quad (13.10)$$

where D_A is the diffusivity of A in the pore water (cm²/sec) and D_C is the diffusivity of the colloids in the pore water (cm²/sec).

In this development the length dimension is assumed to be large with respect to particle diameter in order to invoke the continuum transport expressions. Combining Equations 13.6–13.10 yields

$$\frac{\partial C_{At}}{\partial t} = (\epsilon + \rho_b K_p + \epsilon K_C C_C) \frac{\partial C_A}{\partial t} = \quad (13.11)$$

$$(D_A + D_C K_C C_C)\nabla^2 C_A - (V_w + V_C K_C C_C)\nabla C_A$$

In the derivation we have also assumed constant effective diffusivities for both the chemical and colloids and a uniform colloid concentration in the pore water. If the DOC were treated as a macromolecular chemical species, rigorous modeling would require that a similar expression for DOC be developed and solved simultaneously. However, at bed sediment conditions we expect the uniform colloid concentration assumption to be a good approximation since the net mass of colloids crossing the sediment-water interface will be small.

Equation 13.11 is an approximation of the significant transport phenomena in bed sediments near the sediment-water interface with the exception of bioturbation, and is a starting point in considering nonreactive hydrophobic chemical transport within abiotic bed sediment.

Simplifications can be made in Equation 13.11 when either advection or diffusion processes dominate the transport of the organic pollutant. Here we only consider the diffusion-controlled situation. For this case, Equation 13.11 simplifies, for unidirectional transport, to

$$\frac{\partial C_A}{\partial t} = D_{SC} \frac{\partial^2 C_A}{\partial x^2} \quad (13.12)$$

where D_{SC} is an overall effective diffusion coefficient given by

$$D_{SC} = \frac{D_A + D_C K_C C_C}{\epsilon + \rho_b K_p + \epsilon K_C C_C} \quad (13.13)$$

If the colloid concentration is zero, then the diffusion coefficient expression reduces to that normally used to describe retarded transport of organic pollutants within a sediment bed.[18-21] The overall diffusion coefficient can be increased by several orders of magnitude due to the presence of colloids.

A common modeling approach in aquatic environments considers the sediment to be uniformly contaminated at initial time. The water column is usually considered to be well mixed and to have very low (zero) concentration of the pollutant in question. If the sediment is deep enough so that the chemical concentration remains constant at large depths, the solution to Equation 13.12 is the well-known semi-infinite slab relationship

$$\frac{C_A}{C_{A0}} = erf\left(\frac{x}{\sqrt{4D_{sc}t}}\right) \tag{13.14}$$

Knowing the concentration profile (Equation 13.14), we can use Fick's first law to estimate the flux:

$$n_{AC} = (D_A + D_C K_C C_C) \frac{\partial C_A}{\partial x}\Big|_{x=0} \tag{13.15}$$

where $D_A + D_C K_C C_C$ is the effective diffusivity of species A at the sediment-water interface due to molecular diffusion combined with the amount of A carried across the interface by colloids.

Thus,

$$n_{AC} = C_{At0} \sqrt{\frac{D_{SC}}{\pi t}} \tag{13.16}$$

where C_{At0} is the total initial concentration of the contaminant.

Examination of the flux enhancement is facilitated by noting that for hydrophobic molecules most of the mass of material will be associated with the sediment, and not the water or colloid phases, and the denominator of Equation 13.13 is dominated by the term $\rho_b K_p$. Using this assumption, and taking the ratio of colloidally enhanced flux to that without colloids gives

$$\frac{n_{AC}}{n_A} = \sqrt{\frac{D_A + D_C K_C C_C}{D_A}} \tag{13.17}$$

which is the square root of the enhancement factor for the overall diffusion coefficient. Figure 13.9 shows the calculated enhancement factor as a function of the pollutant hydrophobicity and DOC concentration. For highly hydrophobic compounds, the colloid-mediated flux may be as much as 50 times greater than the flux in the absence of colloids. Flux enhancement is not predicted to be significant for compounds with $\log(K_{oc}) < 4$.

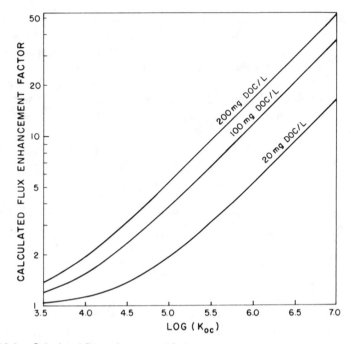

Figure 13.9.   Calculated flux enhancement factors.

CONCLUSIONS

The desorption behavior of dissolved organic carbon from a sediment can be described by a Langmuir-shaped isotherm. On the basis of limited data, the desorption process appears to be reversible. The kinetics of DOC sorption/desorption with the sediment were fast; steady state was reached within minutes. This indicates that pseudo-steady-state assumptions should be acceptable in modeling DOC-mediated transport.

The transport of DOC from a sediment bed to the overlying water, in a system without advection, appears to be a Fickian diffusion process. Estimates of the "binary" DOC diffusivity in water were obtained from three experiments. The average of the estimates was $6.9 \times 10^{-6} \pm 6.4 \times 10^{-6}$ cm^2/sec.

The sediment partition coefficient of naphthalene was shown to be dependent on sediment concentration. Therefore, parameters intended to represent real-world sediments should be evaluated as near to bed sediment conditions as possible.

Modeling considerations show that the presence of DOC may increase the flux of organic pollutants from bed sediments by up to 50 times, depending on the pollutant hydrophobicity and the DOC concentration. At present these are only rough theoretical predictions, and experimental work is planned to test the hypothesis.

REFERENCES

1. Enfield, C. G., G. Bengtsson, and R. Lindqvist. "Influence of Macromolecules on Chemical Transport," *Environ. Sci. Technol.* 23:1278–1286 (1989).
2. McCarthy, J. F., and J. M. Zachara. "Subsurface Transport of Contaminants," *Environ. Sci. Technol.* 23:496–502 (1989).
3. Brownawell, B. J., and J. W. Farrington. "Biochemistry of PCBs in Interstitial Waters of a Coastal Marine Sediment," *Geochim. Cosmochim. Acta* 50:157–169 (1985).
4. Means, J. C., and R. Wijayaratne. "Role of Natural Colloids in the Transport of Hydrophobic Pollutants," *Science* 215:968–70 (1982).
5. Thurman, E. M. *Organic Geochemistry of Natural Waters,* Martinus Nijhoff/Dr. W. Junk, Publishers, Dordrecht, The Netherlands, 1985.
6. Danielsson, L. G. "On the Use of Filters for Distinguishing between Dissolved and Particulate Fractions in Natural Waters," *Water Res.* 16:179–182 (1982).
7. Thibodeaux, L. J., D. D. Reible, and C. S. Fang. "Transport of Chemical Contaminants in the Marine Environment – a Vignette Model," in *Pollutants in a Multimedia Environment,* Y. Cohen, Ed. (New York: Plenum Press, 1986), pp. 49–64.
8. Baron, J. A. "Laboratory Simulation and Diffusion Model for Leaching of Organic Chemicals from Marine Bottom Sediment," Masters Thesis, Louisiana State University, Baton Rouge, LA (1988).
9. Aiken, G. R., O. M. McKnight, and R. L. Wershaw. *Humic Substances in Soil, Sediment and Water* (New York: John Wiley and Sons, 1985).
10. Buffle, J., P. Deladoey, J. Zumstein, and W. Haerdi. "Analysis and Characterization of Natural Organic Matters in Freshwater. I. Study of Analytical Techniques," *Schweiz Z. Hydrol.* 44:325–362 (1982).
11. Hwang, C. P., T. H. Lackie, and R. R. Munch. "Correction for Total Organic Carbon, Nitrate, and Chemical Oxygen Demand When Using the MF-Millipore Filter," *Environ. Sci. Technol.* 13:871–872 (1979).
12. Ogura, N. "Molecular Weight Fractionation of Dissolved Organic Matter in Coastal Seawater by Ultrafiltration," *Mar. Biol.* 24:305–312 (1974).
13. Buffle, J., P. Deladoey, and W. Haerdi. "The Use of Ultrafiltration for the Separation and Fractionation of Organic Ligands in Fresh Waters," *Anal. Chim. Acta* 101:339–357 (1978).
14. Gschwend, P. M., and S. Wu. "On the Constancy of Sediment-Water Partition Coefficients of Hydrophobic Pollutants," *Environ. Sci. Technol.* 19:90–96 (1985).
15. Servos, M. R., and D. C. G. Muir. "The Effect of Suspended Sediment Concentration on the Sediment to Water Partition Coefficient for 1,3,6,8-Tetrachlorodibenzo-*p*-dioxin," *Environ. Sci. Technol.* 23:1302–1306 (1989).
16. Koulermos, A. C. "An Evaluation of the Physico-Chemical Aspects of Natural Colloids with Reference to Their Role in the Transport and Fate of Organic Pollutants in Aquatic Systems," Masters Thesis, Louisiana State University, Baton Rouge, LA (1989).
17. Bird, R. B., W. E. Stewart, and E. N. Lightfoot. *Transport Phenomena* (New York: John Wiley and Sons, 1960).
18. Formica, S. J., J. A. Baron, L. J. Thibodeaux, and K. T. Valsaraj. "PCB Transport into a Lake Sediment. Conceptual Model and Laboratory Simulation," *Environ. Sci. Technol.* 22:1435–1440 (1988).
19. Crank, J. *The Mathematics of Diffusion* (Oxford: Clarendon Press, 1975).

20. Karickhoff, S. W., and K. R. Morris. "Impact of Tubificid Oligochaetes on Pollutant Transport in Bottom Sediments," *Environ. Sci. Technol.* 19:51–56 (1985).
21. Reible, D. D., K. T. Valsaraj, and L. J. Thibodeaux. "Chemodynamics Models of Contaminant Transport from Bed Sediments," in *The Handbook of Environmental Chemistry,* Vol. 2/F, O. Hutzinger, Ed. Springer-Verlag, Germany (1991).
22. Schlicting, H. *Boundary Layer Theory,* 7th ed. (New York: McGraw-Hill Book Company, 1979), pp. 102, 225–230.
23. Sigleo, A. C., and J. C. Means. "Organic and Inorganic Components in Estuarine Colloids: Implications for Sorption and Transport of Pollutants," in *Rev. Environ. Contam. Toxicol.* 112:123–147 (1990).
24. Hassett, J. P., and E. Milicic. "Determination of Equilibrium and Rate Constants for Binding of a Polychlorinated Biphenyl Congener by Dissolved Humic Substances," *Environ. Sci. Technol.* 19:638–643 (1985).
25. Means, J. C., S. G. Wood, J. J. Hassett, and W. L. Banwart. "Sorption of Polynuclear Aromatic Hydrocarbons by Sediments and Soils," *Environ. Sci. Technol.* 14:1524–1528 (1980).

CHAPTER 14

A Thermodynamic Partition Model for Binding of Nonpolar Organic Compounds by Organic Colloids and Implications for Their Sorption to Soils and Sediment

Yu-Ping Chin, Walter J. Weber, Jr., and Cary T. Chiou

INTRODUCTION

Organic colloids, comprised of humic substances and labile macromolecules, are ubiquitous to natural waters. These substances are known to bind effectively with many nonpolar organic contaminants (NOCs) and may significantly alter their fate and transport in the aquatic environment.[1-5] Organic pollutants of this class are usually stable and exhibit sparing solubility in water. Prior research on such binding processes had centered on the development of analytical methodologies used to measure NOC-colloid equilibrium constants and the kinetics of such reactions.[6-14] The specific mechanisms that control the interaction between NOCs and organic colloids of a heterogeneous nature have only recently been better understood.[15,16] In earlier developments, predictions of organic carbon normalized binding constants ($K_{b,oc}$) of NOCs were based on the compounds' water solubilities (or aqueous activity coefficients) or their corresponding octanol-water partition coefficients (K_{ow}). Such empirical models implicitly assume that the equilibrium binding of NOCs with organic colloids was governed by a partition equilibrium similar to that in a solvent-water system, and that the solute's solubility in organic colloids was related to the solute's solubility in a solvent such as octanol. Although the solubility of a NOC in an ordinary solvent (e.g., octanol) can be effectively described by the conventional Raoult's law, this theory failed to account for the solubility in a macromolecular organic geopolymer that comprises most aquatic colloids. The failure in recognizing and explaining the solute solubility in macromolecular substances hinders a proper understanding of the factors controlling the binding of NOCs with natural organic colloids.

An alternative approach to empirical correlations based upon one physicochemical property is to understand and quantify NOC interactions in humic substances as well as in the aqueous phase. Such a modeling approach treats the "binding" of NOCs with macromolecular substances as a partition (solubility) effect. Chiou et al.[15] and recently Chin and Weber[16] were able to develop a more rigorous humic polymer-NOC interaction model by application of the Flory-Huggins theory to quantify solute solubility in polymers, taking into account the effect on solute solubility of the disparity in molecular size between the solute and host humic polymer. Chiou et al. attributed the comparatively small solubility of NOCs in a humic substance (humic acid) relative to that in a nonpolar solvent to the large disparity in polarity between NOCs and the relatively polar humic substrate, consistent with the fact that $K_{b,oc}$ values of NOCs are generally lower than respective K_{ow} values;[17] the difference in $K_{b,oc}$ values among solutes with a given humic material was attributed primarily to the difference in NOC water solubility (or aqueous activity coefficient).

This chapter expands upon earlier efforts and applies $K_{b,oc}$ predictions using the Flory-Huggins partition equation in a simple triphase distribution model to assess the impact of various aquatic organic colloids in affecting NOC sorption by sediments and soils. The modeling approach is conceptually straightforward and provides a better understanding of NOC fate and transport in aqueous systems.

THEORY AND BACKGROUND

The partitioning of a nonionic organic compound between an organic solvent and water can be quantified using a simple linear equilibrium relationship:

$$K_b = C_{op}/C_e \tag{14.1}$$

where C_{op} and C_e are the mass of solute in the organic phase and aqueous phase, respectively. Equation 14.1 can be normalized to different types of organic matter by their fractional organic carbon content, f_{oc}:

$$K_{b,oc} = \frac{K_b}{f_{oc}} \tag{14.2}$$

This equilibrium relationship may be expressed in terms of solute activity coefficients in the respective organic phase (γ_i^{oc}) and water (γ_i^w) by Raoult's law:

$$\ln K_{b,oc} = \ln \gamma_i^w - \ln \gamma_i^{oc} + \ln V_w - \ln V_p - \ln f_{oc} \tag{14.3}$$

where V_w and V_p are the water and amorphous organic substance molar volumes, respectively. Equation 14.3 is applicable only to simple systems where the entropy of mixing follows the ideal Raoult's law in the organic phase. The aqueous-phase solute activity coefficient can be calculated from solubility data or estimated by using any number of linear free-energy relationships such as UNIFAC.[18,19]

Flory[20,21] and Huggins[22] observed that solutions of polymers and liquid organic compounds did not conform to the entropy of mixing as defined by Raoult's law. By using a liquid lattice statistical approach, they were able to define a volume fraction entropy of mixing term:

$$\Delta S_m = R\Sigma(n_i \ln \phi_i) \qquad (14.4)$$

where ϕ_i and n_i are the volume fraction and moles of solute i in solution with polymer p. The NOC volume fraction is defined by the relationship:

$$\phi_i = n_i V_i / (n_p V_p + n_i V_i) \qquad (14.5)$$

where n_p = the moles of polymer in solution with an NOC
V_i = the solute molar volume
V_p = the molar volume of the polymer

In a nonideal binary solution comprised of a NOC and a humic polymer, the free energy of mixing (ΔG_m) for such a system is simply the sum of the enthalpy and entropy components:

$$\Delta G_m = RT \chi n_i \phi_p - T\Delta S_m \qquad (14.6)$$

Combining of Equations 14.4 and 14.6 and differentiating ΔG_m with respect to n_i to express the system in terms of the chemical potential, μ_i, yields

$$\mu_i - \mu_i^\circ = RT\,[\ln \phi_i + \phi_p(1 - V_i/V_p) + \chi\phi_p^2] \qquad (14.7)$$

where μ_i° = the standard state chemical potential
R = the universal gas constant
T = the temperature
χ = the Flory-Huggins interaction parameter, expressing the heat of mixing and the correction for the reduced entropy of mixing caused by inflexibility of polymer segments

Division of both sides of Equation 14.7 by RT results in an expression of the solute in terms of its activity in the polymer phase:

$$\ln a_i = \ln \phi_i + \phi_p(1 - V_i/V_p) + \chi\phi_p^2 \qquad (14.8)$$

The activity of solute can also be defined as the product of its activity coefficient and mole fraction solubility X_i in the polymer:

$$\ln (\gamma_i^{oc} X_i) = \ln \phi_i + \phi_p(1 - V_i/V_p) + \chi\phi_p^2 \qquad (14.9)$$

The mole fraction of component i at dilution simplifies to

$$X_i = n_i/n_p \tag{14.10}$$

and the corresponding volume fraction is defined as

$$\phi_i = n_i V_i/n_p V_p \tag{14.11}$$

Combination of Equations 14.10 and 14.11 and insertion into Equation 14.9 yields an expression for a dilute solution of i in a polymer:

$$\ln \gamma_i^{oc} = \ln V_i - \ln V_p + \phi_p (1 - V_i/V_p) + \chi\phi_p^2 \tag{14.12}$$

In a dilute solution where the solute is a NOC, the colloid volume fraction approaches unity. In addition, NOC molar volumes are much smaller than V_p, and this ratio can be neglected. This simplifies Equation 14.12 to

$$\ln \gamma_i^{oc} = \ln V_i - \ln V_p + 1 + \chi \tag{14.13}$$

Combination of Equations 14.2, 14.3, and 14.13 and conversion to the common logarithm yields

$$\log K_{b,oc} = \log \gamma_i^w + \log (V_w/V_i) - \log \rho - \log (oc) - (1 + \chi)/2.303 \tag{14.14}$$

The term ρ is the density of the polymer, added to express the equilibrium constant in terms of volume per unit mass. Chiou et al. reported this value to be 1.2, based upon the structure of polymers similar to humic acids.[15]

The Flory-Huggins parameter (χ) for a solute in a polymer quantifies deviations from ideal solution behavior caused by the excess heat of mixing and constraints of polymer segments in free mixing. The value χ is comprised of both enthalpic (χ_h) and entropic (χ_s) terms:

$$\chi = \chi_h + \chi_s \tag{14.15}$$

The χ_s value is zero for linear polymers whose chain segments assume nearly complete rotational freedom; for nonlinear polymers, the χ_s value is a positive number, the magnitude of which depends on the degree of segment flexibility and is evaluated empirically. For the interaction between NOC liquids and polar polymers, this term has been determined to be 0.34.[23] The heat of interaction contribution can be estimated a priori using the appropriate regular solution equation.[15,22,24]

Studies show that there exists a critical value of the Flory-Huggins parameter (χ_c), and that $\chi_c = 0.5$. When $\chi < 0.5$, the NOC and polymer are miscible over the entire volume fraction solubility range.[21] Since many humic substances are highly substituted with polar functional groups that can interact with water and each other, nonpolar organic compounds may be expected to exhibit more excess heat of mixing with these polar organic colloids than with less polar homogeneous organic solvents. Hence, χ_h would be numerically large, and the Flory-Huggins parameter would exceed the critical value 0.5. Chiou et al. demonstrated this for NOCs partitioned between soil organic matter and water, where they observed large deviations between predicted and experimental partition coefficients if the heat of mixing term was assumed to be zero.[15]

Evaluation of χ_h involves the application of the regular solution concept. The Scatchard-Hildebrand equation quantifies the heat of mixing of two solution components caused mainly by London forces.[25] More recently, several refinements have been made to account for specific interactions between the solute and solvent. The Blanks and Prausnitz modification of the regular solution model quantifies the heat of mixing between solutions of nonpolar liquids and polar polymers:[23]

$$\chi_h = \frac{V_i}{RT} [(\delta_i - \lambda_p)^2 + \tau_p^2 - 2\Psi] \qquad (14.16)$$

where δ_i = the solute solubility parameter $(cal/cc)^{0.5}$
λ_p = the polymer nonpolar solubility parameter $(cal/cc)^{0.5}$
τ_p = the polymer polar solubility parameters $(cal/cc)^{0.5}$
Ψ = an induction energy term that quantifies permanent dipole-induced dipole forces

This particular approach is well suited for natural aquatic organic colloids and NOCs because it takes into account specific, inductive, and dispersive molecular interactions. The induction parameter is evaluated empirically, and is found to be linearly related to τ_p^2:

$$\Psi = \theta\tau_p^2 \qquad (14.17)$$

where θ is dependent upon the solvent molecular structure. Weimer and Prausnitz observed θ to vary from 0.396 for unsaturated substances, to 0.450 for aromatic compounds.[26]

The heterogeneous nature of organic colloids precludes any attempt to determine directly their solubility parameters by conventional methods (heat of vaporization, refractive index). Chiou et al. were able to indirectly elucidate a total solubility parameter of 13 for soil organic matter.[15] Reported solubility parameters for most NOCs are generally less than 10. This larger value suggests that natural organic substrates are polar by nature, but not as polar as simple alcohols ($\sim$ 15) or water ($>$ 20).

An alternative technique, suggested by Hildebrand et al., is based upon quantifying the molecular interactions between the solute and an adjacent polymer segment.[25] The properties of the segment are approximated by those of the polymer subunit. Although humics and organic colloids are complex, they are comprised of only a few molecular structures. NMR, IR, and pyrolysis followed by GC-MS have revealed acidic functional groups (carboxylic and hydroxyl) attached to a carbon backbone comprised of aromatics, olefins, and aliphatics. Humic acids appeared to be more aromatic than fulvic substances.[27-29] Based upon these observations, methyl salicylate was selected to approximate a typical humic acid subunit. This molecule is an aromatic with methyl, carboxylic, and hydroxyl functionalities — common to all humic acids. The solubility parameters for this compound have been reported by Barton[30] and are comprised of dispersive and polar contribution terms.

Polymaleic acid (PMA) is a substance polymerized by reacting maleic anhydride with pyridine. Several investigators have shown by spectroscopy that the structure of PMA closely resembles fulvic acid.[31,32] The total solubility parameter of maleic anhydride is known,[30] and the polar and nonpolar contributions can be estimated from maleic anhydride's refractive index.[33] This substance is used as a surrogate for more polar organic colloids in estimating binding equilibrium constants using Equation 14.14.

MATERIALS AND METHODS

Solute-Polymer Binding Experiments

Equilibrium binding studies were conducted using the dialysis technique, and the protocol was presented in detail in two previous works.[8,16] The dialysis concept is based upon using a semipermeable membrane that restricts the migration of colloids by size exclusion, while allowing smaller solute molecules to pass through freely. Four organic probes, p-dichlorobenzene (p-DCB), 1,2,4-trichlorobenzene (TCB), 2,5,2'-PCB, and cis-chlordane were chosen on the basis of their potential toxicity and differences in aqueous solubility. Aldrich humic acid (AHA) was chosen to represent a reference organic colloid. Although they are not representative of real-world aquatic organic colloids,[29] they have been well studied, and a large database of pollutant binding constants exists, which is used for evaluating model (Equation 14.14) predictions.

Batch Experiment Sorption Studies

Sediments were obtained from Lake Michigan offshore of South Haven, Michigan. They were wet-sieved to less than 60 μm, freeze-dried, and stored in a sealed container at $-20°C$. The organic carbon content was determined to be 1.42%, using a persulfate oxidation technique. Aqueous total organic carbon (TOC) analyses were performed using a wet-oxidation UV/persulfate analyzer (Dohrmann Co.).

Stock solutions of Aldrich humic acid were made by adding specific amounts of humic acid (400–500 mg) to 1 L of distilled water, and raising the pH to 11 with 1.0 N NaOH. The solution was agitated for 1 hr, titrated with 1.0 N HCL to a pH of 7.0, and filtered through two type A/E prewashed glass-fiber filters (Gelman Science). Working concentrations (12.5 mg/L organic carbon) of humic acid were made by diluting the stock solutions with 10 mM sodium bicarbonate buffered water (pH = 7.5). Sodium azide (1.5 mM) was added to inhibit microbial growth.

Surface water from the Huron River, sampled at Barton Park in Ann Arbor, Michigan, during the spring runoff period, was chosen as a source of natural aquatic organic matter. This sample was filtered through two prewashed type A/E glass-fiber filters, poisoned with sodium azide (1.5 mM), and stored in amber bottles at 4°C. TOC results yielded a concentration of 11 mg/L organic carbon.

Sorption experiments were carried out using 150 cc Hypovials (C. P. Pierce). A 100-cc aliquot of either Huron River water or Aldrich humic acid solution was added to each vial, and the target compound introduced by direct injection in an acetone carrier to give a range of different solute concentrations. Then 100 mg of lacustrine sediment were added to each vial. The vials were sonicated to break up large aggregates and sealed with aluminum crimp seals lined with Tuff-Bond Teflon-lined discs (C. P. Pierce). The sealed bottles were placed on a rotary tumbler, and the systems allowed to equilibrate for 120 hr. A previous study determined this to be sufficient time to attain an apparent equilibrium condition.[34] Sorption experiments for two of the probes (DCB and TCB) were conducted in systems with no headspaces to minimize losses due to volatilization.

A similar set of experiments was conducted using distilled water poisoned with sodium azide to determine the ideal two-phase thermodynamic partition coefficient. Control experiments (i.e., no sorbents added) revealed that solute losses to the walls were 5% or less.

Following equilibration, the vials were centrifuged to separate the solids from the aqueous phase. A portion of the supernatant was pipetted out and extracted with pesticide-free hexane for 30 min. The phases were allowed to separate, and the hexane assayed for the probe on a Hewlett-Packard 5890 gas chromatograph equipped with a [63]Ni electron capture detector and a glass column containing an appropriate stationary phase.

RESULTS AND DISCUSSION

Estimation of NOC-Colloid Binding Constants

Normalized organic carbon NOC–commercial humic acid binding constants for the four compounds and ten other solutes are reported in Table 14.1, along with their octanol-water partition coefficients and aqueous-phase activity coefficients. The fraction organic carbon content of Aldrich humic acid was found to be 0.502, as reported by McCarthy and Jimenez.[7] Figure 14.1 illus-

Table 14.1. Octanol/Water Partition Coefficients and Binding Constants for Selected Organic Solutes and Aldrich or Fluka Humic Acid Polymers

Compound	log (K_{ow})	log ($K_{b,oc}$)obs.	log γ_i^w
TCE	2.53	2.20	3.81
Toluene	2.69	2.27	3.97
Naphthalene	3.38	3.04	4.82
Phenanthrene	4.46	4.00	6.25
Anthracene	4.54	4.21 ± 0.11	6.25
Fluorene	4.18	3.95	6.00
Biphenyl	3.95	3.27	5.72
2,2',5-PCB	6.00	4.57	7.36
2,2',4-PCB	5.67	4.84	7.42
2,2',5,5'-PCB	5.84	4.97 ± 0.35	7.98
α-Chlordane	6.00	4.77	7.78
p,p'-DDT	6.36	5.61 ± 0.11	8.48
1,4-DCB	3.36	2.92	4.79
1,2,4-TCB	3.98	3.11	5.49

Source: All values taken from Chin and Weber,[16] with the exception of p,p'-DDT, which is taken from Chiou et al.[14]

trates the relationship between log $K_{b,oc}$ and log γ_i^w for the target compounds. Linear regression for this relationship reveals

$$\log K_{b,oc} = 0.697 \log \gamma_i^w - 0.451; \ r = .98 \tag{14.18}$$

The high degree of correlation suggests that the binding of NOC by humic acid is strongly dependent on the hydrophobicity of the probe. The large deviation from a slope of unity, however, suggests that there is an increasing incompatibility of solute interactions in the humic polymer as the solute solubility in water decreases.[15]

Since the water solubility or τ_i^w has also been shown to be the dominant

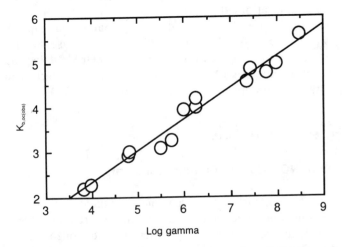

Figure 14.1. Correlation between solute aqueous activity coefficients and observed commercial humic acid binding constants.

factor affecting the magnitude of the octanol-water partition coefficient, a correlation between NOC K_{ow} and $K_{b,oc}$ is expected (Figure 14.2). Regression analysis yields the following equation:

$$\log K_{b,oc} = 0.797 \log K_{ow} + 0.254; \; r = .97 \qquad (14.19)$$

The fit between the two parameters is good, but the experimentally derived binding constants are consistently lower than their respective K_{ow} values, and these differences increase with increasing K_{ow}. Deviations from the slope of unity, while not as large as Equation 14.18, suggest that the activity coefficients of NOCs in the octanol phase are more constant than in humic acids. This is consistent with the findings of Chiou et al. for the correlation of soil organic matter sorption coefficients of NOCs with corresponding K_{ow} values.[15] In both cases the organic phases were more polar than octanol, which resulted in lower partition coefficients for NOCs and a slope less than unity in the correlation of $\log K_{b,oc}$ with $\log K_{ow}$.

While empirical correlations between $K_{b,oc}$ and physicochemical properties of the target compounds reflect consequences of certain thermodynamic principles, they yield little specific information with respect to molecular interactions between NOC and organic colloids. Coefficients of the line derived from linear-regression analysis suggest that NOCs do not form an ideal solution with humic acids, nor does their behavior in the colloidal phase resemble their solubility in octanol. These relationships are useful only for making order-of-magnitude estimates of binding constants in lieu of other methods.

Estimates of NOC–commercial humic acid binding constants using Equation 14.14 and observed values are reported in Table 14.2, along with the required NOC solubility parameters. Correlation between predicted and exper-

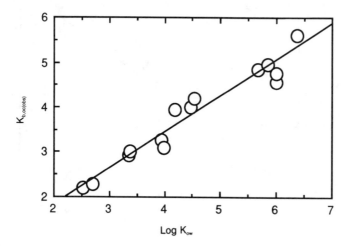

Figure 14.2. Correlation between *n*-octanol–water partition coefficients (K_{ow}) and organic carbon normalized solute–commercial humic acid binding constants ($K_{b,oc}$).

Table 14.2. A Comparison of Predicted and Observed Solute Binding Constants for Commercial Aldrich and Fluka Humic Acids

Compound	δ	log $(K_{b,oc})_{pred.}$	log $(K_{b,oc})_{obs.}$
TCE	9.2	2.28	2.20
Toluene	8.9	2.34	2.27
Naphthalene	9.9	2.89	3.02
Phenanthrene	9.8	3.89	4.00
Anthracene	9.9	3.92	4.21 ± 0.11
Fluorene	9.7	3.87	3.95
Biphenyl	8.3	3.67	3.27
2,2′,5-PCB	9.9	4.85	4.57
2,2′,4-PCB	9.9	4.91	4.84
2,2′,5,5′-PCB	10.4	5.06	4.97 ± 0.35
α-Chlordane	10.4	4.38	4.77
p,p′-DDT	8.8	6.03	5.61 ± 0.11
1,4-DCB	9.7	2.91	2.92
1,2,4-TCB	9.3	3.32	3.11

Source: Barton.[30]
Note: Methyl salicylate parameters: $\lambda = 7.8$ (cal/cc)$^{0.5}$; $\tau = 7.16$ (cal/cc)$^{0.5}$. The value τ incorporates contributions from both polar and hydrogen bonding type interactions.

imental $K_{b,oc}$ for the 14 probes is presented in Figure 14.3. In cases where there are more than one reported literature $K_{b,oc}$ value, the average and standard deviation are presented. Inspection of the data reveals that the equation derived by combining the Flory-Huggins concept for solute solubility in humic polymers with the solute's activity coefficients in water can estimate $K_{b,oc}$ to within less than one-half a log unit for all the compounds studied. Regression analysis reveals the following relationship:

$$\log (K_{b,oc})_{obs.} = 0.945 \log (K_{b,oc})_{pred.} + 0.168; \ r = .98 \qquad (14.20)$$

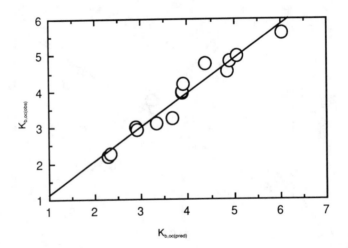

Figure 14.3. Correlation between observed and Flory-Huggins model predicted solute–commercial humic acid binding constants.

Table 14.3. A Comparison of Observed Binding Capacities of the Dispersed Organic Polymer Phases Present in Several Natural Waters and Fulvic Acid to Model Predictions Based upon PMA as a Humic Surrogate Compound

Organic Polymer Source	Log $K_{b,oc}$		
	p,p'-DDT	2,4,4'-PCB	2,2',5,5'-PCB
PMA predictions	5.01	4.21	4.35
Sopchoppy River Water	4.39	3.57	no data
Suwanee River Water	4.39	3.53	no data
Suwanee River F.A.	4.40	3.57	no data
Pakim Pond Water	4.84	no data	no data
Lake Michigan	4.26 ± 0.44	no data	3.88 ± 0.53
Huron River	4.23	no data	3.87

Linearity between the pooled sets of data spans three orders of magnitude, and the coefficients of the correlation for such a small database clearly illustrate the excellent agreement between model output and observed values. Absolute deviations between log $(K_{b,oc})_{pred.}$ and log$(K_{b,oc})_{obs.}$ range from 0.42 to 0.01 log units, with an absolute average deviation of 0.16.

Attempts to predict fulvic acid and natural aquatic colloid $K_{b,oc}$ values using the more polar surrogate, polymaleic acid, are reported in Table 14.3. Agreement between model estimates and the reported values are generally good, with the exception of 2,4,4'-PCB, where a deviation of greater than half an order of magnitude occurs. All the natural water binding constants were lower than their commercial humic acid values, due in large part to the more polar nature of the organic substrate. The model predicts such a decrease in $K_{b,oc}$. This analysis illustrates the sensitivity of the model prediction to the composition of the organic substrate involved in binding. It supports the contention that the commercial humic acid differs in composition (as well as in solubility parameters) from most aquatic organic colloidal materials.[14] Consequently, with respect to NOC–humic polymer interactions, the commercial humic acid is not a good surrogate for real-world aquatic humic substances.

The successful adaptation of the Flory-Huggins concept into a general partition model clearly demonstrated differences in solubility or partitioning behavior of simple nonpolar organic compounds in complex heterogeneous polymers. While estimates of fulvic acid and natural-water organic matter were albeit slightly less successful, trends were observed and predicted where the degree of partitioning decreased with more polar real-world colloids. These findings corroborate earlier work done by Chiou et al.[14] The approach presented here results in an equation that yields both predictive capabilities and insight into the complex nature of molecular interactions between NOCs and humic subphases. In the case of solute binding with natural-water organic matter and fulvic acid, the problem lies in finding a more appropriate surrogate substance rather than in adjusting the physical basis of solute-colloid interaction.

Effects of Colloids on the Sorption of NOCs by Sediments

Humic substances in the aqueous phase have been observed to affect the sorption of NOCs by soils and sediments.[2-4] Such effects can be reasonably explained using a simple triphase model involving equilibrium distribution of a NOC between sorbent, aqueous, and colloid phases. This type of approach is predicated on the following assumptions:

1. The solute residual in water at equilibrium is comprised of both free and colloid-associated forms.
2. Equilibrium exists among all phases.
3. Sorption and binding of the solute to respective solid and colloid phases can be described in terms of partitioning relationships.

The mathematical form of this model is as follows:

$$K_{p,obs.} = \frac{K_p}{(1 + K_{b,oc}X(10^{-6}))} \qquad (14.21)$$

where X = the colloid concentration in mg/L organic carbon
$\quad\quad K_p$ = the sediment partition coefficient in the absence of humic subphases
$\quad\quad K_{p,obs.}$ = the sediment partition coefficient in the presence of humic subphases

The constant 10^{-6} corrects for the proper units of mass and volume. Reasonable estimates of the effect of third-party subphases on the sorption of NOC by sediments/soils can be accomplished if accurate values of K_p, X, and $K_{b,oc}$ can be determined. K_p and X are measured parameters for this work, while $K_{b,oc}$ is estimated from Equation 14.14.

The effects of the humic acid on the sorption of 1,4-DCB, and 1,2,4-TCB by lacustrine sediments are illustrated in Figures 14.4 and 14.5. The solid lines and closed circles represent Equation 14.21 estimates and observed results, respectively, for sorption in the presence of 12.5 mg/L of the commercial humic acid expressed in terms of TOC. The open squares denote sorption data measured only with distilled deionized water and were used to determine K_p. The triphase model predicts essentially no effect at this colloid concentration on sorption by these two substances, and this is verified by the data. Similarly, Huron River water (background organic matter TOC = 11 mg/L) is predicted and observed to have an insignificant effect on the sorption of the probes by the target sorbent, as illustrated in Figures 14.6 and 14.7. These findings corroborate observations made by Caron et al. for sorption of lindane, a compound with K_{ow} slightly lower than that of 1,2,4-TCB.[4] Chiou et al. further support these findings by observing insignificant lindane and trichlorobenzene solubility enhancement by low concentrations of dissolved humic matter.[14]

The relatively low $K_{b,oc}$ values for the two halogenated benzene compounds explain the observations and model predictions presented in Figures 14.4

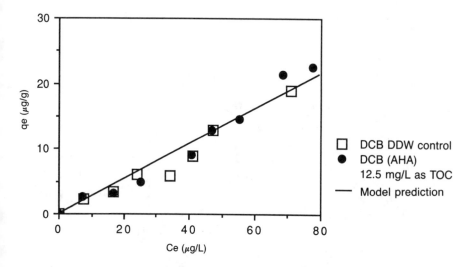

Figure 14.4. Observed and predicted effects of Aldrich humic acid on 1,4-DCB sorption by Lake Michigan sediments (LMS).

through 14.7. A sensitivity analysis was performed using a fixed log ($K_{b,oc}$) value of 3.32 (predicted with Equation 14.14 for TCB binding by AHA) and varying the polymer concentration X in Equation 14.21. The model outputs are illustrated in Figure 14.8. It appears evident from the predictions that K_p is relatively insensitive to the colloid concentration in water, even at TOC levels that are unrealistically high for most natural aqueous systems. A humic acid organic carbon normalized concentration of 40 mg/L affects the partition

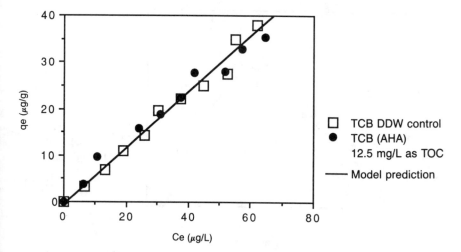

Figure 14.5. Observed and predicted effects of Aldrich humic acid on 1,2,4-TCB sorption by LMS. Reprinted with permission from Chin, Y. P., W. J. Weber, Jr., and B. J. Eadie, *Environ. Sci. Technol.* 24:837 (1990).

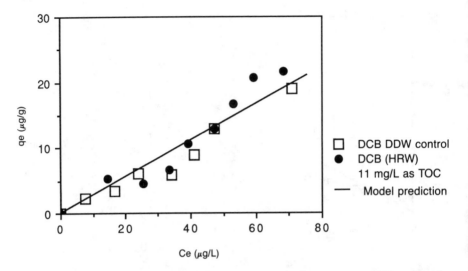

Figure 14.6. Observed and predicted effects of Huron River water on 1,4-DCB sorption by LMS. Reprinted with permission from Chin, Y. P., W. J. Weber, Jr., and B. J. Eadie, *Environ. Sci. Technol.* 24:837 (1990).

coefficient by only 7.8%. Significant changes in K_p are observed only at very high humic substrate concentrations where a 30.2% decrease is observed for a 200 mg/L humic acid solution. Thus, the effects of normal colloid concentrations in natural aquatic systems on the sorption of relatively water-soluble low-molecular-weight organic contaminants by natural solids are relatively small.

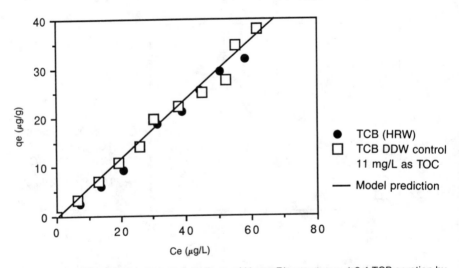

Figure 14.7. Observed and predicted effects of Huron River water on 1,2,4-TCB sorption by LMS. Reprinted with permission from Chin, Y. P., W. J. Weber, Jr., and B. J. Eadie, *Environ. Sci. Technol.* 24:837 (1990).

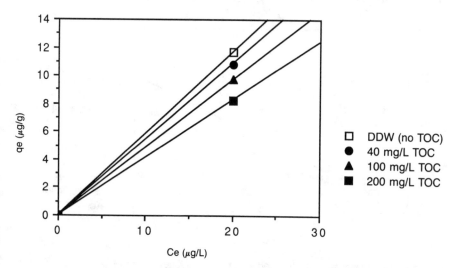

Figure 14.8. $K_{p,obs.}$ predictions for 1,2,4-TCB in the presence of varying amounts of Aldrich humic acid. Reprinted with permission from Chin, Y. P., W. J. Weber, Jr., and B. J. Eadie, *Environ. Sci. Technol.* 24:837 (1990).

The presence of colloids in the aqueous phase has a much more pronounced effect on the two more hydrophobic probes ($K_{ow} > 10^5$). This is illustrated in Figures 14.9 and 14.10 for sorption of 2,5,2'-PCB and α-chlordane, respectively, in the presence of AHA at a concentration of 12.5 mg/L TOC. Agreement between Equation 14.21 predictions and observed values for the effect of a commercial colloid on the sorption of 2,5,2'-PCB is once again good. In the

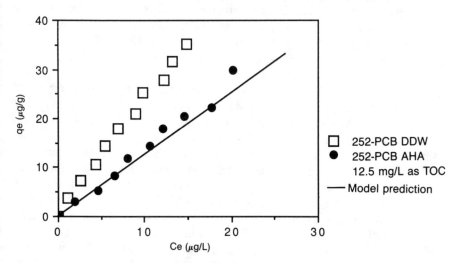

Figure 14.9. Observed and predicted effects of Aldrich humic acid on 2,5,2'-PCB sorption by LMS. Reprinted with permission from Chin, Y. P., W. J. Weber, Jr., and B. J. Eadie, *Environ. Sci. Technol.* 24:837 (1990).

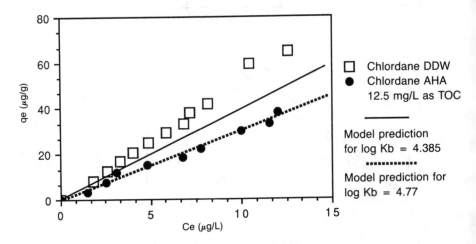

Figure 14.10.   Observed and predicted effects of Aldrich humic acid on α-chlordane sorption by LMS. Reprinted with permission from Chin, Y. P., W. J. Weber, Jr., and B. J. Eadie, *Environ. Sci. Technol.* 24:837 (1990).

case of α-chlordane (solid line), however, $K_{p,obs.}$ model predictions and experimental observations do not corroborate. The $K_{b,oc}$ estimated from Equation 14.14 is approximately a factor of 2.5 lower than the experimental value. By replacing the estimated $K_{b,oc}$ with the observed value, the triphase model fits the experimental data quite well (dashed line). It is rather evident that for these types of nonpolar compounds, accurate values of $K_{b,oc}$ are necessary to provide reliable predictions with Equation 14.21.

The effects of the organic substrates in Huron River water on sorption processes were anticipated to be lower, given their more polar nature and lower binding efficiency as discussed earlier. The effects of Huron River water colloids on the sorption of 2,5,2'-PCB and α-chlordane by Lake Michigan sediments is illustrated in Figures 14.11 and 14.12. Agreement between observed data and model predictions for sorption of 2,5,2'-PCB in the presence of Huron River water is good, using an estimated $K_{b,oc}$ value based upon the structure of PMA, which is similar in properties to the more polar constituents found in natural waters. Estimations for α-chlordane $K_{p,obs.}$ are, however, much less successful, using a hypothetical binding constant predicted from Equation 14.14 with PMA's solubility parameters (solid line). This implies that binding of α-chlordane with Huron River organic matter is much greater than model predictions based upon the surrogate, PMA. As noted previously, estimation of Aldrich humic acid $K_{b,oc}$ for α-chlordane was 0.4 orders of magnitude lower than the experimentally determined value. An inaccurate input function to Equation 14.14 (aqueous-phase solute activity coefficient or solubility parameter value) may be partially responsible for this deviation. It is also plausible that the organic matter present in the Huron River during sampling may have colloidal constituents which could bind hydrophobic substances

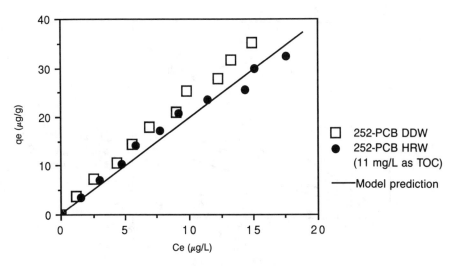

Figure 14.11. Observed and predicted effects of Huron River water on 2,5,2'-PCB sorption by LMS. Reprinted with permission from Chin, Y. P., W. J. Weber, Jr., and B. J. Eadie, *Environ. Sci. Technol.* 24:837 (1990).

more efficiently. This is not unreasonable given the dynamic nature of labile and nonlabile organic macromolecules found in aquatic systems. Rather, such results underscore the need to acquire a reliable database of related physicochemical properties for both target compounds and different types of aquatic colloids.

The effect of the amount of organic colloidal material present in the aqueous phase (X as TOC) on the sorption of 2,5,2'-PCB is illustrated in Figure 14.13. The triphase model is clearly more sensitive to small increases in the dispersed polymer (colloids) concentration for target compounds with large

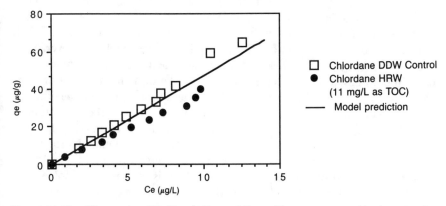

Figure 14.12. Observed and predicted effects of Huron River water on α-chlordane sorption by LMS. Reprinted with permission from Chin, Y. P., W. J. Weber, Jr., and B. J. Eadie, *Environ. Sci. Technol.* 24:837 (1990).

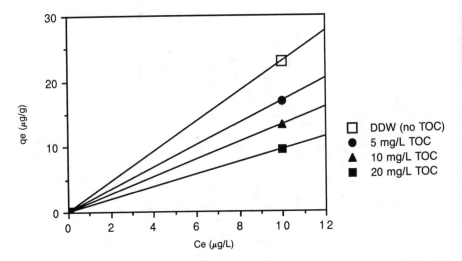

Figure 14.13. $K_{p,obs.}$ predictions for 2,5,2'-PCB in the presence of varying amounts of Aldrich humic acid. Reprinted with permission from Chin, Y. P., W. J. Weber, Jr., and B. J. Eadie, *Environ. Sci. Technol.* 24:837 (1990).

$K_{b,oc}$ values than it is for relatively more water-soluble substances such as DCB and TCB. For 2,5,2'-PCB a realistic surface-water background humic acid concentration of only 5 mg/L (as organic carbon) resulted in a 26.1% decrease in K_p. This is supported by the partition-interaction model of Chiou et al. in explaining the effect of organic substrates on the apparent water solubility of different organic solutes,[14] and by the findings of Brownawell[35] and Brownawell and Farrington[36] for marine systems, in which partition coefficients decreased with increasing TOC levels in near-shore marine sediment cores for various PCB congeners.

It has been suggested by several investigators that $K_{b,oc}$ is approximately equal to the organic carbon normalized solute/sediment partition coefficient, $K_{oc,}$[5,37] while others have shown that $K_{b,oc}$ is generally smaller than K_{oc}.[14,16,17] In the absence of any other data, this assumption is reasonable; however, K_{oc} is expected to be greater than $K_{b,oc}$ because sediment and soil organic matter contain both humic substances and a nonpolar kerogen/humin fraction. This would make the overall sorbent organic matter a more favorable medium for NOC partitioning purposes.[17] To illustrate this, changes in the magnitude of the solute binding constants at a constant colloid concentration (X = 10 mg/L TOC) is examined in Figure 14.14 for α-chlordane, which exhibits the widest variations in $K_{b,oc}$ based on estimations made by Equation 14.14. An 18-fold increase in $K_{b,oc}$ (from $10^{3.5}$ estimated with PMA as the surrogate polymer, to $10^{4.77}$ for Aldrich humic acid) resulted in a 35% change in K_p. Such wide variations again suggest the need for an expanded and more accurate database regarding solute and polymer properties if a thermodynamic modeling approach such as that presented here is to be fully exploited.

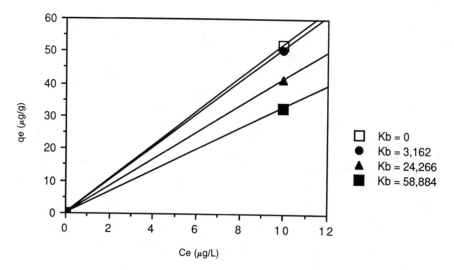

Figure 14.14. $K_{p,obs.}$ predictions for α-chlordane as a function of $K_{b,oc.}$ Reprinted with permission from Chin, Y. P., W. J. Weber, Jr., and B. J. Eadie, *Environ. Sci. Technol.* 24:837 (1990).

ENVIRONMENTAL SIGNIFICANCE

Evidence implicating the binding of organic solutes to a dispersed organic polymer phase as a plausible explanation for the "solids effect" commonly observed in measurements of organic solute sorption by soils and sediments has already been presented.[2,37,38] The work described here supports and helps justify this hypothesis by showing that organic polymers in the aqueous phase can effect significant changes in the sorption characteristics of nonpolar organic contaminants. The magnitude of these effects are dependent upon the nature and amount of dispersed organic polymer present in the aqueous phase and the relative polarity of the solute. The extent to which the solids effect is observed in partitioning experiments is additionally dependent on the desorbability of such polymers from the solids and the effectiveness of the particular solids-liquid separation process employed prior to measurement of "residual" NOC concentration in the aqueous phase.

It is highly unlikely that the amount of organic matter desorbed from the solid phase in most batch experiments would have a major impact on the sorption of relatively soluble low-molecular-weight organic compounds ($K_{ow} < 10^5$). As observed in this chapter, the presence of Aldrich humic acid at very high concentrations had a relatively minor effect on K_p for trichlorobenzene. Gschwend and Wu reported that the amount of TOC desorbed from up to 10 g/L solids concentration was significantly less than 10 mg/L.[37] Thus, one would not expect to see a large change, if any, in the partition coefficient as a function of the solids concentration for weakly hydrophobic substances. Even in subsurface water systems, where these types of contaminants are most prevalent and the ratios of soils to water are high, the solute partition coefficients for these substances should generally remain constant given the low aqueous TOC levels of such systems. Exceptions to this may occur near hazardous waste sites, where cosolvents (e.g., acetone, methanol, etc.), surfactants, or large amounts of heterogeneous polymers in leachates might impact sorption.[39,40] The existence of inorganic colloids may also facilitate transport under such near-site conditions.[41]

Conversely, the amount of polymeric organic matter commonly found in natural aqueous systems is large enough to affect the partitioning of highly water insoluble and recalcitrant organic contaminants such as PCBs, PAHs, and some pesticides, making the apparent K_p values for such compounds highly dependent upon the solids-to-water ratios. These observations are substantiated by the findings of Baker et al., who observed a high dependency of K_p values on solids concentration for sorption of PCBs from the Lake Superior water column.[5] Under such circumstances, the organic geochemistry, fate, and transport of micropollutants become closely dependent on the amount and nature of organic colloidal material present in the aqueous phase.

CONCLUSIONS

The presence of organic macromolecular colloids in aqueous systems complicated predictive equilibrium modeling of hydrophobic organic compound fate and transport in such systems. The research described in this chapter provides insight into the complex thermodynamics that control sorption of such contaminants by sediments and soils, and into the processes that control their interaction with colloidal matter. Observations derived from this work are summarized as follows:

1. Although empirical relationships between organic carbon normalized binding constants and either aqueous-phase activity coefficients or octanol-water partition coefficients were useful, they were limited in their predictive capabilities for solute sorption or binding with organic matter or colloids of varying compositions.
2. Reliable estimates of $K_{b,oc}$ values for binding of hydrophobic organic compounds by a commercial humic acid were made using a partition equation based upon the Flory-Huggins concept. Predictions for NOC binding by a sample of real-world colloids (drawn from surface water samples), using polymaleic acid as a surrogate, were less successful.
3. The effects of organic colloids on the sorption of moderately nonpolar compounds ($K_{ow} < 10^5$) by lacustrine sediments were found to be insignificant, primarily because of their large water solubilities, which caused relatively weak associations with the colloidal phase.
4. Small amounts of colloidal matter were observed to effect large changes on the sorption of hydrophobic NOCs ($K_{ow} > 10^5$) by sediments, due in large part to the high affinity of such compounds for the colloids present in the aqueous phase.

ACKNOWLEDGMENTS

We thank Ingrid Padilla, Kathy Smee, Steven Westenbroek, and Carole Peven for their assistance in the data-acquisition stages of this project. In addition, Phil Gschwend and Bruce Brownawell provided numerous insightful comments during our many discourses with them. This publication is a result of work sponsored by the Michigan Sea Grant College Program, project number R/TS-29 under grant number NA86AA-0-SG043 from the Office of Sea Grant, National Oceanic and Atmospheric Administration, U.S. Department of Commerce, and funds from the State of Michigan. Additional aspects of this research were carried out as part of National Science Foundation project ECE-8503903, Environmental Engineering Program, Dr. Edward H. Bryan, Program Director.

REFERENCES

1. Voice, T. C., C. P. Rice, and W. J. Weber, Jr. "Effect of Solids Concentration on the Sorptive Partitioning of Hydrophobic Pollutants in Aquatic Systems," *Environ. Sci. Technol.* 17:513 (1983).
2. Morel, F. M. M., and P. M. Gschwend. "The Role of Colloids in the Partitioning of Solutes in Natural Waters," in *Aquatic Surface Chemistry,* W. Stumm, Ed. (New York: Wiley Interscience, 1987).
3. Curtis, G. P., M. Reinhard, and P. V. Roberts. "Sorption of Hydrophobic Organic Compounds by Sediments," in *Geochemical Processes at Mineral Surfaces,* J. Davis and K. Hayes, Eds. (Washington, DC: American Chemical Society, 1986).
4. Caron, G., I. H. Suffet, and T. Belton. "Effect of Dissolved Organic Carbon on the Environmental Distribution of Nonpolar Organic Compounds," *Chemosphere* 8:993 (1985).
5. Baker, J. E., P. D. Capel, and S. J. Eisenreich. "Influence of Colloids on Sediment-Water Partition Coefficients of PCB Congeners in Natural Waters," *Environ. Sci. Technol.* 20:1136 (1986).
6. Landrum, P. F., S. R. Nihart, B. J. Eadie, and W. S. Gardner. "Reverse Phase Separation Method for Determining Pollutant Binding to Aldrich Humic Acid and Dissolved Organic Carbon of Natural Waters," *Environ. Sci. Technol.* 18:187 (1984).
7. McCarthy, J. F., and B. D. Jimenez. "Interactions between Aromatic Hydrocarbons and Dissolved Humic Materials," *Environ. Sci. Technol.* 19:1072 (1985).
8. Carter, C. W., and I. H. Suffet. "Binding of DDT to Dissolved Humic Material," *Environ. Sci. Technol.* 16:735 (1982).
9. Landrum, P. F., M. D. Reinhold, S. R. Nihart, and B. J. Eadie. "Predicting the Bioavailability of *P. Hoyi* in the Presence of Humic and Fulvic Materials and Natural Dissolved Organic Matter," *Environ. Toxicol. Chem.* 4:459 (1985).
10. Gauthier, T. D., E. C. Shane, W. F. Guerin, W. R. Seitz, and C. L. Grant. "Fluorescent Quenching Method for Determining the Equilibrium Constants for Polycyclic Aromatic Hydrocarbon Binding to Dissolved Humic Material," *Environ. Sci. Technol.* 20:1162 (1986).
11. Garbarini, D. R., and L. W. Lion. "Evaluation of Sorptive Partitioning of Non-

ionic Pollutants in Closed Systems by Headspace Analysis," *Environ. Sci. Technol.* 19:1122 (1985).

12. Chiou, C. T., R. L. Malcolm, T. I. Brinton, and D. E. Kile. "Water Solubility Enhancement of Some Organic Pollutants by Humic and Fulvic Acids," *Environ. Sci. Technol.* 20:502 (1986).

13. Hassett, J. P., and E. Millicic. "Determination of Equilibrium and Rate Constants for Binding of PCB Congener by Dissolved Humic Substances," *Environ. Sci. Technol.* 19:638-643 (1985).

14. Chiou, C. T., D. E. Kile, T. I. Brinton, R. L. Malcolm, J. A. Leenheer, and P. MacCarthy. "A Comparison of Water Solubility Enhancements of Organic Solutes by Aquatic Humic Materials and Commercial Humic Acids," *Environ. Sci. Technol.* 21:1231 (1987).

15. Chiou, C. T., P. E. Porter, and D. W. Schmedding. "Partitioning Equilibria of Nonionic Compounds between Soil Organic Matter and Water." *Environ. Sci. Technol.* 17:227 (1983).

16. Chin, Y. P., and W. J. Weber, Jr. "Estimating the Effects of Dispersed Organic Polymers on the Sorption of Contaminants by Natural Solids. I. A Predictive Thermodynamic Humic Substance–Organic Solute Interaction Model," *Environ. Sci. Technol.* 23:978 (1989).

17. Chiou, C. T., D. E. Kile, and R. L. Malcolm. "Sorption of Some Organic Liquids on Soil Humic Acid and Its Relation to Partitioning of Organic Compounds in Soil Organic Matter," *Environ. Sci. Technol.* 22:298 (1988).

18. Arbuckle, W. B. "Estimating Activity Coefficients for Use in Calculating Environmental Parameters," *Environ. Sci. Technol.* 17:537 (1983).

19. Campbell, J. R., and R. G. Luthy. "Prediction of Aromatic Solute Partition Coefficients Using the UNIFAC Group Contribution Model," *Environ. Sci. Technol.* 19:980 (1985).

20. Flory, P. J. "Thermodynamics of High Polymer Solutions," *J. Chem. Phys.* 10:51 (1942).

21. Flory, P. J. *Principles of Polymer Chemistry* (Ithaca, NY: Cornell University Press, 1953).

22. Huggins, M. L. "Thermodynamic Properties of Solutions of Long-Chain Compounds," *Ann. NY Acad. Sci.* 43:1 (1942).

23. Blanks, R. F., and J. M. Prausnitz. "Thermodynamics of Polymer Solubility in Polar and Nonpolar Systems," *Ind. Eng. Chem. Fund.* 3:1 (1964).

24. Karickhoff, S. W. "Pollutant Sorption in Aquatic Systems," *J. Hydraul. Eng. ASCE* 10:707 (1984).

25. Hildebrand, J. H., J. M. Prausnitz, and R. L. Scott. *Regular and Related Solutions* (New York: Van Nostrand Reinhold, 1970).

26. Weimer, R. F., and J. M. Prausnitz. "Screen Extraction Solvents This Way," *Hydrocarbon Processing* 44:237 (1965).

27. Schnitzer, M., and S. U. Khan. *Humic Substances in the Environment* (New York: Marcel Dekker, 1972).

28. Aiken, G. R., D. M. McKnight, R. L. Wershaw, and P. MacCarthy. *Humic Substances in Soil, Sediment, and Water* (New York: Wiley Interscience, 1985).

29. Malcolm, R. L., and P. MacCarthy. "Limitations in the Use of Commercial Humic Acids in Water and Soil Research," *Environ. Sci. Technol.* 20:904 (1986).

30. Barton, A. F. M. *Handbook of Solubility Parameters and Other Cohesion Parameters* (Boca Raton, FL: CRC Press, 1983).

31. Welch, D. I. "Studies on the Structure and Behavior of Polymaleic Acid as a Model Soil Organic Polymer," PhD Thesis, Macauley Institute for Soil Research, Aberdeen, U.K. (1981).

32. Spiteller, M., and M. Schnitzer. "A Comparison of the Structural Characteristics of Polymaleic Acid and a Soil Fulvic Acid," *J. Soil Sci.* 34:525 (1983).

33. Karger, B. L., L. R. Snyder, and C. Eon. "An Expanded Solubility Parameter Treatment for Classification and Use of Chromatographic Solvents and Adsorbents," *J. Chromat.* 125:71 (1976).

34. Chin, Y. P., C. S. Peven, and W. J. Weber, Jr. "Estimating Soil/Sediment Partition Coefficients for Organic Compounds by High Performance Reverse Phase Liquid Chromatography," *Water Res.* 7:873 (1988).

35. Brownawell, B. J. "The Role of Colloidal Organic Matter in the Marine Geochemistry of PCBs," PhD Thesis, Woods Hole Oceanographic Institute/MIT, Cambridge, MA (1986).

36. Brownawell, B. J., and J. W. Farrington. "Biogeochemistry of PCBs in Interstitial Waters of a Coastal Marine Sediment," *Geochim. Cosmochim. Acta* 50:157 (1986).

37. Gschwend, P. M., and S. Wu. "On the Constancy of Sediment-Water Partition Coefficients of Organic Pollutants," *Environ. Sci. Technol.* 19:90 (1985).

38. Voice, T. C., and W. J. Weber, Jr. "Solids Effect in Liquid/Solid Partitioning," *Environ. Sci. Technol.* 19:789 (1985).

39. Fu, J., and R. G. Luthy. "Aromatic Compound Solubility in Solvent/Water Mixtures," *J. Environ Eng. ASCE* 112:346 (1986).

40. Nkeddi-Kizza, P., P. S. C. Rao, and A. G. Hornsby. "Influence of Cosolvents on Sorption of Hydrophobic Organic Chemicals by Soils," *Environ. Sci. Technol.* 19:975 (1985).

41. Gschwend, P. M., and M. D. Reynolds. "Monodispersed Ferrous Phosphate Colloids in an Anoxic Groundwater Plume," *J. Contam. Hydrol.* 1:309 (1987).

CHAPTER 15

Applicability of Linear Partitioning Relationships for Sorption of Organic Vapors onto Soil and Soil Minerals

S. K. Ong, S. R. Lindner, and L. W. Lion

INTRODUCTION

It is generally recognized that the transport of volatile organic compounds (VOCs) in the subsurface can be significantly influenced by their movement in the vapor phase of the unsaturated zone. The importance of organic vapor migration in the subsurface is evidenced by recent applications of shallow-soil gas-sampling techniques to predict the extent of groundwater pollution[1] and use of in situ soil-venting techniques to rehabilitate contaminated aquifers.[2] Sorption is a major physical-chemical process that affects the movement of VOCs in the soil matrix. Unlike sorption reactions in the saturated zone, sorption of VOCs in unsaturated soils is complicated by a variable moisture content. Water in the unsaturated zone can vary from fairly dry for surface soils and arid regions, to saturation at the capillary fringe of the water table. Previous studies have shown that sorption of VOCs is substantially influenced by the amount of moisture present in soil.[3-5] Vapor dissolution into pore water and the solid-liquid partitioning of dissolved vapors are linear processes which are likely to govern vapor uptake. However, the variable moisture contents of the soil matrix and the potential range of VOC vapor pressures may result in nonlinear behavior, such as vapor adsorption to soil mineral surfaces and vapor condensation. Although solute competition reportedly does not occur in the sorption of mixtures of nonionic organic compounds under saturated conditions,[6,7] the behavior of vapor mixtures in unsaturated soils has not been extensively evaluated.

In this chapter, we examine the linear processes that affect the sorption of organic vapors onto soil minerals and evaluate their applicability over a range of moisture contents and vapor pressure. Data are presented for sorption of trichloroethylene (TCE) at low (P/P_o < 2%) and high (P/P_o up to 90%)

Table 15.1. Physical-Chemical Properties of Sorbents

	Alumina	Coated Alumina	Iron Oxide	Kaolinite	Montmorillonite	Gila Loam
Particle density (g/cm^3)	2.98	2.57	4.03	2.51	1.8	2.53
pH (H$_2$O extract)	4.5	7.18	6.5	4.2–5.2	8.3	8.5
Organic carbon (%)	0.02	0.45	0.06	0.01	0.02	0.38
Surface area (m^2/g)						
BET nitrogen	143.2	189.3	10.98	8.47	97.42	—
EGME method	183.0	250.0	24.9	21.3	463.3[a]	98.5
CEC (meq/100g)	15.5	15.5	18.5	7.5	113.5	20.6

[a]Surface area computed from water adsorption isotherm.

vapor pressures, and for sorption of vapor mixtures (selected from components of fuels and chlorinated solvents) at low vapor pressures ($P/P_o < 2\%$).

MATERIALS AND METHODS

Sorbents used at low TCE vapor pressures ($P/P_o < 2\%$) were common soil minerals: aluminum oxide (Fisher Scientific, chromatographic grade, mesh size 80–200), ferric oxide (Pfizer, Inc.), Ca^{2+} montmorillonite (Cheto, #SAZ-1, Clay Mineral Society), kaolinite (hydrite flat D type, Georgia Kaolin Co.), and aluminum oxide coated with Aldrich humic acid. Procedures for coating the oxide with humic acid are given by Garbarini and Lion.[8] The sorbents were selected to provide a mix of well-characterized oxides and clay minerals with a range of specific surface areas. Coating of the aluminum oxide permitted controlled addition of organic matter. The aluminum oxide is a microporous material, and montmorillonite is an expanding clay with interlamellar surfaces; the other sorbents are relatively nonporous. Physical-chemical properties of the sorbents are listed in Table 15.1. The minerals were air-dried for ≥ 24 hr before water was applied to obtain the desired moisture contents. The resulting moist sample was then allowed to equilibrate in a closed environment for several days to promote even distribution of water. Alumina and humic-coated alumina were used as sorbents for studies at high TCE relative vapor pressure.

The sorbents used for studies of vapor mixtures were humic-coated aluminum oxide and Gila silt loam. Gila silt loam is an agricultural soil of the Yuma-Wellton area of Arizona-California. It contains 18.4% clay (predominantly montmorillonite) and has an organic carbon content of 0.38%. Twelve sorbates (six fuel components and six chlorinated hydrocarbons) were used to screen for possible mixture effects on the sorption of vapors by coated aluminum oxide.[9] The fuel components were 1,3,5-trimethylbenzene, toluene, cyclohexane, methylcyclohexane, methylcyclopentane, and p-xylene; the chlorinated hydrocarbons were 1,1,1-trichloroethane, perchloroethylene, 1,1-dichloromethylene, trichloroethylene, chloroform, and methylene chloride. Mixtures of organic compounds that exhibited mixture effects in the screening experiments were further evaluated. Two of these studies are reported here: (1)

a ternary vapor study of 1,3,5-trimethylbenzene, toluene, and xylene on humic-coated alumina at 16% moisture content and (2) a binary sorption study involving 1,3,5-trimethylbenzene and toluene on Gila silt loam at 19% moisture content.

The vapor-and liquid-phase partition coefficients (K_d' and K_d, respectively) of nonpolar organic compounds at low concentrations were determined by the headspace analysis method of Garbarini and Lion[8] as modified by Peterson et al.[5] The vapor-phase partition coefficient (K_d') refers to the uptake of the volatile compounds from the vapor phase onto sorbents, which may be either dry or moist. Uptake may therefore involve multiple processes, including dissolution into sorbent-bound water, condensation, adsorption at the mineral-vapor or water-vapor interface, and partitioning into soil organic matter. For analysis of the vapor-phase partition coefficient, fixed volumes of saturated vapor were introduced with a gastight syringe into vials containing sorbent masses ranging from 0.1 to 40 g (dry weight). Moisture contents of the minerals were varied from oven dry to approximately 60% (weight/weight). For each isotherm, twelve to fifteen vials were used, with two or three replicates for each sorbent mass. The nominal volume of the vials was either 60 or 120 mL, depending on the mass of sorbents used. Three to six control vials (without sorbent) received similar additions of the vapor. After introduction of the vapor, each vial was sealed immediately with a Teflon-lined septum and an aluminum crimp cap. For sorption studies of vapor mixtures, two or more organic vapors were introduced through separate gastight syringes before the sample bottles were capped. The bottles were tumbled and allowed to equilibrate for 24–36 hr at 25 ± 1°C. The headspace concentration of the vials was then measured with a gas chromatograph (GC) using a flame ionization detector. Details of analytical procedures are given by Ong[10] and Lindner.[9]

The vapor-phase partition coefficient (K_d') was determined from the following equation, which may be derived from a mass balance of the sample and control vials:[5]

$$\frac{C_B V_B}{C_S V_S} - 1 = K_d' \frac{M}{V_S} \qquad (15.1)$$

where K_d' = the vapor partitioning coefficient (mL/g)
           M = the oven-dry mass of the solids (g)
           C_B = the GC peak area for a control vial
           C_S = the GC peak area for a sample vial
           V_B = the volume of the blank vial (mL)
           V_S = the available gas volume of a sample vial (mL), taking into account the volume occupied by the soil and its moisture

Equation 15.1 is applicable to the linear range of the vapor adsorption isotherm. This is a reasonable assumption for the low vapor pressure range employed in these experiments. Linear regression of a plot of $(V_B C_B / V_S C_S) - 1$ vs M/V_S gives a slope equal to K_d'. A poor fit to Equation 15.1 was taken as evidence of nonlinear behavior.

To determine the liquid-phase partition coefficient (K_d), precise volumes of aqueous solutions of the test solutes were introduced into sample vials (with sorbent) and control vials (without sorbent) containing 20 mL of 0.1 M sodium chloride solution. The vials were equilibrated, and the vapor concentration in the headspace was determined as described above. The aqueous partition coefficient may be determined from Equation 15.2, derived in a manner comparable to Equation 15.1 from a mass balance of the sample and control vials and the assumption that Henry's law and a linear sorption isotherm are obeyed:[5]

$$\left(\frac{C_B}{C_S}\right)\left(\frac{V_{GB} K_H \gamma + V_{LB}}{V_{GS} K_H \gamma + V_{LS}}\right) - 1 = K_d \left(\frac{M}{V_{LS} + V_{GS} K_H \gamma}\right) \quad (15.2)$$

where K_d = the saturated partition coefficient (mL/g)

            C_B = the average GC peak area for control vials

            C_S = the GC peak area for a sample vial

            K_H = the Henry's law constant for the organic vapor (dimensionless)

            γ = the aqueous activity coefficient for the dissolved vapor

V_{GB} and V_{GS} = the gas volumes in the control and sample vials, respectively (mL)

V_{LB} and V_{LS} = the liquid volumes in the control and sample vials, respectively (mL)

            M = the oven dry mass of solids (g)

Sorption studies at higher organic vapor concentrations were performed with a gravimetric sorption apparatus (Figure 15.1) similar to that described by Noll et al.[11] Nitrogen gas saturated with TCE and water vapor was mixed to obtain different relative humidities and TCE relative vapor pressures. When simple mixing of the gas flows was not sufficient to achieve the desired partial pressure of TCE, elevated relative vapor pressures were achieved by heating the gas dispersion bottles containing the organic liquid. The mixed gas stream was thermally equilibrated to 25°C and introduced into a constant temperature (25°C) water-jacketed column containing a glass vessel with a sorbent sample suspended from a quartz spring. Samples were dried in an oven at 105°C for approximately 48 hr before they were transferred to the glass vessel. The samples were then exposed to dry nitrogen gas for 12–24 hr before water and/or TCE vapor were introduced. In binary sorption experiments, the soil sample was first equilibrated with water vapor at a given relative humidity

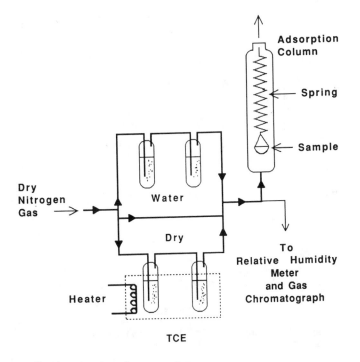

Figure 15.1. Gravimetric adsorption apparatus.

before TCE vapor was introduced. The mass of TCE or water sorbed was determined from the extension of the quartz spring as measured by a cathetometer. The quartz spring had an extension response of 9.94 mg/cm, and the cathetometer has a precision of ±0.00005 cm (or ±5 μg weight change). The column was kept at a constant temperature of 25 ± 1°C using a circulating water bath. Relative humidity was measured with a hygrometer using an optical sensor (General Eastern Model 1500). The relative TCE vapor pressure in the column was computed from the gas flow rates and confirmed by GC analysis.

RESULTS AND DISCUSSION

Low Vapor Concentrations Studies (P/P$_o$ < 2%)

For sorption studies at low vapor concentrations, the amount of TCE sorbed decreased slightly as the moisture content of the sorbent was decreased below saturation. As drying of the solid continued, a moisture content was reached where the mass of TCE sorbed dramatically increased. This behavior is illustrated for TCE sorption onto alumina in Figure 15.2. On this highly porous, high specific surface area sorbent, the sorption of TCE increased greatly when the moisture content was less than 24%. As an aid in the analysis

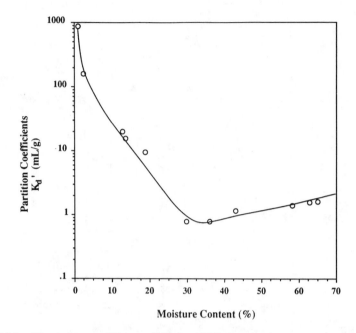

Figure 15.2.   Vapor-phase partition coefficients of TCE onto alumina as a function of moisture content.

of the distribution of TCE vapor in regions of high moisture contents, the uptake of TCE vapor may be attributed to three processes:

1. dissolution of TCE vapor in the soilwater (as governed by its Henry's law constant)
2. sorption of dissolved TCE at the solid-liquid interface (linear partitioning into the organic phase of the sorbent)
3. vapor condensation plus sorption at the water-air interface

Based on this formalism, the vapor-phase partition (K_d') can be expressed in terms of the solid-liquid partitioning coefficient (K_d), Henry's constant (K_H), and the moisture content of the sorbent. This relationship is given by Ong and Lion:[12]

$$K_d' = K_d/K_H + (\text{Moisture Content } \%)/(100 \, K_H \, \gamma \, \rho) + \omega/(C_G M) \quad (15.3)$$

where K_d = liquid-phase partition coefficient (mL/g)
C_G = TCE concentration in the vapor phase (mg/L)
M = mass of sorbent (g)
ρ = density of water (mL/g)
γ = aqueous activity coefficient
K_H = dimensionless Henry's constant

ω = a lumped parameter that includes the effects of water or mineral surface sorption and condensation

If it is assumed that there is negligible condensation or sorption at the water-air interface ($\omega \cong 0$), Equation 15.3 may be used to predict vapor uptake and compared to the experimental data. Results for TCE sorption onto humic-coated alumina (shown in Figure 15.3) indicate that sorption of TCE at high moisture contents can be accounted for by the linear partition processes of dissolution into the aqueous phase plus sorption at the solid-liquid interface. Computation of the predicted vapor-phase partition coefficients were based on an experimentally determined solid-liquid partition coefficient of 0.31 mL/g for humic-coated alumina and a Henry's constant for TCE in distilled water at 25°C of 0.42. In the case of uncoated aluminum oxide (Figure 15.4), where negligible TCE was found to sorb to the mineral surfaces under saturated conditions, vapor-phase uptake of TCE could be completely accounted for by the dissolution of TCE in the aqueous phase. The slight decrease in the vapor partition coefficients as moisture content decreased below saturation is accounted for by the variation in moisture content and the equilibrium capacity for dissolution of TCE in the available water. The processes of vapor condensation and adsorption at the water-air interface did not significantly influence the overall uptake of TCE.

Expression of moisture contents in terms of the number of layers of water

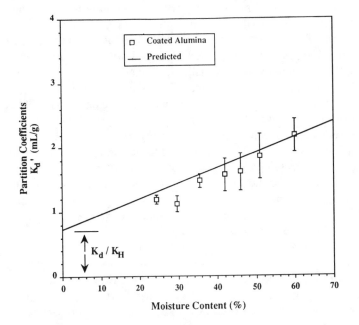

Figure 15.3. Observed and predicted (Equation 15.3) sorption of TCE onto humic-coated alumina.

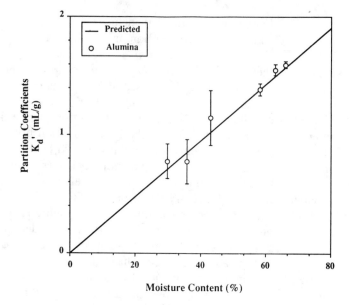

Figure 15.4. Observed and predicted (Equation 15.3) sorption of TCE onto alumina.

molecules on the surface permits comparison of TCE sorption over a wide array of different sorbents. Surface coverage of water molecules can be computed from the surface area of the sorbent by assuming a molecular cross-sectional area of 10.8×10^{-20} m^2 for water[13] and by assuming that water is sorbed uniformly over the surface. The surface areas determined by the Ethylene Glycol Monoethyl Ether (EGME) method without sample pretreatment were used,[14,15] except for montmorillonite, where the surface area was determined from Brunauer, Emmett, Teller (BET) analysis of water vapor adsorption using the gravimetric adsorption apparatus. The partition coefficients for TCE sorption onto several different minerals are plotted against the surface coverage by water molecules, as shown in Figure 15.5. The TCE vapor partition coefficients (K_d') were corrected using Equation 15.3 to remove the contribution of the saturated partition coefficient (K_d); therefore, the y-axis reflects uptake of TCE by dissolution into water on the sorbent plus the contribution of TCE sorption at the air-water interface and condensation, when significant. From Figure 15.5, it can be seen that, at surface coverages exceeding approximately five layers of water molecules, a common process is responsible for TCE uptake on all the experimental surfaces, i.e., vapor dissolution into bound water. Note that application of Equation 15.3 is appropriate only at high moisture contents; the calculated line for TCE dissolution in Figure 15.5 is extended to low moisture contents for purposes of illustrating the point at which the data depart from Equation 15.3. As the moisture content of minerals is reduced below that corresponding to five layers of water molecules (as in semiarid and arid regions or for surface soils), the amount of TCE that is

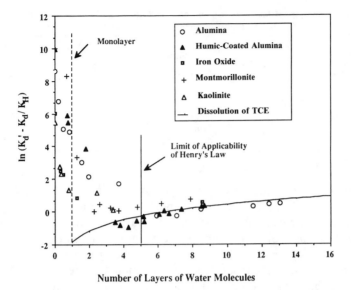

Figure 15.5. Limitation of the applicability of Henry's law constant for dissolution of TCE vapor in soilwater for various minerals.

sorbed increases greatly and cannot be accounted for by dissolution of TCE into condensed water and solid-liquid partitioning. Vapor uptake processes, such as TCE sorption onto the bound water on the mineral surfaces (or possibly direct sorption onto mineral surfaces), may be occurring at low moisture contents. The implication of these results is that transport of organic vapors in drier soils, such as surface soils and soils in semiarid and arid regions, would be retarded relative to the same soils at higher moisture contents.

Behavior of Vapor Mixtures

Partition coefficients of twelve individual hydrocarbons and chlorinated hydrocarbons on humic-coated aluminum oxide at approximately 40% moisture content were found to be linear and correlated well with the aqueous solubility of the sorbate.[9] This is consistent with the observations for TCE noted above, in which vapor sorption can be accounted for by the processes of vapor dissolution into water associated with the solid phase and sorption at the solid-liquid interface. However, the sorption of mixtures of six hydrocarbons or six chlorinated hydrocarbons at lower moisture contents than 40% was, in some cases, found to exhibit mixture effects that were dependent on mixture composition.[9]

The behavior of toluene in a ternary vapor mixture with 1,3,5-trimethylbenzene and *p*-xylene provides an illustration of these mixture effects. The sorbent used was humic-coated aluminum oxide at 16% moisture content (approximately three layers of water molecules). In this study, parti-

Table 15.2. Sorption of Mixtures of Toluene, *p*-Xylene, and 1,3,5-Trimethylbenzene on Humic-Coated Alumina at 16% Moisture Content

	Vapor	$K_d'(\pm 95\%\ Cl)$ (mL/g)
Single Vapor	toluene	55.9 ± 8.9
	p-xylene	173.4 ± 39.1
	trimethylbenzene	342.6 ± 100.7
Ternary Vapor	toluene	37.9 ± 2.0
	p-xylene	171.2 ± 19.1
	trimethylbenzene	470.4 ± 215

tion coefficients of toluene were reduced from 55.9 ± 8.9 mL/g to 37.9 ± 2.0 mL/g (see Table 15.2). Note that the vapor-phase partition coefficients (within a 95% confidence interval) of *p*-xylene and 1,3,5-trimethylbenzene showed no significant change. Since the moisture content of the humic-coated alumina was less than five layers of water, it is conceivable that the observed competitive effects could result from vapor competition for sites on the surface of sorbent-bound water. However, this is considered unlikely since the isotherms were linear (implying an excess of surface sites if vapor uptake was at an interface).

A similar effect was observed (Figure 15.6) for a binary vapor study with toluene and 1,3,5-trimethylbenzene on Gila silt loam at a moisture content of

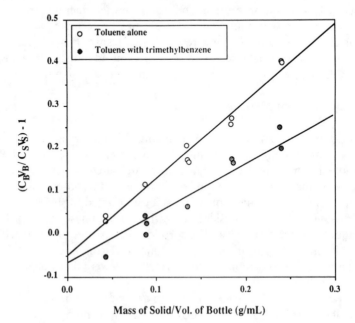

Figure 15.6. Suppressed sorption of toluene onto Gila silt loam (at 19% moisture content) in the presence of trimethylbenzene.

19% (approximately seven layers of water). The partition coefficient of toluene was found to decrease from 1.46 ± 0.12 mL/g to 0.70 ± 0.35 mL/g in the presence of 1,3,5-trimethylbenzene. The results obtained with TCE (see above discussion) suggest that Henry's law and the saturated partition coefficient should adequately predict uptake from the vapor phase at this moisture content. Therefore, vapor adsorption at (and competition for) surfaces is not deemed likely.

The results showing competition of vapors do not correspond to previous studies under saturated conditions, where a lack of competition between dissolved solutes has been observed.[7,16] Presumably the differences in mixture effects observed under saturated vs unsaturated conditions may be attributed to the relative amounts of water and nonionic organic compound present. At low moisture contents, water on the sorbent has its structure influenced by polar interaction with charged mineral surfaces and the air-water interface.[17,18] Vapor dissolution, as governed by Henry's law, is influenced by the aqueous activity coefficient, and it is conceivable that, at low moisture contents, dissolution of one vapor may influence the structure of the aqueous phase sufficiently to alter the aqueous activity coefficient of other nonionic solutes. At higher moisture contents, addition of small amounts of one VOC would be anticipated to be less likely to influence the activity of other VOCs. Under such conditions, linear isotherms without competition should occur. The frequent occurrence of contaminated sites with mixtures of volatile hydrocarbons would suggest a need for further research on VOC mixture effects.

High Vapor Concentration Studies ($0 < P/P_0$ up to 90%)

Sorption isotherms for TCE onto alumina over a wide vapor range are shown in Figure 15.7. As observed in the above studies at low vapor concentration, sorption of TCE was affected by the presence of moisture and decreased with an increase in moisture content. Instead of the linear isotherms observed at low vapor pressures, sorption isotherms at higher vapor pressures were nonlinear over the RH studied (0 to 70%). Isotherm curvature of this type is consistent with that which characterizes multilayer vapor sorption.[19] When the isotherms are expressed in terms of the total volume of water plus TCE adsorbed (Figure 15.8), the high vapor pressure regions of the isotherm are coincident regardless of the sorbent moisture content. Condensation of TCE and water vapor in the pores of the sorbent is most likely to explain this observation. Similar studies were conducted with alumina coated with humic acid. As shown in Figure 15.9, sorption isotherms were again nonlinear over the range of RH studied, and the presence of water reduced the amount of TCE sorbed. Expressing these isotherms on a total volume basis, as before, results in a similar coincidence in the high vapor pressure region of the isotherms (Figure 15.10). This result, obtained on a porous solid with an organic phase, would appear to further implicate vapor condensation as a mechanism

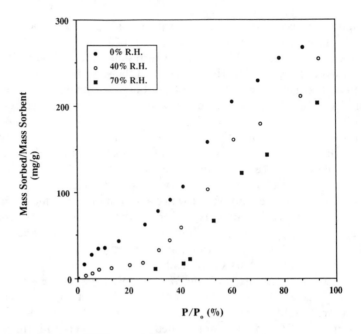

Figure 15.7.   Sorption of TCE onto alumina.

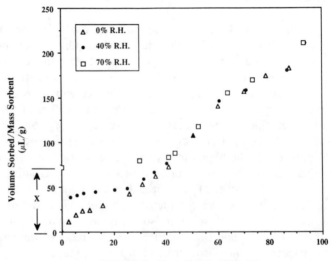

Figure 15.8.   Sorption of TCE and water (on a volume basis) onto alumina; X, volume of water sorbed at 70% RH.

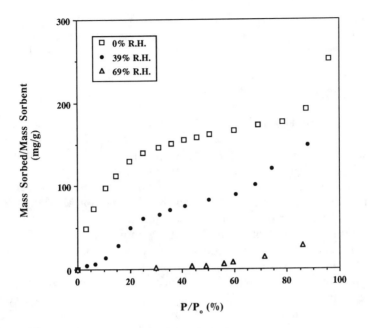

Figure 15.9. Sorption of TCE onto humic-coated alumina.

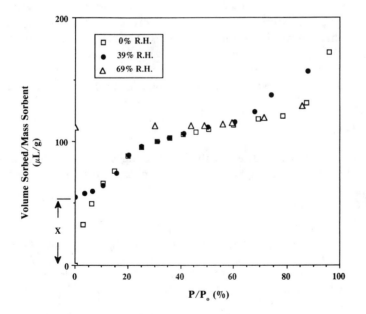

Figure 15.10. Sorption of TCE and water (on a volume basis) onto humic-coated alumina; X, volume of water sorbed at 39% RH.

in the uptake of TCE at high vapor pressures. Vapor condensation would be most likely to occur in soils that are close to vapor sources, such as leaking storage tanks, or in the presence of non-aqueous-phase liquids (NAPLs).

CONCLUSIONS

For sorption of TCE at low vapor pressure on sorbents with high moisture contents, vapor uptake can be accounted for by two linear partition processes: dissolution of TCE in soil water and sorption at the solid-liquid interface. However, at moisture contents below the amount equivalent to five layers of surface coverage by water, the processes of vapor dissolution and partitioning cannot account for the mass of TCE which is sorbed. Sorption of TCE at the water-air interface or direct sorption to mineral surfaces are likely explanations for the increased uptake of TCE at low moisture contents. For organic vapors at high vapor pressures, condensation of vapors may occur in the pore space of the soil that is not occupied by water.

Suppressed sorption relative to the linear isotherm for single vapors was observed for several mixtures of organic vapors at low moisture contents. Decreased sorption may imply that vapor competition was occurring and could reflect activity effects in surface-bound water. At the moisture contents considered in this research, the sorption of nonionic vapor mixtures and water appears to reflect their relative affinities for the surface and relative concentration in the vapor phase.

The complicating effects of a variable moisture content, high vapor pressures, and vapor mixture effects may alter sorption in the unsaturated zone relative to that observed in saturated systems. The successful use of saturated linear partition coefficients and the Henry's constant to predict vapor uptake will depend on the moisture content present, the presence of other organic vapors, and the relative vapor pressure of the organic compounds of concern.

ACKNOWLEDGMENT

This research was supported by the Engineering Services Laboratory of the U.S. Air Force Engineering Services Center under contract F08635–85–C-003.

REFERENCES

1. Marrin, D. L., and H. B. Kerfoot. "Soil-Gas Sampling Techniques," *Environ. Sci. Technol.* 22(7):740–745 (1988).
2. Baehr, A. L., G. E. Hoag, and M. C. Marley. "Removing Volatile Contaminants from the Unsaturated Zone by Inducing Advective Air-Phase Transport," *J. Contam. Hydrol.* 4:1–26 (1989).
3. Chiou, C. T. "Theoretical Considerations of the Partition Uptake of Nonionic Organic Compounds by Soil Organic Matter," in *Reactions and Movement of*

Organic Chemicals in Soils, B. L. Sawhney and K. Brown, Eds., Soil Science Society of America Special Pub. No. 22 (Madison, WI: Soil Science Society of America, 1989), pp. 1–29.

4. Hance, R. J. "The Adsorption of Atratron and Monuron by Soils at Different Water Contents," *Weed Res.* 17:137–201 (1977).

5. Peterson, M. S., L. W. Lion, and C. A. Shoemaker. "Influence of Vapor Phase Sorption and Diffusion on the Fate of Trichloroethylene in an Unsaturated Aquifer System," *Environ. Sci. Technol.* 22:571–578 (1988).

6. Chiou, C. T., T. D. Shoup, and P. E. Porter. "Mechanistic Roles of Soil Humus and Minerals in the Sorption of Nonionic Organic Compounds from Aqueous and Organic Solutions," *Org. Geochem.* 8:9–14 (1985).

7. Chiou, C. T., P. E. Porter, and D. W. Schmedding. "Partition Equilibria of Nonionic Organic Compounds between Soil Organic Matter and Water," *Environ. Sci. Technol.* 17:227–231 (1983).

8. Garbarini, D. R., and L. W. Lion. "Evaluation of Sorptive Partitioning of Nonionic Pollutants in Closed System by Headspace Analysis," *Environ. Sci. Technol.* 19:1122–1128 (1985).

9. Lindner, S. R. "Sorption of Volatile Organic Compounds and Their Mixtures on Synthetic and Natural Soils," PhD Thesis, Cornell University, Ithaca, NY (1990).

10. Ong, S. K. "Sorption Equilibria of Trichloroethylene Vapor onto Moist Soils and Minerals," PhD Thesis, Cornell University, Ithaca, NY (1990).

11. Noll, K. E., A. A. Aguwa, Y. P. Fang, and P. T. Boulanger. "Adsorption of Organic Vapors on Carbon and Resin," *ASCE J. Environ. Eng.* 111(4):487–500 (1985).

12. Ong, S. K., and L. W. Lion. "Mechanisms of Trichloroethylene Vapor Sorption onto Soil Minerals," *J. Env. Qual.* 20:180–188 (1991).

13. Livingston, H. F. "The Cross Sectional Areas of Molecules Adsorbed on Solid Surfaces," *J. Colloid Sci.* 4:447–458 (1949).

14. Carter, D. L., M. M. Mortland, and W. D. Kemper. "Specific Surface," in *Methods of Soil Analysis, Part 1,* A. Klute, Ed. (Madison, WI: Soil Science Society of America, 1986), pp. 413–423.

15. Cihacek, L. J., and J. M. Bremner. "A Simplified Ethylene Glycol Monoethyl Ether Procedure for Assessment of Soil Surface Area," *Soil Sci. Soc. Am. J.* 43:821–822 (1979).

16. Karickhoff, S. W., D. S. Brown, and T. A. Scott. "Sorption of Hydrophobic Pollutants on Natural Sediments," *Water Res.* 13:241–248 (1979).

17. Grim, R. E. *Clay Mineralogy* (New York: McGraw-Hill Book Company, 1953).

18. Sposito, G., and R. Prost. "Structure of Water Adsorbed on Smectites," *Chem. Rev.* 82(6):554–573 (1982).

19. Brunauer, S., P. H. Emmett, and E. Teller. "Adsorption of Gases in Multimolecular Layers," *J. Am. Chem. Soc.* 60:309 (1938).

CHAPTER 16

Competitive Effects in the Sorption of Nonpolar Organic Compounds by Soils

Joseph J. Pignatello

INTRODUCTION

Hydrophobic phase-partitioning to natural soil organic matter is widely regarded to be the mechanism of sorption of nonpolar organic compounds to soils and sediments, provided water is present and the organic matter content is above trace levels.[1,2] According to this model, the organic matter behaves like a hydrophobic liquid phase that incorporates the sorbate homogeneously in the organic solid network, and sorption is driven primarily by entropy (exclusion from water) rather than by enthalpy (formation of bonds with the sorbent).

Among the major lines of evidence presented in support of the hydrophobic partitioning model for soils is the apparent absence of competition between sorbates.[1] Competition is characteristic of surface adsorption. For example, uptake of solutes by activated carbon[3-5] and inorganic oxides[5] as well as uptake of vapors by dry soils[6] are subject to competitive effects. Competition is not characteristic of phase partitioning in dilute systems, such as partitioning between water and an immiscible organic solvent. The lack of competition in soils is posited by unchanged sorption isotherms, compared to the single solute isotherms, of the bisolute pairs 1,3-dichlorobenzene and 1,2,4-trichlorobenzene,[7] and parathion and lindane.[8] In these two studies, the sorbed concentrations of the cosolutes were adjusted to approximately 1:1. Also, Karickhoff et al. stated, without giving data, that they observed no competition between pyrene and phenanthrene on a sediment.[9]

Recently, however, reports of multisolute effects in aqueous systems involving natural sorbents have appeared. Sorption of tetrachloroethene on a soil containing 1.14% organic carbon (OC) was inhibited by 1,4-dichlorobenzene.[10] Mutual suppression of sorption on inactive biomass from sewage sludge was observed between chlorobenzene and ethylbenzene and between various substituted phenols.[11] Desorption rates of a hexachlorobiphenyl from a sediment were altered in the presence of other PCB conge-ners.[12] Sorption of polychlorodibenzodioxins (PCDDs) in soil-methanol-water systems was reduced in the presence of small concentrations of pentach-lorophenol or chlorobenzene.[13]

Bisolute effects are important both for their mechanistic implications and for their potential impacts on contaminant transport and fate. The present study was prompted by observations of bisolute effects on nonequilibrium sorption behavior of halogenated hydrocarbons (see below). It is reported here that the sorption of several nonpolar halogenated aliphatic and aromatic hydrocarbons to four soils is moderately suppressed in the presence of a sec-ond solute.

EXPERIMENTAL

Materials

Three soils were collected from the upper 15-cm horizon of agricultural fields. They will be referred to by town location in Connecticut: Warehouse Point, Windsor, and Granby. All three are Merrimac sandy loams, in which the dominant clays are illitic. The soils were sieved (2 mm) at field moisture and stored at 4°C. OC was measured as CO_2 released from pulverized and acidified samples (Galbraith Laboratories, Knoxville, TN). A fourth soil is a high organic peat soil (International Humic Substances Society reference peat) kindly provided by C. T. Chiou (U.S. Geological Survey, Denver, CO).

Humic acid was prepared from topsoil collected at the Windsor site. The soil was first treated for 1 hr with 0.1 N HCl and then extracted with degassed 0.1 N NaOH overnight with shaking. After settling, the supernatant was centri-fuged at 8600 x g for 20 min and acidified with HCl to pH 1.5. The precipitate was carried through two additional dissolution-centrifugation-precipitation steps and then dialyzed with distilled water in SpectraPor-6 dialysis bags (molecular weight cutoff, 1000; Spectrum Medical Industries) until chloride was removed. The freeze-dried solid was crushed to pass through a 500 μm sieve. Elemental analysis (Galbraith) gave C, 51.1; H, 4.66; N, 6.35; S, 0.83; Cl, 0.03; ash, 8.63%.

All compounds (liquids at room temperature) were reagent grade and veri-fied to be free of contaminants that would interfere with the gas chromato-graphic (GC) quantitation of the paired solute, which could cause misinterpre-tation. The compounds are listed along with their abbreviations in Table 16.1.

Procedures

Sorption experiments were carried out in screw-cap vials with Teflon-lined septa or Teflon Mininert valves (Pierce Chemical).

"Nonequilibrium" sorption experiments were carried out as described previ-

Table 16.1. Compounds Used and Their Molecular Free Surface Areas

Compound	Abbreviation	S_M, A^2
chlorobenzene	Cl-Bz	123.9
1,2-dibromoethane	EDB	109.8
1,2-dibromo-3-chloropropane	DBCP	155.9
1,2-dibromobutane	DBBu	157.3
1,4-dibromopentane	DBPen	182.3
1,2-dibromopropane	DBPr	136.4
1,2-dichlorobenzene	Cl_2-Bz	141.0
dichloromethane	DCM	78.9
1,2,4-trichlorobenzene	Cl_3-Bz	158.1
1,1,2-trichloroethane	TCA	118.8
trichloroethene	TCE	110.7

Note: Molecular free surface areas tabulated in Sabljic[18] or calculated according to Gavezzotti.[17]

ously.[14,15] Briefly, EDB was equilibrated with the soil suspension for 24 hr. TCE was added simultaneously or 24 hr prior to EDB. The solutes were then purged from aqueous suspension for 96 hr in the presence of Tenax GC polymeric adsorbent beads, which acted as a sink for desorbed solute. The residual EDB remaining in the soil was determined by hot acetone extraction and GC quantitation.

"Equilibrium" sorption experiments were carried out by combining sorbent and aqueous solution and then adding the solutes simultaneously as methanol solutions. The aqueous solution consisted of distilled water containing 0.01 M $CaCl_2$ plus 200 mg/L $HgCl_2$ as biocide — except when the sorbent was humic acid particles the solution consisted of 0.001 N HCl. Soil to water ratios were adjusted from 1:1 to 1:10 to result in 50 ± 20% of the test solute in solution. All samples in a given experiment had the same soil-to-water ratio and contained the same total volume of methanol to ensure that all replicates were treated the same except for the concentrations of the solutes. The methanol did not exceed 0.5% of solution. Control experiments indicated the order of treatment (single-solute vs bisolute samples) was unimportant.

The vials were tumbled on a hematology mixer (Fisher Scientific) at 21 ± 1°C during the equilibration period. The vials were then centrifuged at $750 \times g$ (IEC model UV) to visual clarity. The effect of colloidal particles on partition constants was expected to be insignificant since all compounds, with the exception of Cl_3-Bz, have organic carbon-based partition coefficients (K_{oc}) below 500 L/kg. The K_{oc} of Cl_3-Bz was not of direct concern here.

Aqueous concentrations were determined by withdrawing 0.1–1 mL supernatant and extracting with hexane for GC analysis on a 0.53 mm × 30 m DB-624 capillary column (J & W Scientific, Folsom, CA). Within experimental error, controls containing only water lost no solute during the equilibration period. Sorbed concentrations were determined by mass balance in some experiments. In other experiments, sorbed concentrations were determined independently by decanting the supernatant, extracting the solids with acetone overnight, and phase-transferring the solutes from the extracting solution into hexane after dilution with nine parts water; the assay was corrected for solute in the soil interstitial water.

RESULTS

Bisolute Effects on Nonequilibrium Sorption

Recently we investigated mechanisms of slowly reversible (nonequilibrium) sorption of low-molecular-weight nonpolar compounds.[14-16] As part of these studies, a "competition" experiment between EDB and TCE was carried out to differentiate between opposing mechanistic hypotheses for retarded desorption, namely, between slow molecular diffusion through organic matter phases, which was assumed to be insensitive to the presence of a second solute, and surface adsorption or other specific interactions which are likely to be subject to competition. The slowly reversible concentration was defined as the sorbed concentration remaining after a purge scheme.[14,15] The slowly reversible concentration of EDB in the Warehouse Point soil was found to be markedly suppressed by TCE in solution (Figure 16.1). Suppression of this residual by TCE was most pronounced at low TCE concentrations and was practically independent of whether TCE was added to the soil simultaneously with EDB or 24 hr prior to EDB in the prepurge incubation period. Additional experiments were necessary at this point to determine whether TCE was directly

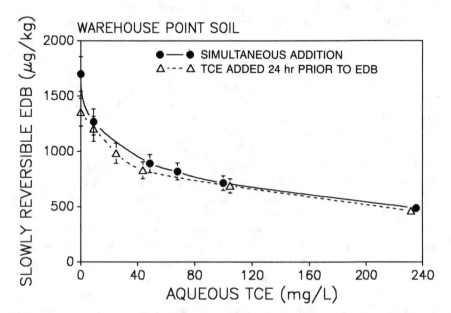

Figure 16.1.   Effect of TCE on the slowly reversible fraction of EDB in the Warehouse Point soil. Prepurge sorption equilibration period (24 hr) for EDB. Initial sorbed EDB, 10 mg/kg. Error bars, standard deviation of triplicates.

affecting EDB sorption/desorption kinetics or indirectly affecting the total prepurge sorbed EDB concentration.

Bisolute Effects on Equilibrium Sorption

In initial experiments, the 24-hr single-point OC-based sorption constant (K_{oc}) of EDB in the Warehouse Point soil was evaluated as a function of sorbed TCE concentration. The single-point K_{oc} was taken simply as the ratio of sorbed to aqueous concentrations using the same mass of EDB in each replicate. The results (Figure 16.2) confirm that TCE decreases the equilibrium sorption of EDB. At maximum TCE, K_{oc} was reduced by a factor of about 3, which is comparable to the effect of TCE on the slowly reversible concentration of EDB in Figure 16.1. Noticeable effects on K_{oc} require sorbed TCE concentrations to be little more than the sorbed EDB concentration, which was initially 0.5 g/(kg OC). The sorption isotherm of TCE was linear (inset to Figure 16.2). At maximum, TCE was only about 10% of its aqueous solubility and about 5% of the reported limiting sorption capacity of soil organic matter for TCE.[1] Solution-phase EDB was around 1% of its solubility. This indicates that the experiment was conducted under conditions of regular (dilute) sorption behavior of both EDB and TCE.[1]

In contrast to TCE, Cl_2-Bz has much less effect on the single-point K_{oc} of EDB (Figure 16.3). The decline in K_{oc} is evident only at the highest concentrations and reaches a maximum factor of 1.4. The sorption isotherm of Cl_2-Bz (inset) is linear and its solubility (0.15 g/L) was not approached.

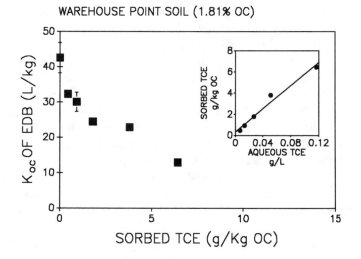

Figure 16.2.   Effect of TCE on the 24-hr K_{oc} of EDB in the Warehouse Point soil. Total EDB added, 110 µg to 5 g soil and 6.8 mL solution. Error bars (shown when larger than the symbol size) indicate standard deviation of triplicates. *Inset,* sorption isotherm of TCE.

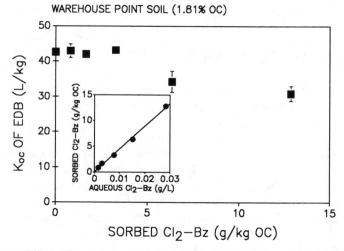

Figure 16.3.   Effect of Cl_2-Bz on the 24-hr K_{oc} of EDB in the Warehouse Point soil. Legend to Figure 16.2 applies. *Inset,* sorption isotherm of Cl_2-Bz.

Figures 16.4 to 16.7 indicate that bisolute effects occur in several soils and between a variety of nonpolar solute pairs. The sorption isotherms of EDB in the presence of a cosolute (TCE or DBPr) in Granby, peat, and Windsor soils are presented in Figure 16.4. For these experiments, sorbed concentration was determined independently by acetone extraction to eliminate the uncertainties associated with calculating sorbed concentration by mass balance. ^{14}C-EDB was used in the Granby and peat soils and unlabeled EDB in the Windsor soil. In all cases, the K_{oc} of EDB decreased with increasing concentration of the sorbed cosolute (cosorbate). Bisolute effects are evident even when the cosorbate concentration is only about threefold higher than EDB.

Addition of a second solute can displace a previously sorbed compound into solution. Replicates of peat soil (Figure 16.5A) were equilibrated with EDB for a 48-hr period and then either spiked with TCE (16-fold sorbed excess over EDB) or an amount of methanol equivalent to that in the TCE spike solution. Over the next several days, the apparent K_{oc} of the samples that received only methanol remained the same, while the TCE-spiked replicates showed displacement of EDB into solution resulting in an apparent decrease in K_{oc}. Analogous results were obtained with the Windsor soil (data not shown).

A similar displacement experiment performed on humic acid particles from the Windsor soil showed little or no evidence of bisolute effects (Figure 16.5B). Assuming that all sorption occurs to the organic matter fraction of the soils, these results imply that bisolute effects depend on the nature of the organic matter.

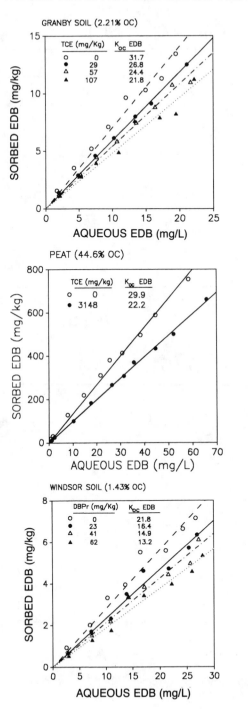

Figure 16.4. Sorption isotherms (48 hr) of EDB showing decreasing K_{oc} (slopes) with increasing sorbed TCE *(a and b)* or DBPr *(c)*.

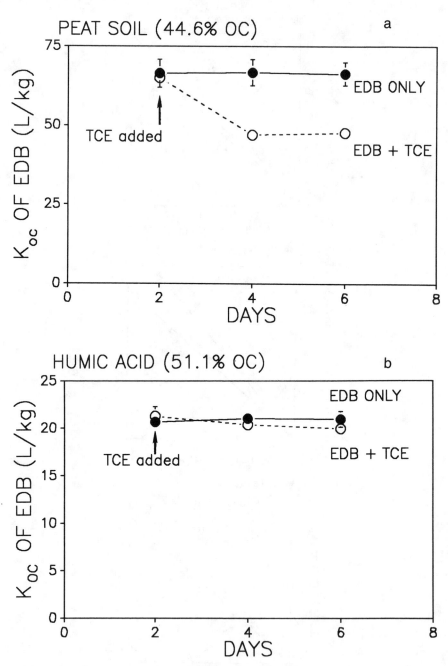

Figure 16.5. Displacement of presorbed ^{14}C-EDB by addition of TCE in peat soil *(a)* and humic acid particles *(b)*.

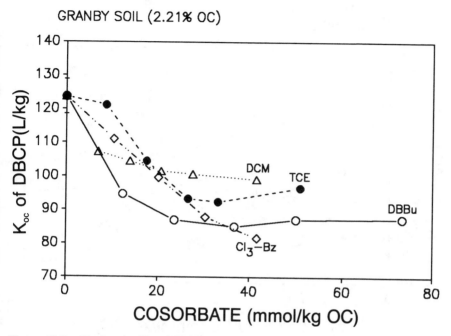

Figure 16.6. Single-point K_{oc} of DBCP in binary mixture with various compounds. Initial DBCP, 2 mmol/(Kg OC).

Dependence on Cosolute Properties

The ability of cosolute to suppress the sorption of a particular compound is dependent on its sorptive strength. The K_{oc}s of DBCP (Figure 16.6) and TCE (Figure 16.7) declined, in similar fashion, with increasing molar cosorbate concentration. The greatest rate of change in K_{oc} occurred below a molar ratio of cosorbate to test compound of approximately 10:1 (i.e., below ~20 mmol/ kg cosorbate). Changes in K_{oc} appear to flatten at levels above about 40 mmol/ kg cosorbate.

Cosolute ability to depress K_{oc} is also slightly dependent on compound structure, indicating a weak molecular sieving behavior of the sorbent. This can be visualized by relating a quantitative measure of the bisolute (competitive) effect with a steric parameter that reflects the difference in size between the two competing compounds. Steric effects are likely to play a role in sorbate competition, since all of these compounds are nonpolar and have no interacting functional groups. Lacking a mathematic model, we are restricted to an empirical measure of competition, which may be taken as the K_{oc} suppression in the plateau region, i.e., above 20 mmol/(kg OC) cosorbate concentration (refer to Figures 16.6 and 16.7). An appropriate steric parameter is the molecular free surface area, S_M, defined by Gavezzotti to be the sum of atomic surface areas calculated from van der Waals' radii minus the surfaces between the atoms that are unavailable due to bonding (Table 16.1).[17,18] Other steric param-

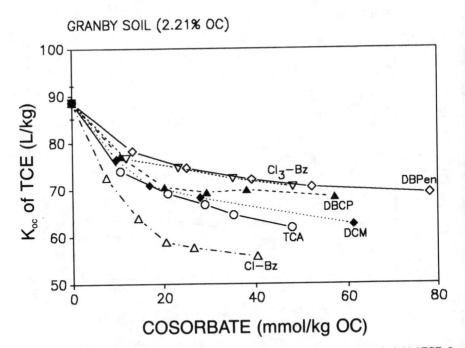

Figure 16.7. Single-point K_{oc} of TCE in binary mixture with various compounds. Initial TCE, 2 mmol/(kg OC).

eters, such as molecular volume[19] or the "total surface area" of Yalkowski and coworkers[20,21] may be equally suitable but were not examined.

Figure 16.8 shows the percent decrease in K_{oc} of each of three test compounds (using data on EDB, TCE, and DBCP from Figures 16.2, 16.3, 16.6, and 16.7) as a function of the absolute value of the difference in molecular free surface area, $|\Delta S_M|$, between the competing solutes. All three data sets indicate that "competition" is greater as the cosolutes become closer in surface area. This suggests that sorption at these competitive "sites" is based in part on molecular size and shape.

DISCUSSION

We may consider the following causes for the bisolute effects:

1. cosolute enhancement of aqueous solubility
2. cosorbate effects on the bulk properties of the organic matter sorbent leading to diminished "solubility" of the test compound in the organic phase
3. a minor amount of sorption in the soils attributable to the presence of an adsorbent
4. deviation from ideal behavior of the soil organic matter as a hydrophobic partitioning medium

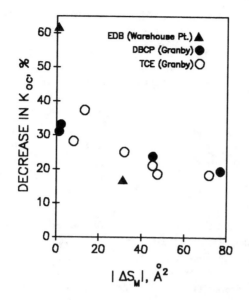

Figure 16.8. Magnitude of competitive sorption (decrease in K_{oc} at cosorbate of 40 mmol/kg OC) as a function of the difference in molecular free surface area between test compound (EDB, DBCP, or TCE in indicated soil) and cosorbate.

Solubility

Cosolute enhancement of aqueous solubility can be rejected because the solution concentrations of the cosolutes were much too low to affect solubility.[22] Furthermore, no decrease in the sorption constant was observed when methanol alone was added, even though methanol (at 0.2–0.5% of solution volume) greatly exceeded the cosolute concentration. Finally, enhancement of solubility in mixed solvent systems is known to be exponential in cosolvent mole fraction of the liquid phase;[22,23] here, however, the cosolute effect on K_{oc} plateaus at higher cosolute concentrations. It is noteworthy that pentachlorophenol and chlorobenzene inhibited PCDD sorption when the solution phase consisted of 50–75% methanol in water;[13] it is highly unlikely that PCDD solubility was affected by the cosolute at such high methanol levels.

Sorbent Properties

Cosorbate effects on the bulk properties of the sorbent (organic matter phase) deserves careful consideration in view of the known or postulated effects of organic solvents on both natural organic matter and synthetic polymers. Enhanced desorption (extraction) of hydrophobic compounds from natural organic matter by organic solvents has been attributed to gel swelling of the organic matter phase by the solvent, coupled with the high dissolving powers of the solvent.[24] Small organic molecules are capable of swelling ("plasticizing") many synthetic polymers by weakening polymer-polymer interactions; for this reason, their solubility within the polymer increases exponentially with concentration.[25] On this basis, and assuming that aqueous solubility is unaffected, a cosorbate might be predicted to, if anything, increase the solubility of a compound in soil organic matter, and hence increase the sorption constant. In the soils, however, the bisolute effects are in the opposite direction (solubility of the test compound in the organic polymer decreases) and, moreover, are stronger at relatively low cosorbate concentration. It should be noted that the maximum bisolute effects occur at cosorbate concentrations of less than 0.30% on a weight basis of the soil organic matter. For these reasons, cosorbate effects on bulk sorbent properties are unlikely.

Adsorption

Controversy continues over the issue of surface adsorption in soils.[1,26,27] The likelihood of adsorption is greater in soils of very low organic carbon ($<$ ~0.1%), such as aquifer sediments,[28,29] or under very dry conditions,[6] neither of which apply here. The presence of a minor amount of adsorbent, however, could result in competitive effects. The leveling of K_{oc} at high cosorbate concentrations suggests bisorbent behavior of the soils — one type that is subject to bisolute effects and one that is not. The absence of bisolute effects in the humic acid particles is further support for this. Thus, the humic acid component of the soil may behave as a partition medium and another component as an adsorbent.

The nature of this adsorbent can only be speculative at this time. Murphy et al. examined sorption of hydrophobic organics on humic-coated clays and hematite.[30] They proposed that the organic matter can act as a hydrophobic surface rather than a hydrophobic phase. Smith et al. studied sorption of halogenated hydrocarbons on clays exchanged with tetraalkylammonium cations.[31] Some sorbate-organocation systems displayed nonlinear isotherms and competitive effects typical of adsorption, which was suggested to occur on the hydrophobic surface created by the organocation. Other systems gave more linear isotherms and no competition — behavior that is characteristic of partitioning.

It is worth noting that "competition" between sorbates in this study is clearly much weaker than on some known adsorbents; for example, substitution on activated carbon is often nearly one for one.[5] In the soils, however, generally less than 0.03 moles are displaced for every mole of cosorbate sorbed. Molecular sieving behavior is not always strong in adsorption. Molecular sieving by activated carbon is obscured by its relatively open structure.[32] Zeolites are well-known molecular sieves.

Nonideal Behavior

Finally, we may consider the competitive field to be the soil organic matter phase itself. Soil organic matter is heterogeneous,[33,34] and it would be surpris-

Table 16.2. Freundlich Power Coefficients for EDB Sorption Isotherms in the Presence and Absence of Cosolute

Soil	Cosolute	Cosolute Conc. (mg/kg soil)	Freundlich n	Regression r^2
Peat	TCE	0	0.85	0.9986
		3148	0.98	0.9996
Windsor	DBPr	0	0.88	0.9884
		23	0.97	0.9887
		41	0.98	0.9925
		62	1.01	0.9778
Granby	TCE	0	0.87	0.9958
		29	0.92	0.9972
		57	0.93	0.9981
		107	0.94	0.9855

ing if it behaved as an ideal hydrophobic liquidlike partition medium where sorbate becomes homogeneously dispersed. Its three-dimensional polymeric structure may contain internal sites (compartments, pockets) that have a higher sorption potential than the bulk organic phase. Competition is possible if these internal sites are of molecular dimensions and restrictive of the number and size of sorbate molecules. Molecular sieving effects are feasible in this context, especially if one considers a gradation of internal site dimensions, each with an optimal affinity for a sorbate of a given size. Such structures may be associated with the more rigid regions in the organic polymers. The absence of competition in humic acid particles is consistent with its more homogeneous, gel quality compared to, say, humin, which is more rigid and hydrophobic.[33,34]

Adsorption and nonideal hydrophobic partitioning both should impart nonlinearity to single-solute isotherms. There are numerous examples in the literature where weak-to-moderate curvature is observed in the sorption isotherm. By either mechanism, the isotherm is predicted to be steeper at low concentrations as sorbate preferentially binds to sites of higher sorption potential, whether on surfaces or at internal sites. The observed nonlinearity will depend on the fraction of the total sorption attributable to nonpartitioning. These points are illustrated for EDB in the peat soil. Although the single-solute isotherm of EDB (Figure 16.4B) appears to be linear, closer inspection shows that the isotherm has curvature and that the soil has a higher affinity for EDB at low EDB concentrations. The log S vs log C isotherm, where S and C are sorbed and aqueous concentrations, respectively, is linear with slope of 0.85 (Table 16.2). This slope is equivalent to the power coefficient n in the Freundlich equation

$$S = K_F \cdot C^n$$

where K_F is the Freundlich sorption constant. A slope of less than 1 indicates curvature in the S vs C isotherm. Illustrated another way, Figure 16.9

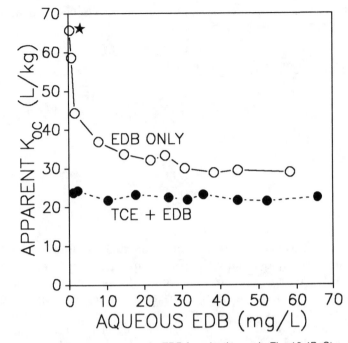

Figure 16.9.   Ratio of sorbed to aqueous EDB from isotherms in Fig. 16.4B. Starred (*) point was taken from the displacement experiment of Fig. 16.5A.

shows that the apparent single-point K_{oc} along the isotherm is considerably higher at low EDB concentrations, approaching a constant value at high concentrations. [This behavior explains the high single-point K_{oc} of EDB observed in the displacement experiment of Fig. 16.5A (starred datum in Figure 16.9) compared to the K_{oc} calculated from the (assumed linear) sorption isotherm for the peat soil.]

TCE, however, completely suppresses the tendency of EDB to sorb more strongly at low concentrations. The power coefficient n becomes 0.98 (Table 16.2), and the single-point K_{oc} is unchanged over the entire concentration range (Figure 16.9). Table 16.2 shows a similar trend in n for EDB in the other two soil systems for which isotherms were available. Figure 16.9 also indicates that a compound is more susceptible to competitive displacement at low concentrations. Taken together, these results support the bisorbent behavior of the soils in aqueous suspension, regardless of the mechanism of the competitive effect.

SUMMARY

Nonpolar halogenated aliphatic and aromatic compounds show bisolute effects (competitive sorption) in soils of moderate to high OC. Inclusion of a second solute decreased the single-solute partition constant by up to a factor of 3. Displacement of a presorbed compound into solution by the addition of excess cosolute was observed. Cosolute also reduced the kinetically slow sorbed fraction, but in retrospect this could be attributed to a reduction in the total initial sorbed concentration. Competition is more effective when the solutes are close in molecular free surface area. A particular compound appears to be more susceptible to displacement at low concentrations. Humic acid particles show no competition. The results are consistent with either a small amount of adsorption occurring in the soil, or the soil organic matter behaving as a nonideal hydrophobic partitioning medium. In the latter case, it is proposed that the organic matrix contains internal sites that have a higher sorption potential than the bulk organic phase. A trend toward greater sorption at low concentrations, which is diminished or erased in the presence of cosorbate, is consistent with either mechanism. The internal site mechanism is more consistent with the widespread experience that sorption from solution to natural minerals is very weak. However, further work is needed to differentiate them.

ACKNOWLEDGMENT

I thank Marta Day for her enthusiasm and diligence in providing technical assistance. This work was supported in part by the United States Department of Agricultural, National Pesticide Impact Assessment Program.

REFERENCES

1. Chiou, C. T. "Theoretical Considerations of the Partition Uptake of Nonionic Organic Compounds by Soil Organic Matter," in *Reactions and Movement of Organic Chemicals in Soils*, B. L. Sawhney and K. Brown, Eds. (Madison, WI: Soil Science Society of America, 1989), pp. 1–29.
2. Karickhoff, S. W. "Organic Pollutant Sorption in Aquatic Systems," *J. Hydraul. Eng.* 110:707–735 (1984).
3. Rosene, M. R., and M. Manes. "Application of the Polanyi Adsorption Potential Theory to Adsorption from Solution on Activated Carbon. 7. Competitive Adsorption of Solids from Water Solution," *J. Phys. Chem.* 80(9):953–959 (1976).
4. Rosene, M. R., and M. Manes. "Application of the Polanyi Adsorption Potential Theory to Adsorption from Solution on Activated Carbon. 9. Competitive Adsorption of Ternary Solid Solutes from Water Solution," *J. Phys. Chem.* 81(17):1646–1650 (1977).
5. Kipling, J. J. *Adsorption from Solutions of Non-electrolytes* (New York: Academic Press, 1965), pp. 160–190.
6. Chiou, C. T., and T. D. Shoup. "Soil Sorption of Organic Vapors and Effects of Humidity on Sorptive Mechanism and Capacity," *Environ. Sci. Technol.* 19(12):1196–1200 (1985).
7. Chiou, C. T., P. E. Porter, and D. W. Schmedding. "Partition Equilibria of Non-

ionic Organic Compounds between Soil Organic Matter and Water," *Environ. Sci. Technol.* 17(4):227–231 (1983).

8. Chiou, C. T., T. D. Shoup, and P. E. Porter. "Mechanistic Roles of Soil Humus and Minerals in the Sorption of Nonionic Organic Compounds from Aqueous and Organic Solutions," *Org. Geochem.* 8:9–14 (1985).

9. Karickhoff, S. W., D. S. Brown, and T. A. Scott. "Sorption of Hydrophobic Pollutants on Natural Sediments," *Water Res.* 13:241–248 (1979).

10. McGinley, P. M., L. E. Katz, and W. J. Weber, Jr. "Multi-solute Effects in the Sorption of Hydrophobic Organic Compounds by Aquifer Soils," in *Preprint Extended Abstracts,* Division of Environmental Chemistry, American Chemical Society 198th Meeting, Miami, FL (1989), pp. 146–149.

11. Selvakumar, A., and H-N. Hsieh. "Competitive Adsorption of Organic Compounds by Microbial Biomass," *J. Environ. Sci. Health* A23(8):729–744 (1988).

12. Coates, J. T., and A. W. Elzerman. "Desorption Kinetics for Selected PCB Congeners from River Sediments," *J. Contam. Hydrol.* 1:191–210 (1986).

13. Walters, R. W., Z. Yousefi, A. L. Tarleton, S. A. Ostazeski, and D. C. Barry. "Assessment of the Potential for Transport of Dioxins and Codisposed Materials to Groundwater," EPA Technical Report EPA/600/6-89/002 (1989).

14. Pignatello, J. J. "Slowly Reversible Sorption of Aliphatic Halocarbons in Soils. I. Formation of Residual Fractions," *Environ. Toxicol. Chem.* 9:1107–1115 (1990).

15. Pignatello, J. J. "Slowly Reversible Sorption of Aliphatic Halocarbons in Soils. II. Mechanistic Aspects," *Environ. Toxicol. Chem.* 9:1117–1126 (1990).

16. Steinberg, S. M., J. J. Pignatello, and B. L. Sawhney. "Persistence of 1,2-Dibromoethane in Soils: Entrapment in Intraparticle Micropores," *Environ. Sci. Technol.* 21:1201–1208 (1987).

17. Gavezzotti, A. "Molecular Free Surface: A Novel Method of Calculation and Its Uses in Conformational Studies in Organic Chemistry," *J. Am. Chem. Soc.* 107:962–967 (1985).

18. Sabljic, A. "On the Prediction of Soil Sorption Coefficients of Organic Pollutants from Molecular Structure: Application of Molecular Topology Model," *Environ. Sci. Technol.* 21:358–366 (1987).

19. Gavezzotti, A. "The Calculation of Molecular Volumes and the Use of Volume Analysis in the Investigation of Structured Media and Solid-State Organic Reactivity," *J. Am. Chem. Soc.* 105:5220–5225 (1983).

20. Yalkowski, S. H., and S. C. Valvani. "Solubilities and Partitioning: II. Relationship between Aqueous Solubilities, Partition Coefficients, and Molecular Surface Areas of Rigid Aromatic Hydrocarbons," *Chem. Eng. Data* 24:127–129 (1979).

21. Yalkowski, S. H., R. J. Orr, and S. C. Valvani. "Solubility and Partitioning: III. The Solubility of Halobenzenes in Water," *Ind. Eng. Chem. Fundam.* 18:351–353 (1979).

22. Pinal, R., P. S. C. Rao, L. S. Lee, P. V. Cline, and S. H. Yalkowski. "Cosolvency of Partially Miscible Organic Solvents on the Solubility of Hydrophobic Organic Chemicals," *Environ. Sci. Technol.* 24:639–647 (1990).

23. Yalkowski, S. H., S. C. Valvani, and G. L. Amidon. "Solubility of Nonelectrolytes in Polar Solvents: IV. Nonpolar Drugs in Mixed Solvents," *J. Pharm. Sci.* 65:1488–1494 (1976).

24. Freeman, D. H., and L. S. Cheung. "A Gel Partition Model for Organic Desorption from a Pond Sediment," *Science* 214:790–792 (1981).

25. Rogers, C. E. "Solubility and Diffusivity," in *Physics and Chemistry of the Organic*

Solid State, D. Fox, M. M. Labes, and A. Weissberger, Eds. (New York: Interscience Publ., 1965), pp. 509–635.

26. Mingelgrin, U., and Z. Gerstl. "Reevaluation of Partitioning as a Mechanism of Nonionic Chemicals Adsorption in Soils," *J. Environ. Qual.* 12:1–11 (1983).

27. Hassett, J. J., and W. L. Banwart. "The Sorption of Nonpolar Organics by Soils and Sediments," in *Reactions and Movement of Organic Chemicals in Soils,* B. L. Sawhney and K. Brown, Eds. (Madison, WI: Soil Science Society of America, 1989), pp. 31–44.

28. Stauffer, T. B., W. G. MacIntyre, and D. C. Wickman. "Sorption of Nonpolar Organic Chemicals on Low-Carbon-Content Aquifer Materials," *Environ. Toxicol. Chem.* 8:845–852 (1989).

29. Schwarzenbach, R. P., and J. Westall. "Transport of Nonpolar Organic Compounds from Surface Water to Groundwater: Laboratory Sorption Studies," *Environ. Sci. Technol.* 15:1360–1367 (1981).

30. Murphy, E. M., J. M. Zachara, and S. C. Smith. "Influence of Mineral-Bound Humic Substances on the Sorption of Hydrophobic Organic Compounds," *Environ. Sci. Technol.* 24:1507–1516 (1990).

31. Smith, J. A., P. R. Jaffé, and C. T. Chiou. "Effect of Ten Quaternary Ammonium Cations on Tetrachloromethane Sorption to Clay from Water," *Environ. Sci. Technol.* 24:1167–1172 (1990).

32. Manis, M. "The Polanyi Adsorption Potential Theory and Its Application to Adsorption from Water Solution onto Activated Carbon," in *Activated Carbon Adsorption of Organics from the Aqueous Phase,* Volume 1, I. H. Suffet and M. J. McGuire, Eds. (Ann Arbor, MI: Ann Arbor Science, 1980), pp. 43–64.

33. Hayes, M. H. B., and F. L. Himes. "Nature and Properties of Humus Mineral Complexes," in *Interaction of Soil Minerals with Natural Organics and Microbes,* P. M. Huang and M. Schnitzer, Eds. (Madison, WI: Soil Science Society of America, 1986), pp. 103–158.

34. Hayes, M. H. B., and R. S. Swift. "The Nature of Soil Organic Colloids," in *The Chemistry of Soil Constituents,* D. J. Greenland and M. H. B. Hayes, Eds. (New York: John Wiley and Sons, 1978), pp. 179–320.

PART III

BIODEGRADATION OF ORGANIC CONTAMINANTS IN SOILS AND SEDIMENTS

Biodegradation of PCBs by Aerobic Microorganisms

Peter Adriaens, Chi-Min Huang, and Dennis D. Focht

INTRODUCTION

The current method for decontaminating soils of PCBs is solvent extraction and subsequent thermal destruction.[1] In addition to the high energy costs incurred from incineration, variable amounts of polychlorinated dibenzo-*p*-dioxins (PCDDs) and dibenzofurans (PCDFs) are generated as toxic by-products. Thus, alternative methods, such as in situ (no displacement of the soil matrix) and onsite (with displacement) bioremediation processes, would be advantageous economically and environmentally.[2] For these methods to be applicable and effective, information is required on the degradation kinetics of PCBs (both in pure culture and in soil), and on the fate and inhibitory effects of products generated from cometabolism of PCBs.

The potential of both aerobic and anaerobic microbial degradation of PCBs represents a conceptually valid, but untested, process. Although neither aerobic nor anaerobic bacteria can use PCBs as sources of carbon and energy, aerobic bacteria are able to oxidize and split the aromatic ring, yet cannot dehalogenate any of the products (e.g., chlorobenzoic acids) that are produced. Anaerobes, on the other hand, use the PCB molecule as an electron acceptor and dechlorinate higher chlorinated congeners without destroying the PCB nucleus.[3]

PCBs AND THEIR DEGRADATION PRODUCTS IN THE SOIL

The role of microbial activity vis-à-vis metabolism of xenobiotic compounds is governed by the same basic principles as soil organic matter decomposition. The influence of environmental parameters, such as temperature, pH, aeration, nutrient availability, depth, diffusion, and spatial variability, on bacterial metabolism in soils is described elsewhere.[4]

The partitioning of PCBs in different compartments of the environment depends to a large extent on the sorption reactions. Generally, sorption increases with increasing chlorine content of the chlorobiphenyl, surface area, and organic carbon content of the sorbent.[5] The relationship between the organic carbon content and hydrophobicity of organic compounds has been described by Karickoff et al.[6] as

$$K_{oc} = 0.63 \; K_{ow} \qquad (r^2 = 0.96) \qquad (17.1)$$

which relates the octanol-water partition coefficient (K_{ow}) to the organic carbon–water partition coefficient (K_{oc}). Incorporation of ^{14}C-labeled Aroclor 1242 and its degradation products into humus has been shown to occur,[7] although the extent is much less than that observed with more polar xenobiotic compounds such as ^{14}C-labeled chlorocatechols (intermediates from chlorobenzoic acid degradation) and 2,4-dichlorophenoxy acetic acid (ring-^{14}C and 2-^{14}C).[8]

Mineralization of PCBs in soil is normally a very slow process. When soil was amended with straw and sludge, and followed with anaerobic or aerobic incubations, or a combination of both, no more than 3% mineralization of [14]C-[U]d Aroclor 1242 was observed over 30 weeks.[9] In marked contrast, the addition of biphenyl to soil resulted in over 40% mineralization over a 49-day period.[7] Enrichment of soil with biphenyl was commensurate with the growth of biphenyl-utilizing bacteria (Figure 17.1), which fortuitously metabolized PCBs by enzymes of low substrate specificity to products requiring enzymes of high specificity in accordance with the scheme shown in Figure 17.2. Inoculating the soil at different inocula densities of *Acinetobacter* sp. strain P6, a PCB-cometabolizing strain,[10] decreased the apparent lag phase in CO_2 production from biphenyl and [14]CO_2 production from PCBs (Figure 17.3).

Acinetobacter sp. strain P6 effectively cometabolizes PCBs as long as biphenyl is present but has little effect in soil in its absence.[9] As soon as the substrate is depleted, the numbers of biphenyl-oxidizing bacteria drop dramatically (Figure 17.1), and the rates of CO_2 production begin to decline (Figure 17.3). CO_2 evolution from the mineralization of biphenyl could be described by the 3/2-order kinetic model derived by Brunner and Focht.[11] The results further validated that exponential growth kinetics are more realistic at lower cell densities ($< 10^6$/g soil), where substrate diffusion through the biofilm absorbed on the soil colloids is not limiting, while linear growth kinetics are more applicable at higher cell densities ($> 10^6$/g), where diffusion is limiting.[7]

The mineralization of PCBs, as measured from [14]CO_2 evolution, was delayed in all cases until after the mineralization of biphenyl (Figure 17.3). The [14]CO_2 evolution data did not fit the 3/2-order kinetic model for two reasons: first, Aroclor 1242 contains a mixture of 30 to 50 different compounds;[12] second, *Acinetobacter* sp. strain P6 can neither dehalogenate nor use any of the congeners in this mixture for growth.[10,13] Hence, indigenous commensals affect direct production of [14]CO_2 by mineralizing the products formed from

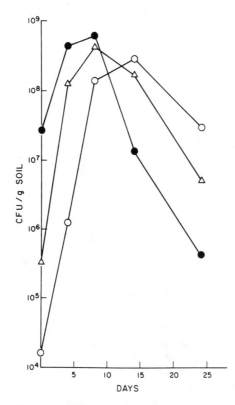

Figure 17.1. Growth and death of biphenyl-utilizing bacteria in soil treated with biphenyl to enrich for cometabolism of PCBs. Treatments represent no inoculation (○), and inoculation with *Acinetobacter* sp. strain P6 at 10^5 (△) and 10^9 (●) cells/g. Reprinted from Focht and Brunner.[7] Copyright 1985, American Society for Microbiology.

cometabolism of PCBs. Thus, the rate-limiting step in $^{14}CO_2$ evolution from PCBs is the cometabolic one. Further evidence of this effect can be seen from the concurrence of the inflection points in Figure 17.3B, where the rate is maximal, with the maximum population densities of the biphenyl oxidizers at the same time in Figure 17.1. This concept was kinetically consistent with a fit to a typical first-order sequential reaction model for the highest inoculum treatment in which growth was absent.[7]

Although chlorobenzoic acids are a major degradation product of PCBs, there is little information on their longevity in soil. 3-Chlorobenzoate (3-CB) was refractive to attack in soil by the indigenous microflora but was completely mineralized upon inoculation with *P. alcaligenes* C-O, which was isolated from sewage by enrichment culture with 3-CB.[14] The maximum growth rate μ_{max} (0.32 hr^{-1}) and cell yield coefficient Y (34 g cells/mol 3-CB) were the same in soil and pure culture, but the saturation constant K_s was considerably higher in soil (6.0 vs 0.18 mM). Monod kinetics demonstrated, in this case,

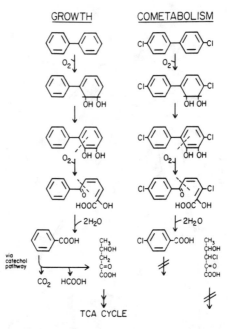

Figure 17.2. Fortuitous catabolic pathways of biphenyl and 4,4′-dichlorobiphenyl in pure culture of *Acinetobacter* P6. Reprinted from Focht.[4] Copyright 1988, Plenum Publishing Corporation.

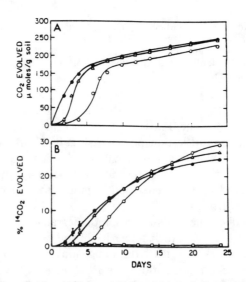

Figure 17.3. *(A)* Mineralization of biphenyl (3.3 mg/g) in soil. Symbols represent the same treatments described in Figure 17.1. *(B)* Mineralization of ^{14}C-labeled Aroclor 1242. Symbols represent the same treatments described in Figure 17.1 with the addition of a control not treated with biphenyl (□). Reprinted from Focht and Brunner.[7] Copyright 1985, American Society for Microbiology.

that physical factors (diffusional effects upon the estimate of K_s), and not biological factors (μ_{max} and Y), governed the lower rate of 3-CB metabolism in soil. Moreover, the issue of competition between *P. alcaligenes* C-O and indigenous microflora was moot since the latter could not metabolize the substrate.

METABOLISM OF PCBs AND THEIR PRODUCTS IN CULTURE

Only monochlorobiphenyls are known to be used as sole sources of carbon and energy because growth is supported by metabolism of products originating from the nonchlorinated ring since chlorobenzoic acids are not further metabolized.[10,15-17] In only a few cases, have mineralization and dechlorination from 4-chlorobiphenyl been observed with a single isolate.[18-20] Biphenyls chlorinated on both rings do not serve as carbon and energy sources because the ring fission products that are formed after a dioxygenase attack in the 2,3-position are not dehalogenated in accordance with the example shown in Figure 17.2. Hence, PCBs can be cometabolized aerobically only when an additional carbon source is provided for the growth of the bacteria.[9,11,21-24] A newly discovered degradation pathway of 4-chlorobiphenyl leads to the accumulation of 4-chloroacetophenone.[25]

Cells actively growing on biphenyl appear to be far more active in cometabolism of PCBs than washed resting cell suspensions, as noted in comparisons of resting cells vs growing cells of *Acinetobacter* sp. strain P6 and *Arthrobacter* sp. strain B1B with Aroclor 1254. The cells transformed 32 and 23% of the [^{14}C]Aroclor 1254, respectively; resting cells of the same respective cultures transformed only 17 and 8%.[13] Moreover, this study showed that *Acinetobacter* sp. strain P6 was one of the most versatile PCB-transforming organisms reported — it transformed the components of 25 of the 40 largest peaks (tetra- and pentachlorobiphenyls) of Aroclor 1254. Some of these congeners, however, lacked the open 2,3-positions required for the established 2,3-dioxygenase attack prior to ring fission. Transformation of these isomers to chlorobenzoic acids via the meta-fission pathway (Figure 17.2) was not evident, but mono-and dihydroxylated PCBs have been reported in this strain.[10] Dioxygenation of PCBs at the 3,4-site has also been shown to occur in *Alcaligenes eutrophus* H850.[26]

The fate of hydroxy-PCBs is unclear. *Pseudomonas* sp. strain HBP1, which utilized 2-hydroxybiphenyl (2-HBP) as a sole carbon source, oxidized it by an NADH-dependent 2-monooxygenase to 2,3-dihydroxybiphenyl, which was metabolized through the standard biphenyl pathway (Figure 17.2). A similar conversion of 2,2'-dihydroxybiphenyl to a 2,3,2'-trihydroxybiphenyl led to the formation of salicylic acid through the same pathway. Although the monooxygenase hydroxylated other phenolic substrates with specific structural requirements, several chlorinated hydroxybiphenyls (HBPs) served only as pseudosubstrates for the monooxygenase by affecting consumption of NADH and oxygen without being hydroxylated.[27] Metabolism of 3-HBP by *Pseudomonas* sp. strain FH23 also proceeded via a monooxygenase attack to form 2,3-dihydroxybiphenyl, which was also metabolized through the standard biphenyl pathway.[28] However, 4-hydroxybiphenyl was not attacked at the 3-position by a monooxygenase. Instead, the unsubstituted ring was attacked by a dioxygenase to form 2,3,4'-trihydroxybiphenyl; this was then metabolized through the standard biphenyl pathway giving 4-hydroxybenzoic acid, rather than benzoic acid, as an intermediate. Except for 3-and 4-chlorobiphenyl, no significant oxygen uptake activity could be observed on PCBs. Based on the reports by Kohler et al.[27] and Higson and Focht,[28] it would appear unlikely that hydroxybiphenyl degraders hydroxylated PCBs.

Unlike the metabolism of 3-chlorobenzoic acid, which loses its halogen after ring fission,[29] 4-chlorobenzoic acid is converted to 4-hydroxybenzoic acid under either aerobic or anaerobic conditions, even though the bacteria are obligate aerobes, namely *Arthrobacter*,[30] *Acinetobacter*,[31] and *Pseudomonas*.[32] An NADH-dependent monooxygenase in *Acinetobacter* sp. strain 4-CB1 was found to convert 4-hydroxybenzoic acid to 3,4-dihydroxybenzoic acid (protocatechuate), which was further degraded by ortho-(3,4-) and meta-(4,5-) fission enzymes of the protocatechuate pathway (Figure 17.4). 3,4-Dichlorobenzoic acid (3,4-DCB) could not be used as a growth substrate but was converted to 3-chloro-4-hydroxybenzoic acid (3-C-4-OHB) when 4-chlorobenzoic acid was supplied as the growth substrate. Growth on 3-C-4-OHB (μ_{max} = 0.15 hr^{-1}) was comparable to growth on 4-chlorobenzoic acid (μ_{max} = 0.12 hr^{-1}). 3-C-4-OHB was hydrolytically dehalogenated to 4-carboxy-1,2-benzoquinone, which was further metabolized via ortho-ring fission. Why strain 4CB-1 will not grow on 3,4-DCB, even though it transforms it to a growth substrate, remains unclear.

COCULTURE MINERALIZATION OF SELECTED PCB CONGENERS

An axenic consortium of *Acinetobacter* sp. strain P6, which transforms PCBs to chlorobenzoic acids, and *Acinetobacter* sp. strain 4-CB1, a commensal that grows on chlorobenzoic acids, was used to mineralize 4,4'-dichlorobiphenyl (Figure 17.4). Mineralization (e.g., the conversion of organochlorine substrate to CO_2 and HCl) can be inferred from chloride release because strain 4CB-1 grows on 4-chlorobenzoic acid, which is produced from 4,4'-dichlorobiphenyl, and releases chloride. The suspended coculture incubations (20 days) of both *Acinetobacter* spp. with 0.6 mM 4,4'-dichlorobiphenyl resulted in 50% dehalogenation and only 5% accumulation of 4-chlorobenzoic acid. When *Acinetobacter* sp. strain P6 was cultured alone, 47% of the cometabolized 4,4'-dichlorobiphenyl accumulated as 4-chlorobenzoic acid and 18% as inorganic chloride. When *Acinetobacter* sp. strain P6 was incubated alone with 4-chlorobiphenyl (1.2 mM), 4-chlorobenzoic acid accumulated stoichiometrically. In contrast, 4,4'-dichlorobiphenyl was completely mineralized by this coculture.[31]

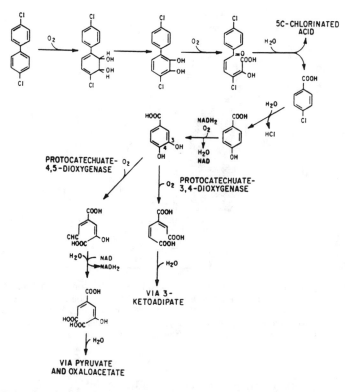

Figure 17.4. Axenic coculture metabolism of 4,4'-dichlorobiphenyl (4,4'-DCBP) by *Acinetobacter* sp. strain P6 and *Acinetobacter* sp. strain 4CB1. Strain P6, while growing on biphenyl, fortuitously metabolizes 4,4'-DCBP to 4-chlorobenzoic acid, which is then utilized as a growth substrate by strain 4CB1. Reprinted from Adriaens et al.[31] Copyright 1989, American Society for Microbiology.

The potential of coculture applications for continuous degradation of PCBs was demonstrated in an aerobic fixed-film reactor system (Figure 17.5).[33] Both bacteria developed a microbial film on a polyurethane foam bed, using benzoic acid (500 mg/L) as the growth substrate. After 14 days, the biofilm was in steady state (bacterial growth in equilibrium with the substrate added) and a Chromosorb column, coated with the PCB congeners, was inserted in the medium supply line. The benzoic acid medium was supplied with 13.7 and 5.6 μmol of 4,4'-and 3,4-dichlorobiphenyl, and 1 μmol of 3,3',4,4'-tetrachlorobiphenyl per day. Of the total 4,4'-dichlorobiphenyl transformed, 52% was cometabolized to ring fission product and 11% to 4-chlorobenzoic acid. All of the added 3,4-dichlorobiphenyl was either cometabolized to 3,4-dichlorobenzoate (83%) or further degraded by *Acinetobacter* sp. strain 4-CB1 to 3-chloro-4-hydroxybenzoic acid and 4-carboxy-1,2-benzoquinone. About 40% of 3,3',4,4'-tetrachlorobiphenyl was transformed to 3,4-dichlorobenzoate.

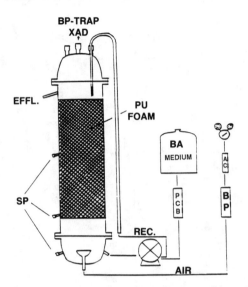

Figure 17.5.   Schematic of a continuous-flow fixed-bed reactor containing *Acinetobacter* sp. strain P6 and *Acinetobacter* sp. strain 4-CB1 immobilized onto a polyurethane (PU) foam: *AC*, activated carbon; *BP*, biphenyl; *BA*, benzoic acid; *PCB*, polychlorinated biphenyls; *REC.*, recycling flow; *SP*, sampling ports; *XAD*, amberlite resin. Reprinted from Adriaens and Focht.[33] Copyright 1990, American Chemical Society.

Although the chlorobiphenyls were degraded to various intermediates, the extent of dehalogenation was low: 6.5% and 10% of 4,4'-dichlorobiphenyl and 3,4-dichlorobiphenyl, respectively. The concentration of chlorobenzoic acids in the reactor was very low, considering the amount of PCB added per day. When substrate concentrations decreased below the saturation constant (K_s, mg/L) for growth, the Monod equation simplifies to a rate of utilization which is first order with respect to both organism (X, mg/L) and substrate (S, mg/L) concentration (i.e., $dS/dt = kXS/K_s$). This implies that the rate of degradation of the chlorinated cosubstrates is slowed down compared to the rate of benzoic acid utilization. Chlorobenzoic acid persistence could be due to insufficient exposure time between bacteria and substrates in the continuous system. Moreover, it was observed from bacterial counts of sloughed-off cells in the effluent that the ratio of *Acinetobacter* sp. strain 4-CB1 to *Acinetobacter* sp. strain P6, which was originally 1:1, had dropped to 1:100 after 4 weeks. These results are consistent with Monod kinetics and the competitive exclusion principle since strain P6 was a better competitor than strain 4-CB1: $\mu_{max} = 0.28$ vs 0.18 hr^{-1}; $K_s = 42$ vs 75 μM, respectively. Aside from microbial kinetics, the physicochemical characteristics of substrate diffusion play an important role. The organic carbon-water partitioning coefficient K_{oc} increases, and molecular diffusivity in the biofilm decreases, with increased chlorination and molecular volume of the compounds. Although the bulky molecular volumes of the PCB congeners render them less diffusive than the chlorobenzoic acids, the latter need to diffuse between both bacteria to be degraded by *Acinetobacter* sp. strain 4-CB1.

STRAIN CONSTRUCTION STRATEGIES FOR PCB MINERALIZATION

Aside from using axenic consortia of at least two bacteria to mineralize PCBs, we have directed our effort toward the construction of recombinant strains.[34-36] The advantages of the latter approach over coculture systems have been summarized by Focht as follows:[4]

1. No cosubstrate analog would be required for growth and enzyme induction.
2. There would be no competition from indigenous bacteria for the growth substrate.
3. There would be no diffusion limitations of PCB cometabolites (i.e., chlorobenzoic acids) between bacteria.
4. Only the growth rate of the recombinant rather than the two cocultures would have to be managed.

The involvement of degradative plasmids in biphenyl and chlorobiphenyl metabolism has been demonstrated. Either the plasmid contains genes for complete mineralization of 4-chlorobiphenyl[18,19] or partial metabolism to 4-chlorobenzoic acid.[37] The greater flexibility of plasmid vs chromosomal DNA in terms of replication and genetic transfer has led to speculation that bacteria in stressed environments might evolve novel catabolic abilities more rapidly than those in nonstressed environments. Noting that enrichment cultures of chlorobenzene utilizers took 7–13 months to isolate,[38-42] Krockel and Focht used the rationale of accelerated evolution (or enhanced recombination) in a continuous amalgamated column chemostat system (Figure 17.6).[34] Using two parental strains with complementary degradation pathways for 3-chlorobenzoic acid and benzene metabolism, a recombinant with chlorobenzene-and, fortuitously, 1,4-dichlorobenzene-degrading abilities was produced in 2 weeks. The recombinant strain, *P. putida* CB1–9, harbored a 33-kb plasmid, which was formed by insertion of a chromosomal DNA fragment from *P. alcaligenes* into a 57-kb TOL plasmid of *P. putida* R5–3.[35] The smaller size of the recombinant plasmid coincided with the loss of xylene and methylbenzoate metabolism, which is consistent with the observation that these substrates strongly induce for the meta-fission pathway, which would be suicidal for chlorobenzene catabolism.[38]

Using the same trichemostat apparatus (Figure 17.6) with the same rationale behind fortuitous selection on chlorobenzene, Huang constructed a *m*-dichlorobenzene utilizer, *Pseudomonas* sp. strain CB35 from parental matings with *P. putida* strain R5–3, the chlorobenzene cometabolizer, with *Pseudomonas* sp. strain HF1, a 2,4-dichlorophenoxyacetate utilizer.[36] The focal point for redirection to a productive pathway is the common intermediate 3,5-dichlorocatechol (Figure 17.7).

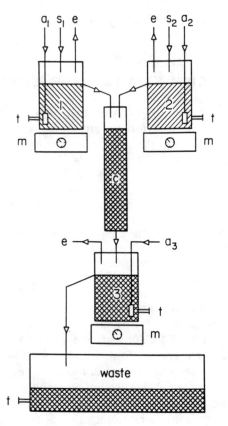

Figure 17.6. Schematic of the trichemostat used by Krockel and Focht for construction of a chlorobenzene-utilizing recombinant. The same apparatus was used for construction of a *m*-dichlorobenzene utilizer by growing the benzene utilizer *P. putida* strain R5-3 with toluene in chemostat 1 and the 2,4-D utilizer *Pseudomonas* sp. strain HF1 with benzoic acid in chemostat 2. Chlorobenzene-utilizing recombinants were selected in chemostat 3, which was saturated with chlorobenzene vapors. Reprinted from Krockel and Focht.[34] Copyright 1987, American Society for Microbiology.

A parental mating of *Pseudomonas* sp. strain HF1, which grew on 3-chlorobenzoic acid, and *Acinetobacter* sp. strain P6, which cometabolized PCBs, was conducted by Huang to construct a 3-chlorobiphenyl-degrading recombinant, *Acinetobacter* sp. strain CB15 (Figure 17.8).[36] The mechanism of DNA rearrangement has not been elucidated at this time. Strain CB15 grows on either 3-chlorobiphenyl or 3-chlorobenzoic acid and produces inorganic chloride from both substrates, but it does not grow on 3,3'-dichlorobiphenyl. The ability to cometabolize 3,3'-dichlorobiphenyl indicates that product inhibition may prevent growth. Moreover, the possession of two potentially incompatible ring fission enzymes, namely, meta fission of the chlorobiphenyl catechol and ortho fission of the 3-chlorocatechol, also may be problematical.

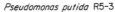

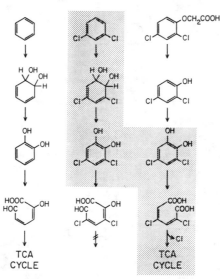

Figure 17.7.   Rationale for construction of a *m*-dichlorobenzene-utilizing recombinant strain from the mating pair described in Figure 17.5. Strain R5–3 fortuitously metabolizes *m*-dichlorobenzene via the unproductive meta-fission pathway. The focal point for redirection to a productive pathway is at 3,5-dichlorocatechol. From Huang.[36]

Growth on 3-chlorobiphenyl is slow and reaches an optical density of only 0.20 after 20 days (Figure 17.9). Nevertheless, the potential exists for constructing strains that grow on higher chlorinated biphenyls using the trichemostat method if the requisite gene pools of the parental strains are compatible and complementary.

CONCLUSIONS AND FUTURE APPLICATIONS

Whether cocultures of bacteria or recombinant strains are eventually developed for degrading PCBs or other chlorinated aromatic hydrocarbons (CAHs) will depend on integration and optimization of microbiological and physicochemical characteristics. PCBs and most CAHs do not serve as growth substrates but are degraded. This requires knowledge of appropriate growth substrates, such as nonchlorinated analogs. Additionally, chlorinated intermediates which are more refractive than the starting compound may accumulate and may act as metabolic inhibitors to the bacteria. Lastly, the microbial reaction rates will determine the engineering constraints, such as reactor size, application conditions for soil inocula, and length of time required to obtain a given degree of degradation.

Based on the information provided above, the application of cocultures for

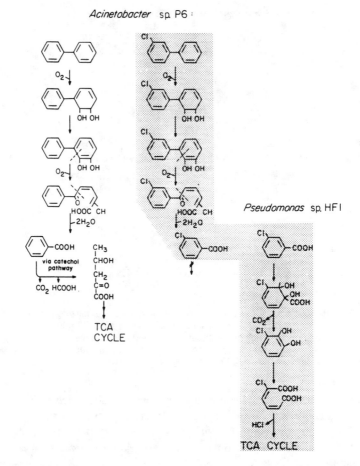

Figure 17.8.   Rationale for construction of a 3-chlorobiphenyl utilizer: *Acinetobacter* sp. strain P6, a PCB cometabolizer, is mated with *Pseudomonas* sp. strain HF1, a 3-chlorobenzoic acid utilizer, to produce *Acinetobacter* sp. strain CB15, which grows on 3-chlorobiphenyl. From Huang.[36]

the mineralization of specific PCB congeners is possible. Expanding the attack on PCB congeners will involve isolating and screening a vast array of biphenyl utilizers for cometabolic activity and the direct isolation of microorganisms capable of utilizing polychlorinated benzoic acids. Considerable effort on the first part has been undertaken here and elsewhere,[43] but to our knowledge, very little attention has been devoted to the second aspect. Work is currently under way in isolating and studying cultures that grow on various isomers of di-and trichlorobenzoic acids since these are more certain than monochloro-benzoic acids to be the products generated from cometabolism of PCBs. The lack of commercial chlorinated aliphatic acids of the type produced by cometabolism of PCBs places an additional restriction on the enrichment culture

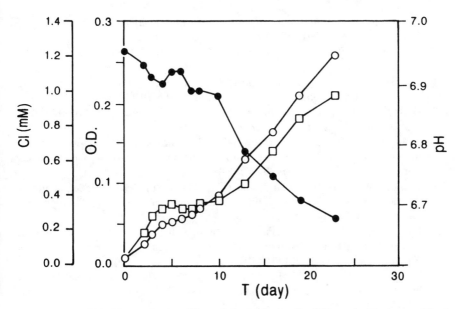

Figure 17.9. Chloride production (○), cell (optical) density (□), and pH change (●) of *Acinetobacter* sp. strain CB15 grown on 3-chlorobiphenyl. From Huang.[36]

procedure for isolating the microorganisms that contain the complete complementary catabolic pathway. Until this problem is resolved, we will be working with either an incomplete microbial consortia or with recombinants that do not contain all the genes coding for the complete catabolic pathway.

ACKNOWLEDGMENTS

We would like to acknowledge partial funding received for this work from Occidental Chemical Corporation, the National Science Foundation ECE 8419315, and the University of California Toxic Substances Teaching and Research Program.

REFERENCES

1. Electric Power Research Institute. "PCB Disposal Manual," EPRI CS-4098, Project 1263-14 (1985).
2. McCormick, D. "One bug's meat . . ." *Bio/Technology* 3:429–435 (1985).
3. Quensen, J. F., J. M. Tiedje, and S. A. Boyd. "Reductive Dechlorination of Polychlorinated Biphenyls by Anaerobic Microorganisms from Sediments," *Science* 242:752–754 (1988).
4. Focht, D. D. "Performance of Biodegradative Microorganisms in Soil: Xenobiotic Chemicals as Unexploited Metabolic Niches," in *Environmental Biotechnology:*

Reducing Risks from Environmental Chemicals through Biotechnology, G. S. Omenn, Ed. (New York: Plenum Press, 1988), pp. 15–29.

5. Waid, J. S. *PCBs and the Environment,* Vol. I, II, and III (Boca Raton, FL: CRC Press, 1986).

6. Karickoff, S. W., D. S. Brown, and T. A. Scott. "Sorption of Hydrophobic Pollutants on Natural Sediments," *Wat. Res.* 13:241–248 (1979).

7. Focht, D. D., and W. Brunner. "Kinetics of Biphenyl and Polychlorinated Biphenyl Metabolism in Soil," *Appl. Environ. Microbiol.* 50:1058–1063 (1985).

8. Stott, D. E., J. P. Martin, D. D. Focht, and K. Haider. "Biodegradation, Stabilization in Humus, and Incorporation into Soil Biomass of 2,4-D and Chlorocatechol Carbons," *Soil Sci. Soc. Am. J.* 47:66–70 (1983).

9. Brunner, W., F. H. Sutherland, and D. D. Focht. "Enhanced Biodegradation of Polychlorinated Biphenyls in Soil by Analog Enrichment and Bacterial Inoculum," *J. Environ. Qual.* 14:324–328 (1985).

10. Furukawa, K., N. Tomikuza, and A. Kamibayashi. "Effect of Chlorine Substitution on the Bacterial Metabolism of Various Polychlorinated Biphenyls," *Appl. Environ. Microbiol.* 38:301–310 (1979).

11. Brunner, W., and D. D. Focht. "Deterministic Three-Half-Order Model for Microbial Degradation of Added Carbon Substrates in Soil," *Appl. Environ. Microbiol.* 47:167–172 (1984).

12. Hutzinger, O., S. Safe, and V. Zitka. *The Chemistry of PCB's* (Boca Raton, CRC Press, 1974).

13. Kohler, H.-P. E., D. Kohler-Staub, and D. D. Focht. "Cometabolism of PCBs: Enhanced Transformation of Aroclor 1254 by Growing Bacterial Cells," *Appl. Environ. Microbiol.* 55:1940–1945 (1988).

14. Focht, D. D., and D. Shelton. "Growth Kinetics of *Pseudomonas alcaligenes* C-0 Relative to Inoculation and 3-Chlorobenzoate Metabolism in Soil," *Appl. Environ. Microbiol.* 53:1846–1849 (1987).

15. Furukawa, K., K. Tonomura, and A. Kamibayashi. "Effect of Chlorine Substitution on the Biodegradability of Polychlorinated Biphenyls," *Appl. Environ. Microbiol.* 35:223–227 (1978).

16. Kong, H.-L., and G. S. Sayler. "Degradation and Total Mineralization of Monohalogenated Biphenyls in Natural Sediment and Mixed Bacterial Culture," *Appl. Environ. Microbiol.* 46:666–672 (1983).

17. Masse, R., F. Messier, L. Peloquin, C. Ayotte, and M. Sylvestre. "Microbial Degradation of 4-Chlorobiphenyl, a Model Compound of Chlorinated Biphenyls," *Appl. Environ. Microbiol.* 47:947–951 (1984).

18. Shields, M. S., S. W. Hooper, and G. S. Sayler. "Plasmid Mediated Mineralization of 4-Chlorobiphenyl," *J. Bacteriol.* 163:882–889 (1985).

19. Hooper, S. W., T. C. Dockendorff, and G. S. Sayler. "Characteristics and Restriction Analyses of the 4-Chlorobiphenyl Catabolic Plasmid, pSS50," *Appl. Environ. Microbiol.* 55:1286–1288 (1989).

20. Parsons, J. R. "Biodegradation of Chlorinated Aromatic Compounds in Chemostat Cultures," PhD Thesis, University of Amsterdam (1988).

21. Ahmed, M., and D. D. Focht. "Degradation of Polychlorinated Biphenyls by Two Species of *Achromobacter,*" *Can. J. Microbiol.* 19:47–52 (1973).

22. Bedard, D. L., R. L. Wagner, M. J. Brennan, M. L. Haberl, and J. F. Brown, Jr. "Extensive Degradation of Aroclors and Environmentally Transformed Polych-

lorinated Biphenyls by *Alcaligenes eutrophus* H 850," *Appl. Environ. Microbiol.* 53:1094–1102 (1987).

23. Furukawa, K. "Microbial Degradation of Polychlorinated Biphenyls (PCB's)," in *Biodegradation and Detoxification of Environmental Pollutants,* A. M. Chakrabarty, Ed. (Boca Raton, FL: CRC Press, 1982), pp. 33–57.

24. Furukawa, K., N. Tomikuza, and A. Kamibayashi. "Metabolic Breakdown of Kanechlors (Polychlorobiphenyls) and Their Products by *Acinetobacter* Sp.," *Appl. Environ. Microbiol.* 46:140–145 (1983).

25. Barton, M. R., and R. L. Crawford. "Novel Biotransformation of 4-Chlorobiphenyl by a *Pseudomonas* Sp.," *Appl. Environ. Microbiol.* 54:594–595 (1988).

26. Bedard, D. L., M. L. Haberl, R. May, and M. J. Brennan. "Evidence for Novel Mechanisms of Polychlorinated Biphenyl Metabolism in *Alcaligenes eutrophus* H850," *Appl. Environ. Microbiol.* 53:1103–1112 (1987).

27. Kohler, H.-P. E., D. Kohler-Staub, and D. D. Focht. "Degradation of 2-Hydroxybiphenyl and 2,2'-Dihydroxybiphenyl by *Pseudomonas* Sp. Strain HBP1," *Appl. Environ. Microbiol.* 54:2683–2688 (1988).

28. Higson, F. K., and D. D. Focht. "Bacterial Metabolism of Hydroxylated Biphenyls," *Appl. Environ. Microbiol.* 55:946–952 (1989).

29. Reineke, W. "Microbiological Degradation of Halogenated Aromatic Compounds," in *Microbial Degradation of Halogenated Aromatic Compounds,* D. T. Gibson, Ed. (New York: Marcel Dekker, 1984), pp. 319–360.

30. Muller, R., J. Thiele, U. Klages, and F. Lingens. "The Origin of [^{18}O] Water into 4-Hydroxybenzoic Acid in the Reaction of 4-Chlorobenzoate Dehalogenase from *Pseudomonas* Sp. CBS3," *Biochem. Biophys. Res. Commun.* 124:178–182 (1984).

31. Adriaens, P., H.-P. E. Kohler, D. Kohler-Staub, and D. D. Focht. "Bacterial Dehalogenation of Chlorobenzoates and Co-culture Biodegradation of 4,4'-Dichlorobiphenyl," *Appl. Environ. Microbiol.* 55:887–892 (1989).

32. Marks, T. S., R. Wait, A. R. W. Smith, and A. V. Quirk. "The Origin of the Oxygen Incorporated during the Dehalogenation/Hydroxylation of 4-Chlorobenzoate by an *Arthrobacter* Sp.," *Biochem. Biophys. Res. Comm.* 124:669–674 (1984).

33. Adriaens, P., and D. D. Focht. "Continuous Coculture Degradation of Selected Polychlorinated Biphenyl Congeners by *Acinetobacter* Spp. in an Aerobic Reactor System," *Environ. Sci. Technol.* (in press).

34. Krockel, L., and D. D. Focht. "Construction of Chlorobenzene – Utilizing Recombinants by Progenitive Manifestation of a Rare Event," *Appl. Environ. Microbiol.* 53:2470–2475 (1987).

35. Carney, B. F., L. Krockel, J. V. Leary, and D. D. Focht. "Identification of *Pseudomonas alcaligenes* Chromosomal DNA in the Plasmid DNA of the Recombinant Chlorobenzene-Degrading *Pseudomonas putida* Strain CB1-9," *Appl. Environ. Microbiol.* 55:1037–1039 (1989).

36. Huang, C.-M. "Strain Construction Strategies for Chlorinated Aromatic Hydrocarbon-Utilizers by Multiple Chemostat," PhD Thesis, University of California, Riverside, CA (1989).

37. Khan, A., and S. Walia. "Cloning of Bacterial Genes Specifying Degradation of 4-Chlorobiphenyl from *Pseudomonas putida* OU83," *Appl. Environ. Microbiol.* 55:798–805 (1989).

38. Reineke, W., and H.-J. Knackmuss. "Microbial Metabolism of Haloaromatics:

Isolation and Properties of a Chlorobenzene-Degrading Bacterium," *Appl. Environ. Microbiol.* 47:395–402 (1984).

39. DeBont, J. A. M., M. J. A. W. Vorage, S. Hartmans, and W. J. J. van den Tweel. "Microbial Degradation of 1,3-Dichlorobenzene," *Appl. Environ. Microbiol.* 52:677–680 (1986).

40. Shraa, G., M. L. Boone, M. S. M. Jetten, A. R. W. van Neerven, P. J. Colberg, and A. J. B. Zehnder. "Degradation of 1,4-Dichlorobenzene by *Alcaligenes* Sp. Strain A175," *Appl. Environ. Microbiol.* 52:1374–1381 (1986).

41. Haigler, B. E., S. F. Nishino, and J. C. Spain. "Degradation of 1,2-Dichlorobenzene by a *Pseudomonas* Sp.," *Appl. Environ. Microbiol.* 54:294–301 (1988).

42. Spain, J. C., and S. F. Nishino. "Degradation of 1,4-Dichlorobenzene by a *Pseudomonas* sp." *Appl. Environ. Microbiol.* 53:1010–1019 (1987).

43. Bedard, D. L., R. Unterman, L. H. Bopp, M. J. Brennan, M. L. Haberl, and C. Johnson. "Rapid Assay for Screening and Characterizing Microorganisms for the Ability to Degrade Polychlorinated Biphenyls," *Appl. Environ. Microbiol.* 51:761–768 (1986).

CHAPTER 18

Microbial Oxidation of Natural and Anthropogenic Aromatic Compounds Coupled to Fe(III) Reduction

Debra J. Lonergan and Derek R. Lovley

INTRODUCTION

Aromatic compounds are important constituents of naturally occurring organic matter, and they are prevalent anthropogenic contaminants in aquatic sediments and groundwater. Many of these sedimentary environments have anaerobic zones. With the development of anaerobic conditions, Fe(III) is, in general, the most abundant potential electron acceptor for organic matter oxidation.[1-3] This suggests that the oxidation of aromatic compounds coupled to the reduction of Fe(III) could be an important mechanism for the mineralization of aromatic compounds. This conclusion is supported by the frequent observation of the accumulation of Fe(II) in groundwater contaminated with aromatics[4-6] and the finding that aromatic compounds are readily oxidized in sediments in which Fe(III) reduction is the predominant terminal electron accepting process.[5]

Both microbiological[5,7] and abiotic[8] mechanisms for the reduction of Fe(III) by aromatics have been proposed. However, several factors indicate that abiotic oxidation of aromatic compounds coupled to Fe(III) reduction cannot bring about significant mineralization of aromatic compounds in most environments. There are few aromatic compounds that react with Fe(III) at the circumneutral pH typical of most natural environments.[8] Furthermore, for those aromatic compounds that do abiotically react with Fe(III), the abiotic oxidation with Fe(III) is typically only a two electron transfer and does not result in the breakage of the aromatic ring or evolution of carbon dioxide. For example, hydroquinone, an aromatic compound which can reduce Fe(III) at pH values below pH 6, transfers two electrons to Fe(III)[8] according to the reaction:

$$\text{hydroquinone} + 2\,\text{Fe(III)} \rightarrow p\text{-benzoquinone} + 2\,\text{Fe(II)}$$

Table 18.1. Oxidation of Aromatic Compounds Coupled to the Reduction of Fe(III)

Substrates metabolized with the reduction of Fe(III) to Fe(II) by the Fe(III)-reducing isolate, GS-15:
Toluene
Benzoate, Benzaldehyde, Benzylalcohol
p-Hydroxybenzoate, p-Hydroxybenzaldehyde, p-Hydroxybenzylalcohol
Phenol
p-Cresol

Substrates metabolized with the reduction of Fe(III) to Fe(II) by Fe(III)-reducing enrichment cultures:
m-Hydroxybenzoate
2,5-Dihydroxybenzoate, 3,4-Dihydroxybenzoate
m-Cresol
Nicotinic Acid
o-Phthalic acid
Syringic Acid, Ferulic Acid
Phenylacetate
Tyrosine

Substrates tested but not metabolized with the reduction of Fe(III) to Fe(II) either by GS-15 or enrichment cultures:
Benzene, o-Xylene, m-Xylene, p-Xylene, Ethylbenzene
o-Cresol
o-Hydroxybenzoate
m-Phthalic acid, p-Phthalic acid
o-Toluic acid, m-Toluic acid, p-Toluic acid
Naphthalene

In contrast, microbial oxidation of monoaromatic compounds can result in the complete conversion of the aromatic compound to carbon dioxide with 20–40 moles of Fe(III) reduced for each mole of the aromatic compound oxidized.[5,7] More evidence for the necessity for microbial activity to bring about the oxidation of aromatics with the reduction of Fe(III) is the finding that if microbial activity is inhibited in Fe(III)-reducing sediments, then the oxidation of typical aromatic compounds such as benzoate and toluene is also inhibited.[5]

The dissimilatory Fe(III)-reducing microorganism GS-15 has provided a model microorganism for the oxidation of aromatic compounds coupled to Fe(III) reduction.[5,7] This organism uses Fe(III) to oxidize a variety of aromatic compounds, including important contaminants such as toluene, phenol, and p-cresol (Table 18.1). Toluene metabolism by GS-15 has added significance because it was the first example of a microorganism of any kind in pure culture that could oxidize an aromatic hydrocarbon under anaerobic conditions. However, there are also many aromatic compounds which GS-15 cannot oxidize (Table 18.1).

Given the potential for widespread significance of microbial oxidation of aromatic compounds with the reduction of Fe(III), we investigated whether there might be other Fe(III)-reducing microorganisms which could oxidize aromatic compounds that GS-15 did not metabolize. The results demonstrate that Fe(III)-reducing microorganisms, or microbial consortia containing

Fe(III)-reducers, can oxidize a wide range of aromatic compounds with Fe(III) as the sole electron acceptor.

MATERIALS AND METHODS

The enrichment cultures were initiated with an inoculum of freshwater sediments that had been collected from the site in the Potomac River which yielded the aromatic-oxidizing, Fe(III)-reducing microorganism, GS-15.[9] Previous investigations have indicated that Fe(III) reduction can be an important mechanism for organic matter oxidation in these sediments.[9-11]

Sediments (1.0 g) were added under a stream of N_2-CO_2 (80:20) to 10 mL of the previously described anaerobic medium[12] in anaerobic pressure tubes. The medium contained (in grams per liter of deionized water): $NaHCO_3$, 2.5; $CaCl_2 \cdot 2H_2O$, 0.1; KCl, 0.1; NH_4Cl, 1.5; and $NaH_2PO_4 \cdot H_2O$, 0.6; as well as a mixture of vitamins and trace minerals. Approximately 100 mmol of Fe(III) in the form of a poorly crystalline Fe(III) oxide was provided per liter of medium. The gas phase was N_2-CO_2 (80:20). The pH of the sterile autoclaved medium was 6.7. Individual aromatic compounds were added to the enrichment culture medium from anaerobic sterile stock solutions (5 or 10 mM) to provide an initial concentration of ~ 0.5 mM.

Fe(III) reduction in the tubes was initially monitored by observing the production of magnetite which accumulated as a black precipitate in the bottom of the tube.[13] After at least 18 repeated transfers (10% inoculum) in a medium with the compound under study as the sole electron donor, the cultures were transferred into fresh medium with the appropriate aromatic compound as the sole electron donor and Fe(III) oxide as the sole electron acceptor.

The cultures were anaerobically sampled over time with a syringe and needle. To measure aromatic compounds, the aromatics in culture filtrates were separated with HPLC as previously described.[7] The aromatics were quantified with a variable wavelength UV detector that was set at 254 nm, except for the studies with m-cresol when the absorbance was monitored at 275 nm.

Fe(II) production from Fe(III) reduction was measured as previously described.[10] Subsamples were extracted in 0.5 N HCl for 15 minutes to dissolve Fe(II) minerals, and Fe(II) was determined with ferrozine.

RESULTS AND DISCUSSION

Lignin is the most abundant natural source of aromatic compounds in sedimentary environments, and lignin oligomers and monomers are metabolized under methanogenic conditions.[14] In order to examine the potential for microbial oxidation of lignin monomers with Fe(III) as the electron acceptor, the metabolism of two common lignin monomers, syringic acid and ferulic acid, was examined. Both compounds were metabolized over time with the reduction of Fe(III) in enrichment cultures that had been developed with these compounds (Figures 18.1 and 18.2). No intermediates were detected during the HPLC analysis of the parent compounds. There was no consistent loss of syringic acid or ferulic acid over time in sterile controls. This indicated that, in the absence of Fe(III)-reducing microorganisms, these compounds did not react with poorly crystalline Fe(III) oxide at pH 6.7. The stoichiometry of Fe(II) accumulation and syringic acid and ferulic acid loss indicated that, in the enrichment cultures, syringic acid was oxidized to carbon dioxide via

$$C_9H_{10}O_5 + 36 \ Fe(III) + 22 \ H_2O \rightarrow 9 \ HCO_3^- + 36 \ Fe(II) + 45 \ H^+$$

and ferulic acid was oxidized according to

$$C_{10}H_{10}O_4 + 42 \ Fe(III) + 26 \ H_2O \rightarrow 10 \ HCO_3^- + 42 \ Fe(II) + 52 \ H^+$$

These results demonstrate that either Fe(III)-reducing microorganisms or microbial consortia that include Fe(III) reducers can completely oxidize ferulic acid and syringic acid with Fe(III) as the sole electron acceptor.

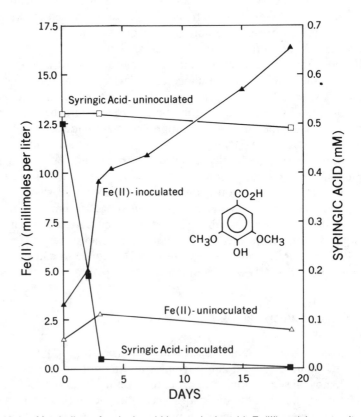

Figure 18.1.   Metabolism of syringic acid in a syringic acid–Fe(III) enrichment culture and a sterile control.

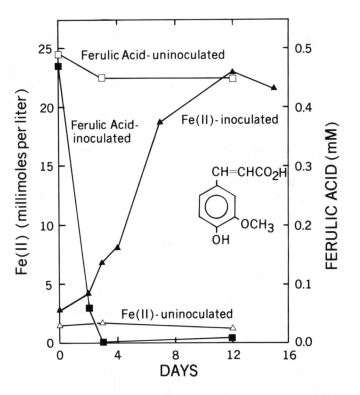

Figure 18.2. Metabolism of ferulic acid in a ferulic acid–Fe(III) enrichment culture and a sterile control.

Although there was no apparent abiotic oxidation of syringic and ferulic acid at pH 6.7, there was an initial accumulation of a small amount of Fe(II) in the 0.5 N HCl extractions in the presence of ferulic and syringic acid (Figures 18.1 and 18.2). This suggested that, at low pH, syringic and ferulic acid could reduce Fe(III) oxides without the aid of Fe(III)-reducing microorganisms. This is consistent with the general finding that, although most aromatic compounds do not react with Fe(III) at the circumneutral pHs typical of most aquatic sediments, some do react at lower pHs.[8] Even under the acidic conditions of the Fe(II) extraction procedure, the extent of abiotic Fe(III) reduction was much less than the microbial Fe(III) reduction during complete oxidation of syringic acid or ferulic acid.

Nitrogen-containing heterocyclic aromatic compounds are common in nature as well as being frequent components of waste materials.[15] Nicotinic acid is frequently regarded as a model heterocyclic compound for decomposition studies. Under anaerobic conditions, *Clostridium barkeri* ferments nicotinic acid to acetate and carbon dioxide,[16,17] and *Desulfococcus niacini* completely oxidizes nicotinic acid to carbon dioxide with sulfate as the electron acceptor.[18]

When nicotinic acid was added to sterile Fe(III) oxide medium, the concentration of dissolved nicotinic acid was only half of that measured when nicotinic acid was added to the same volume of medium without the Fe(III) oxide (Figure 18.3). There was no Fe(III) reduced in the sterile Fe(III) oxide medium with added nicotinic acid. These results suggested that half of the nicotinic acid had adsorbed onto the Fe(III) oxide. The nicotinic acid enrichment culture removed the dissolved nicotinic acid with the concomitant reduction of Fe(III) (Figure 18.3). No intermediates were observed during nicotinic acid metabolism. The nicotinic acid that was initially adsorbed, as well as the dissolved nicotinic acid, appeared to be available for microbial metabolism since the amount of Fe(II) produced was consistent with $\sim$ 0.5 mM of nicotinic acid being completely oxidized to carbon dioxide according to

$$C_6H_5NO_2 + 22\ Fe(III) + 16\ H_2O \rightarrow 6\ HCO_3^- + 22\ Fe(II) + NH_4^+ + 27H^+$$

Hydroxylated aromatic compounds are often important contaminants and are often intermediates in the anaerobic metabolism of other aromatic compounds.[14,19,20] The pathways for the anaerobic oxidation of various hydroxyl-

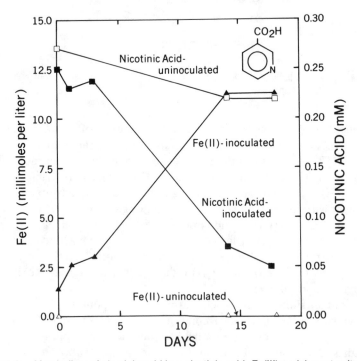

Figure 18.3.   Metabolism of nicotinic acid in a nicotinic acid–Fe(III) enrichment culture and a sterile control. The concentration of dissolved nicotinic acid was 0.62 mM when an equivalent amount of nicotinic acid was added to the same volume of medium with the Fe(III) oxide omitted.

ated aromatics may be isomer specific and may also be different with different electron acceptors.[20,21] Previous studies have indicated that, in addition to phenol, aromatic compounds with a hydroxyl group in the *para* position are readily oxidized with Fe(III) as the electron acceptor.[7] The Fe(III) reducer GS-15 was found to oxidize *p*-cresol, *p*-hydroxybenzoate, *p*-hydroxy-benzylalcohol, and *p*-hydroxybenzaldehyde. GS-15 did not oxidize the *meta* and *ortho* isomers of hydroxybenzoate and cresol (Table 18.1). Enrichment cultures which could couple the oxidation of *m*-hydroxybenzoate and *m*-cresol to the reduction of Fe(III) were readily established (Figures 18.4 and 18.5). No intermediates were detected, and there was no oxidation of the substrates under sterile conditions. The stoichiometry of substrate loss and Fe(II) accumulation was consistent with *m*-hydroxybenzoate oxidation according to

$$C_7H_6O_3 + 28\ Fe(III) + 18H_2O \rightarrow 7HCO_3^- + 28\ Fe(II) + 35\ H^+$$

and *m*-cresol oxidation with this stoichiometry:

$$C_7H_8O + 34\ Fe(III) + 20\ H_2O \rightarrow 7HCO_3^- + 34\ Fe(II) + 41H^+$$

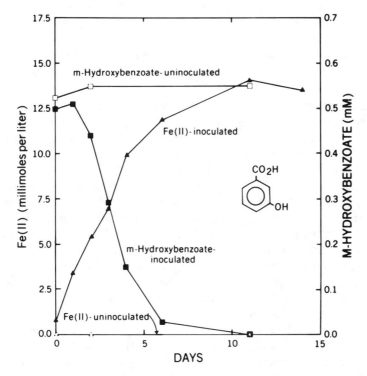

Figure 18.4.   Metabolism of *m*-hydroxybenzoate in a *m*-hydroxybenzoate–Fe(III) enrichment culture and a sterile control.

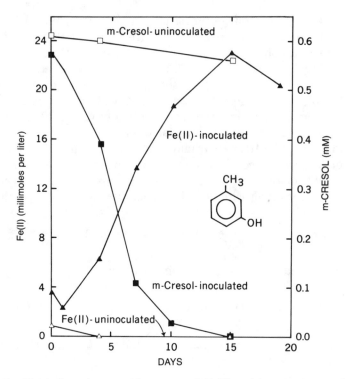

Figure 18.5. Metabolism of *m*-cresol in a *m*-cresol–Fe(III) enrichment culture and a sterile control.

Attempts to obtain Fe(III)-reducing enrichments that would oxidize *o*-cresol or *o*-hydroxybenzoate were unsuccessful. Difficulty in developing enrichments to degrade *o*-cresol under anaerobic conditions have been reported previously. There was no degradation of *o*-cresol after 56 days of incubation with methanogenic sewage sludge in one study,[22] and there was a lag period of more than 90 days before *o*-cresol was metabolized in incubations of aquifer sediments under sulfate-reducing or methanogenic conditions in another.[23] An enrichment culture with *o*-cresol as the electron donor and sulfate as the electron acceptor yielded a gas vesicle–containing sulfate reducer.[24] However, *o*-cresol was toxic to the sulfate reducer, and *o*-cresol metabolism was much slower than *p*-cresol metabolism in comparable *p*-cresol enrichments.

Our inability to obtain an enrichment that would oxidize *o*-hydroxybenzoate is less easily explained since *o*-hydroxybenzoate was metabolized in incubations of aquifer sediments under nitrate-reducing, sulfate-reducing, and methanogenic conditions.[20] Furthermore, *Desulfobacterium phenolicum* can oxidize *o*-hydroxybenzoate with the reduction of sulfate.[24]

Phthalic acids, especially the *ortho* isomer, are ubiquitous environmental contaminants that have previously been shown to be degraded under anaerobic conditions by denitrifying bacteria.[25-27] As was noted for nicotinic acid above, half of the *o*-phthalic acid that was added to sterile Fe(III) oxide medium adsorbed to the Fe(III) oxide (Figure 18.6). The concentration of *o*-phthalic acid did not decrease over time in sterile controls. In the enrichment culture, there was a steady decline in *o*-phthalic acid concentrations over time that was associated with an increase in Fe(II) (Figure 18.6). Assuming that, in addition to the dissolved *o*-phthalic acid, the *o*-phthalic acid that was initially adsorbed was metabolized, then the loss of *o*-phthalic acid and the production of Fe(II) was in accordance with the reaction:

$$C_8H_6O_4 + 30\ Fe(III) + 20\ H_2O \rightarrow 8\ HCO_3^- + 30\ Fe(II) + 38\ H^+$$

Attempts to develop Fe(III)-reducing enrichment cultures capable of oxidizing *m*-phthalic acid or *p*-phthalic were unsuccessful — perhaps because these compounds were insoluble in the culture medium. Growth could be restricted when both the electron donor and the electron acceptor are insoluble.

Stable Fe(III)-reducing enrichment cultures were also established with tyrosine, phenylacetate, 2,5-dihydroxybenzoate, and 3,4-dihydroxybenzoate. Although detailed studies on the stoichiometry of the oxidation of these compounds with the reduction of Fe(III) are yet to be completed, preliminary data indicate that these compounds are also completely oxidized to carbon dioxide with Fe(III) as the sole electron acceptor. Attempts to establish Fe(III)-reducing enrichment cultures with a number of other aromatic compounds have been unsuccessful (Table 18.1). However, this is not definitive evidence

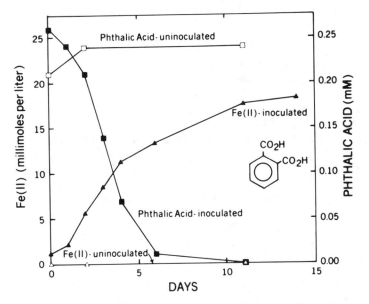

Figure 18.6. Metabolism of *o*-phthalic acid in an *o*-phthalic acid–Fe(III) enrichment culture and a sterile control. The concentration of dissolved *o*-phthalic acid was 0.62 mM when an equivalent amount of *o*-phthalic acid was added to the same volume of medium with the Fe(III) oxide omitted.

that these compounds cannot be oxidized under Fe(III)-reducing conditions. For example, GS-15 and Fe(III)-reducing enrichments did not oxidize xylenes or benzene, but geochemical evidence has indicated that these aromatic compounds were oxidized in the Fe(III)-reducing zone of a contaminated aquifer.[5,28]

In summary, these results demonstrate that in the presence of the appropriate Fe(III)-reducing microorganisms, a wide variety of aromatic compounds can be oxidized with Fe(III) as the electron acceptor (Table 18.1). As has been previously suggested,[5,7,8] the studies summarized here further indicate that abiotic oxidation of aromatics with Fe(III) is unlikely to be an important mechanism for the mineralization of most aromatic compounds, especially at the circumneutral pH typical of most natural aquatic environments.

Previous investigations have demonstrated Fe(III)-reducing microorganisms may play an important role in the mineralization of aromatic contaminants in groundwater,[5] and Fe(III) reduction is considered to be an important process for the oxidation of naturally occurring organic matter in aquatic sediments and groundwater.[29-32] Given the widespread availability of Fe(III) in many soils and sediments and the wide range of aromatic compounds that can be oxidized by microbial Fe(III) reduction, it is clear that oxidation of aromatic compounds coupled to Fe(III) reduction has the potential to be an important process in the mineralization of both naturally occurring and contaminant aromatic compounds whenever anaerobic conditions develop in sedimentary environments.

REFERENCES

1. Bostrom, K. "Some pH-Controlling Redox Reactions in Natural Waters," in *Equilibrium Concepts in Natural Waters*, R. E. Gould, Ed. (Washington, DC: American Chemical Society, 1967), pp. 286–311.

2. Ponnamperuma, F. N. "The Chemistry of Submerged Soils," *Adv. Agron.* 24:29–96 (1972).

3. Van Breeman, N. "Effects of Seasonal Redox Process Involving Iron on the Chemistry of Periodically Reduced Soils," in *Iron in Soils and Clay Minerals*, J. W. Stucki, B. A. Goodman, and U. Schwertmann, Eds. (Boston: D. Reidel Publishing Company, 1988), pp. 797–809.

4. Ehrlich, G. G., E. M. Godsy, D. F. Goeritz, and M. F. Hult. "Microbial Ecology of a Creosote-Contaminated Aquifer at St. Louis Park, Minnesota," *Dev. Ind. Microbiol.* 24:235–245 (1983).

5. Lovley, D. R., M. J. Baedecker, D. J. Lonergan, I. M. Cozzarelli, E. J. P. Phillips, and D. I. Siegel. "Oxidation of Aromatic Contaminants Coupled to Microbial Iron Reduction," *Nature* 339:297–299 (1989).

6. Schwille, F. "Anthropogenically Reduced Groundwaters," *Hydrol. Sci. Bull.* 21:629–645 (1976).

7. Lovley, D. R., and D. J. Lonergan. "Anaerobic Oxidation of Toluene, Phenol, and *p*-Cresol by the Dissimilatory Iron-Reducing Organism, GS-15," *Appl. Environ. Microbiol.* 56:1858–1864 (1990).

8. LaKind, J. S., and A. T. Stone. "Reductive Dissolution of Goethite by Phenolic Reductants," *Geochem. Cosmochim. Acta* 53:961–971 (1989).
9. Lovley, D. R., and E. J. P. Phillips. "Organic Matter Mineralization with Reduction of Ferric Iron in Anaerobic Sediments," *Appl. Environ. Microbiol.* 51:683–689 (1986).
10. Lovley, D. R., and E. J. P. Phillips. "Availability of Ferric Iron for Microbial Reduction in Bottom Sediments of the Freshwater Tidal Potomac River," *Appl. Environ. Microbiol.* 52:751–757 (1986).
11. Lovley, D. R., and E. J. P. Phillips. "Competitive Mechanisms for Inhibition of Sulfate Reduction and Methane Production in the Zone of Ferric Iron Reduction in Sediments," *Appl. Environ. Microbiol.* 53:2636–2641 (1987).
12. Lovley, D. R., and E. J. P. Phillips. "Novel Mode of Microbial Energy Metabolism: Organic Carbon Oxidation Coupled to Dissimilatory Reduction of Iron or Manganese," *Appl. Environ. Microbiol.* 54:1472–1480 (1988).
13. Lovley, D. R., J. F. Stolz, G. L. Nord, and E. J. P. Phillips. "Anaerobic Production of Magnetite by a Dissimilatory Iron-Reducing Microorganism," *Nature* 330:252–254 (1987).
14. Young, L. Y., and A. C. Frazer. "The Fate of Lignin and Lignin-Derived Compounds in Anaerobic Environments," *Geomicrobiol. J.* 5:261–293 (1987).
15. Berry, D. F., A. J. Francis, and J. Bollag. "Microbial Metabolism of Homocyclic and heterocyclic Aromatic Compounds under Anaerobic Conditions," *Microbial Rev.* 51:43–59 (1987).
16. Pastan, I., L. Tsai, and E. R. Stadtman. "Nicotinic Acid Metabolism," *J. Biol. Chem.* 239:902–906 (1964).
17. Stadtman, E. R., T. C. Stadtman, I. Pastan, and L. D. Smith. "*Clostridium barkeri sp. n.*," *J. Bacteriol.* 110:758–760 (1972).
18. Imhoff-Stuckle, D., and N. Pfennig. "Isolation and Characterization of a Nicotinic Acid-Degrading Sulfate-Reducing Bacterium, *Desulfococcus niacini Sp. nov.*," *Arch. Microbiol.* 136:194–198 (1983).
19. Evans, W. C., and G. Fuchs. "Anaerobic Degradation of Aromatic Compounds," *Ann. Rev. Microbiol.* 42:289–317 (1988).
20. Kuhn, E. P., J. M. Suflita, M. D. Rivera, and L. Y. Young. "Influence of Alternate Electron Acceptors on the Metabolic Fate of Hydroxybenzoate Isomers in Anoxic Aquifer Slurries," *Appl. Environ. Microbiol.* 55:590–598 (1989).
21. Tschech, A., and B. Schink. "Fermentative Degradation of Monohydroxybenzoates by Defined Syntrophic Cocultures," *Arch. Microbiol.* 145:396–402 (1986).
22. Boyd, S. A., D. R. Shelton, D. Berry, and J. M. Tiedje. "Anaerobic Biodegradation of Phenolic Compounds in Digested Sludge," *Appl. Environ. Microbiol.* 46:50–54 (1983).
23. Smolenski, W. J., and J. M. Suflita. "Biodegradation of Cresol Isomers in Anoxic Aquifers," *Appl. Environ. Microbiol.* 53:710–716 (1987).
24. Bak, F., and F. Widdel. "Anaerobic Degradation of Phenol and Phenol Derivatives by *Desulfobacterium phenolicum Sp. Nov.*," *Arch. Microbiol.* 146:177–180 (1986).
25. Aftring, R. P., B. E. Chalker, and B. F. Taylor. "Degradation of Phthalic Acids by Denitrifying Mixed Cultures of Bacteria," *Appl. Environ. Microbiol.* 41:1177–1183 (1981).
26. Aftring, R. P., and B. F. Taylor. "Aerobic and Anaerobic Catabolism of Phthalic Acid by a Nitrate-Respiring Bacterium," *Arch. Microbiol.* 130:101–104 (1981).

27. Nozawa, T., and Y. Maruyama. "Denitrification by a Soil Bacterium with Phthalate and Other Aromatic Compounds as Substrates," *J. Bacteriol.* 170:2501–2505 (1988).

28. Baedecker, M. J., D. I. Siegel, P. Bennett, and I. M. Cozzarelli. "The Fate and Effects of Crude Oil in a Shallow Aquifer. I. The Distribution of Chemical Species and Geochemical Facies," in *U.S. Geological Survey Water Resources Division Report 88-4220*, Mallard and Ragone, Ed. (1989), pp. 13–20.

29. Aller, R. C., J. E. Macklin, and R. T. Cox, Jr. "Diagenesis of Fe and S in Amazon Inner Shelf Muds: Apparent Dominance of Fe Reduction and Implications for the Genesis of Ironstones," *Cont. Shelf Res.* 6:263–289 (1986).

30. Champ, D. R., J. Gulens, and R. E. Jackson. "Oxidation-Reduction Sequences in Ground Water Flow Systems," *Can. J. Earth Sci.* 16:12–23 (1979).

31. Lovley, D. R. "Organic Matter Mineralization with the Reduction of Ferric Iron: A Review," *Geomicrobiol. J.* 5:375–399 (1987).

32. Lovley, D. R., F. H. Chapelle, and E. J. P. Phillips. "Fe(III)-Reducing Bacteria in Deeply Buried Sediments of the Atlantic Coastal Plain," *Geology* 18:954–957 (1990).

CHAPTER 19

Occurrence and Speciation of Naturally Produced Organohalogens in Soil and Water

A. Grimvall, H. Borén, and G. Asplund

INTRODUCTION

It has long been known that there are naturally produced organohalogens. As early as the 1930s, lichenologists and mycologists noted that certain organisms produced specific halometabolites.[1,2] Somewhat later, pharmacologists became interested in the antibiotic properties of halometabolites,[3] and biochemists started systematic studies of haloperoxidases.[4] Marine biologists have shown that chlorinated, brominated, and iodinated organic compounds can be produced by algae,[5] and botanists have compiled lists of higher plants known to produce organohalogens.[6]

So far, more than 700 naturally produced organohalogens have been identified.[7,8] Despite this substantial and detailed knowledge about naturally produced organohalogens, their presence in the environment has rarely been discussed in quantitative terms. Without investigating the matter, it has been assumed that practically all organohalogens in air, water, and soil are of industrial origin, and group parameters such as adsorbable organic halogens (AOX) have been suggested as suitable indicators of industrial pollution.[9,10]

The first calculations showing that natural halogenation is far from negligible were made in the early and mid-1980s, when several estimates of the production of halogenated organic compounds in marine environments were presented.[11-14] During the past few years, studies in Sweden and Denmark have drawn attention to halogenation processes in terrestrial environments and given additional evidence of a large natural production of organohalogens.[15-17] In this chapter these studies will be reviewed. Some results from ongoing investigations will be added, and the occurrence, origin, and speciation of organohalogens in aquatic and terrestrial environments will be discussed.

OCCURRENCE OF ADSORBABLE ORGANIC HALOGENS IN AQUATIC AND TERRESTRIAL ENVIRONMENTS

The procedure for measuring AOX in water was originally developed to monitor the total amount of organohalogens in drinking water.[18,19] Since then, this procedure has also been applied to wastewater and surface water, and it has been shown that organohalogens are ubiquitous in aquatic environments.[20-23] Recently, this was further documented in surveys of AOX in rivers in Sweden[24] and groundwaters in Denmark.[16] In the latter studies, AOX was present in measurable concentrations in all analyzed surface water samples and in all but one of the groundwater samples (Figure 19.1). Substantial amounts of AOX have also been found in marine environments: A survey of AOX in the Baltic Sea resulted in AOX concentrations from 5 to 15 μg/L.[17]

Concentrations of AOX in terrestrial environments are less extensively documented. In a study performed in Sweden, all analyzed soil samples contained measurable amounts of organohalogens, and the AOX content was particularly high in organic soils.[15]

NEW EVIDENCE OF A LARGE NATURAL PRODUCTION OF ORGANOHALOGENS

The difficulties in assigning observed AOX concentrations in surface waters to known industrial sources of organohalogens were rather recently discussed in two independent reports.[23,25] Continued studies at Linköping University have shown that in Sweden seemingly unpolluted surface waters may have AOX concentrations approaching 200 μg Cl/L, a level equal to or exceeding the concentrations observed in some of the most polluted rivers in Europe.[21] More solid evidence of a large natural production of AOX was presented by Asplund et al., who showed that organohalogens are present (0.2–0.4 mg Cl/g

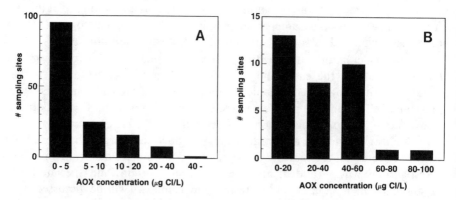

Figure 19.1. Frequency distribution of AOX in *(A)* groundwater in Denmark (modified from Grön[16]) and *(B)* river water in Sweden (modified from Enell et al.[24]).

fulvic acid, FA) even in old groundwaters (1300–5200 years).[15] Provided that no artifacts had affected the results of the AOX analyses (see below), this proves that there are considerable amounts of organohalogens of preindustrial origin in the environment. The same conclusion was drawn by Grøn in his study of AOX in Danish groundwaters.[16]

The existence of a large natural production of organohalogens was further confirmed by mass balance calculations for a raised bog in Sweden.[15] Measurements of AOX in organic matter leached from peat showed that the total storage of AOX in the bog was several hundred times greater than the annual wet deposition onto the bog. Consequently, air pollution from industrialized society could not account for more than a fraction of the detected organohalogens. Neither could the presence of AOX in the bog be explained by local discharges to water, since the bog area is not populated and does not receive water from surrounding areas. The total pool of leachable AOX in peat in Sweden was estimated to be approximately 300,000 metric tons.

The evidence of natural production of organohalogens presented above was entirely based on AOX determinations. Therefore, it is important to discuss potential artifacts in the analytical procedure used. Two problems are of particular interest: (1) incomplete removal of inorganic halides prior to the final detection of AOX and (2) interference from substances other than halogens.

The separation of organohalogens from inorganic halides has been subject to several investigations. Wigilius et al. tested and rejected the hypothesis that the simultaneous presence of humic substances and moderate concentrations of chloride (≤ 100 mg/L) may affect the result of an AOX analysis.[23] In brackish water and seawater, interference by inorganic halides represents a more serious problem. However, Borén et al.[26] have recently shown that one of the standard methods for analyzing AOX in water[27] may be sufficiently improved to permit reproducible determinations in brackish water. Interference by substances other than halogens has been thoroughly studied by Grøn.[16] By utilizing neutron activation analysis, he showed that the sum of the chlorine, bromine, and iodine content was approximately equal to the AOX content, as determined by the standard microcoulometric titration. In summation, there is no indication that analytical artifacts could undermine the conclusion that there is a large pool of naturally produced AOX in soil and water.

ORIGIN OF ORGANOHALOGENS IN AQUATIC AND TERRESTRIAL ENVIRONMENTS

The presence of naturally produced AOX in surface water, groundwater, and soil does not necessarily imply that the organohalogens are formed at these sites. Theoretically, organohalogens could originate in the sea or in the atmosphere and then be transported to other environments. However, a closer examination of existing data shows that such long-range transport cannot be responsible for more than a small part of the observed spatial distribution of AOX.

The AOX concentrations in Figure 19.2 were observed in streams that were little affected by local sources of pollution. If long-range atmospheric deposition made the major contribution, it is unlikely that there would be large differences in AOX concentration between adjacent rivers. Neither can atmospheric deposition be responsible for certain tributaries of the studied lake having an average AOX concentration of almost 100 μg Cl/L, even though the AOX concentration in precipitation is only about 10 μg Cl/L in this part of Sweden. This implies that organohalogens must be produced naturally either in aquatic or terrestrial environments.

Evidence that the net production of AOX in lakes is small or nonexistent is

Figure 19.2.   Average AOX concentration (μg Cl/L) in the tributaries of Lake Vättern, Sweden. Mean values of 3 to 4 samples. *Source:* Grimvall et al.[17]

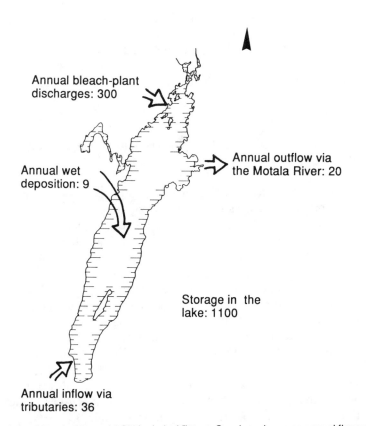

Annual bleach-plant
discharges: 300

Annual outflow via
the Motala River: 20

Annual wet
deposition: 9

Storage in the
lake: 1100

Annual inflow via
tributaries: 36

Figure 19.3.   Mass balance of AOX for Lake Vättern, Sweden, shown as annual fluxes (metric tons Cl/year) and as storage (metric tons Cl) in the lake. *Source:* Grimvall et al.[17]

provided by the results presented in Figures 19.2 and 19.3, as well as by the fact that only one freshwater organism, a cyanobacterium, is known to produce a halometabolite.[28] As shown in Figure 19.3, the export of AOX from Lake Vättern, Sweden, is much smaller than the input from the atmosphere, tributaries, and direct industrial discharges. Furthermore, the lowest AOX values in the tributaries shown in Figure 19.2 were observed at sampling sites immediately downstream from lakes. Altogether, these results indicate that in limnic environments a large net production of AOX is very unlikely. The only remaining possibility is a substantial natural production of AOX in soil or vegetation.

At the present, it is not known whether the largest contribution of organohalogens can be traced to processes in soil or vegetation. Analyses of precipitation and canopy drip from Norwegian spruce (*Picea abies*) have shown that the AOX content in canopy drip is so much higher than in precipitation that the difference can hardly be explained by evaporation or by dry deposition.[29] This indicates that natural production in or on vegetation may be an important source of AOX. However, it should not be overlooked that there are also soil processes that may give substantial contributions of AOX (see below). The origin of the AOX found in precipitation and marine environments is outside the scope of this chapter. It is presently not even known whether the major fraction of these organohalogens is of natural or anthropogenic origin.

SPECIATION OF ADSORBABLE ORGANIC HALOGENS IN THE ENVIRONMENT

The halometabolites identified so far are relatively low-molecular-weight, specific compounds.[8] The organohalogens accounting for the major fraction of the AOX content in soil and water are of a different character, and several results indicate that the major fraction of the organically bound halogens are incorporated into humic substances.

Surveys of surface water and groundwater in Scandinavia have shown that samples with a high content of organic matter usually have the highest AOX concentrations,[16,24] and humic substances isolated from water contain considerable amounts of AOX.[15,23] Using size-exclusion chromatography, Wigilius et al. showed that AOX and TOC had approximately the same molecular weight distribution in the investigated surface water samples; a small shift toward lower molecular weights for the organohalogens was the main difference.[23] In Danish groundwaters, fractionation of organic matter by ultrafiltration and ion exchange techniques gave similar results; the amount of AOX in the different fractions was almost proportional to the TOC value.[16]

Until now, few investigators have measured the distribution of the different halogens that may contribute to the occurrence of AOX in organic matter from soil and water. DeLeer analyzed some Dutch soil samples by neutron activation analysis and found organically bound chlorine, but no bromine or iodine.[30] Analyzing Danish groundwaters by the same technique, Grön found all three halogens.[16] In particular, there seemed to be a marked contribution of organically bound bromine and iodine in groundwaters that could have been affected by seawater intrusion.

Attempts to identify specific organohalogens in unpolluted water or soil samples have been moderately successful. Swedish and Dutch soil samples with a high AOX-to-TOC ratio have been subjected to pyrolysis followed by gas chromatographic analysis of hexane extracts. The results of constant neutral loss scans and negative chemical ionization with selected ion monitoring have indicated the presence of several chloroalkanes or alkenes. Furthermore, monochlorophenol has been tentatively identified in a soil pyrolysate.[31] Another chlorophenol, 2,4,6-trichlorophenol, and its methylated analogue, 2,4,6-trichloroanisole, have been detected in Swedish surface waters not exposed to any industrial emissions.[17,32] This indicates that these compounds are of natural origin (see also below).

NATURAL HALOGENATION OF HUMIC MATTER

The speciation of AOX in water and soil indicates that one of the major natural halogenation processes is a rather unspecific halogenation of humic matter. It has long been known that low-molecular-weight organohalogens of industrial origin can be adsorbed to humic matter.[33] However, halogens can also be incorporated into humic substances by other mechanisms. Sakar et al. showed that 2,4-dichlorophenol was incorporated into stream fulvic acid by enzymatic coupling,[34] and in a series of experiments, Behrens found that iodide was probably incorporated into soil organic matter by similar reactions.[35,36]

Some recently compiled reports have drawn further attention to enzymatically mediated halogenation processes in soil. Laboratory studies of the fungal enzyme chloroperoxidase (EC 1.10.1.11), from *Calderiomyces fumago*, have shown that in the presence of low concentrations of chloride and hydrogen peroxide, this enzyme may chlorinate humic substances. The curves in Figure 19.4 show that fulvic acids may incorporate up to 5% chlorine and that the velocity of the halogenation reaction reaches a maximum at pH 3. Experiments with peroxidase-mediated iodination of humic acids gave similar results.[38] Complementary measurements showed that extracts obtained from soil samples by means of a procedure for peroxidase isolation had a measurable chlorinating capacity.[37] This means that enzymatically catalyzed halogenation of humic substances may comprise a plausible explanation of the large natural production of organohalogens in terrestrial environments.

When experiments were performed with different enzyme concentrations, the velocity of the halogenation reaction was almost proportional to the enzyme concentration. However, in particular at very low pH values, the combined presence of bromide and hydrogen peroxide resulted in a measurable halogenation of humic acids, even in the absence of peroxidases (Figure 19.5). It is not clear whether there is also an abiotic production of organically bound chlorine.

Further experiments with chloroperoxidase strongly indicated that hypochlorous acid is an intermediate product in chloroperoxidase-mediated chlorination of surface waters.[39] The reaction mixture in these experiments had a characteristic smell of chlorine, and chlorine gas could be stripped from the solution. Furthermore, both chlorination with sodium hypochlorite and chloroperoxidase-mediated chlorination resulted in practically identical ECD (electron capture detector) chromatograms of stripping extracts. Analysis of chlorophenols showed that for both chlorination methods, 2,4,6-trichlorophenol was formed in comparatively high concentration (Figure 19.6). This strengthens the conclusions that hypochlorous acid is an intermediate in chloroperoxidase-mediated chlorination of natural organic matter and that 2,4,6-trichlorophenol may be produced naturally.

The natural occurrence of a chloroperoxidase-like catalyst in soil and the occurrence of active chlorine as an intermediate in chloroperoxidase-mediated

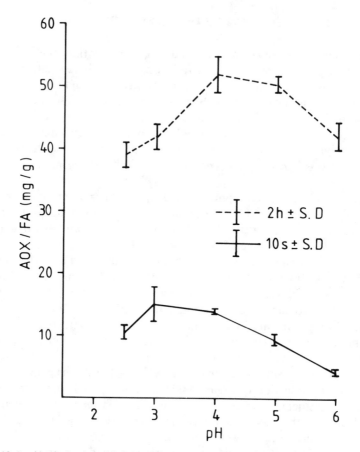

Figure 19.4.   Net formation of AOX by chloroperoxidase-mediated chlorination of fulvic acid (0.5 mg dissolved in 20 mL phosphate buffer) at different pH values. Reaction time 2 h and 10 s, respectively. *Source:* Asplund et al.[37]

chlorination of humic substances indicate that organohalogens other than 2,4,6-trichlorophenol are also produced naturally. In particular, one could expect to find several organohalogens known from chlorine disinfection of drinking water and chlorine bleaching of pulp.

If 2,4,6-trichlorophenol is naturally produced, one could also expect dimeric products such as corresponding dioxins. Upon storage of 2,4,6-trichlorophenol in a neutral water solution, several reaction products were formed. One of them was tentatively identified as 4-(2,4,6-trichlorophenoxy)-2,6-trichlorophenol by means of full scan mass spectroscopy.[39] The results just cited are preliminary, but raise several questions regarding the traditional division of trace organics into biogenic and xenobiotic compounds.

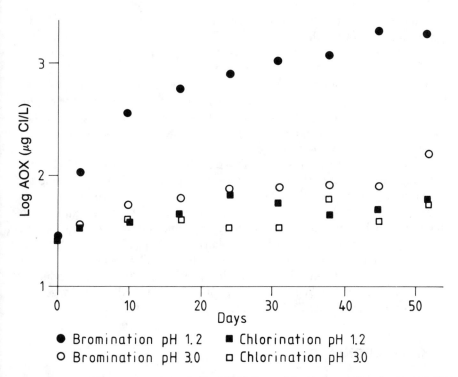

Figure 19.5.   AOX concentrations obtained by H_2O_2-mediated halogenation of organic matter in a surface water sample. *Source:* Hodin et al.[39]

CONCLUDING REMARKS

Due to their persistence and xenobiotic character, a great number of organohalogens have long been focal points in environmental research. The discovery

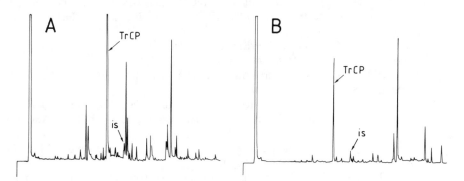

Figure 19.6.   ECD chromatograms from analysis of chlorophenols in surface water chlorinated *(A)* by chloroperoxidase-mediated chlorination and *(B)* by sodium hypochlorite. *Source:* Hodin et al.[39]

of a large natural production of organohalogens could perhaps lead to the conclusion that organohalogens are less hazardous to the environment than previously assumed. However, for several reasons, this conclusion cannot be justified by the presented results. A great number of organohalogens are, most probably, truly xenobiotic. Other compounds, previously assumed to be xenobiotic, may also be naturally produced, but anthropogenic sources are responsible for the highest concentrations in the environment. Furthermore, the mere fact that a compound is naturally produced does not imply that it is not hazardous to the environment.

The significance of the results reviewed in this chapter is strongly related to the question of whether there is a temporal trend in the natural production of organohalogens. Some of the organohalogens found in the environment are apparently of preindustrial origin, but it cannot be excluded that natural production of such compounds may partly be an indirect effect of general changes in the environment. One of the plausible halogenation mechanisms, chloroperoxidase-mediated chlorination of humic substances, seems to be favored by a low pH value. This makes it urgent to discuss natural production of organohalogens in relation to soil acidification and other large-scale changes in the terrestrial environment.

ACKNOWLEDGMENTS

The authors are grateful to the Swedish Environmental Protection Agency for financial support.

REFERENCES

1. Hardiman, J., J. Keane, and T. J. Nolan. "The Chemical Constituents of Lichens Found in Ireland," *Scient. Proc. Royal Dublin Soc.* 21:141–145 (1934).
2. Raistrick, H., and G. Smith. "Studies in the Biochemistry of Micro-Organisms. The Metabolic Products of *Aspergillus terreus thom,*" *Biochem. J.* 30:1315–1322 (1936).
3. Iwata, K., and I. Yosioka. "Terricin, a New Antibiotic Substance Produced by *Aspergillus terreus,*" *J. Antibiotics* ser. B 3:193–197 (1950).
4. Shaw, P. D., and L. P. Hager. "An Enzymatic Chlorination Reaction," *J. Am. Chem. Soc.* 81:1011–1012 (1961).
5. Fenical, W. "Natural Halogenated Organics," in *Marine Organic Chemistry,* Elsevier Oceanography Series 31, E. K. Duursma and R. Dawson, Eds. (Amsterdam: Elsevier, 1981), pp. 375–393.
6. Engvild, K. "Chlorine-Containing Natural Compounds in Higher Plants," *Phytochemistry* 25:781–791 (1986).
7. Siuda, J. F., and J. F. deBernardis. "Naturally Occurring Halogenated Organic Compounds," *Lloydia* 36:107–143 (1973).
8. Neidleman, S. L., and J. Geigert. *Biohalogenation—Principles, Basic Roles and Applications* (Chichester, Eng.: Ellis Horwood, 1986).

9. Hoffman, H.-J., G. Bühler-Neiens, and D. Laschka. "AOX in Schlämmen und Sedimenten – Bestimmungsverfahren und Ergebnisse," *Vom Wasser* 71:125–134 (1988).

10. Martin, J. T., and Y. Takahashi. "Total, Extractable and Leachable Organic Halides in Soil and Sediments," in *Chemical and Biological Characterization of Municipal Sludges, Sediments, Dredge Spoils and Drilling Muds,* J. J. Lichtenberg, J. A. Winter, C. I. Weber, and L. F. Fradkin, Eds. (Philadelphia: Special Technical Publications, 1988).

11. Kocher, D. C. "A Dynamic Model of the Global Iodine Cycle and Estimations of Dose to the World Population from Release of Iodine-129 to the Environment," *Environ. Int.* 5:15–31 (1981).

12. Whitehead, D. C. "The Distribution and Transformation of Iodine in the Environment," *Environ. Intern.* 10:321–339 (1984).

13. Gschwend, P. M., J. K. MacFarlane, and K. A. Newman. "Volatile Halogenated Organic Compounds Released to Seawater from Temperate Marine Microalgea," *Science* 227:1033–1035 (1985).

14. Harper, D. B. "Halomethane from Halide Ion – a Highly Efficient Fungal Conversion of Environmental Significance," *Nature* 315:55–57 (1985).

15. Asplund, G., A. Grimvall, and C. Pettersson. "Naturally Produced Adsorbable Organic Halogens (AOX) in Humic Substances from Soil and Water," *Sci. Total Environ.* 81/82:239–248 (1989).

16. Grön, C. "Organic Halogens in Danish Groundwaters," PhD Thesis, Institute of Applied Geology, Technical University of Denmark, Copenhagen (1989).

17. Grimvall, A., H. Borén, S. Jonsson, S. Karlsson, and R. Sävenhed. "Organohalogens of Natural and Industrial Origin in Large Recipients of Bleach Plant Effluents," Third IAWPRC Symposium on Forest Industry Wastewaters, Tampere, Finland, June 5–8, 1990.

18. Kühn, W., and H. Sontheimer. "Zur analytischen Erfassung organischer Chlorverbindungen mit der temperatur-programmierten Pyrohydrolyses," *Vom Wasser* 43:327 (1974).

19. Takahashi, Y., R. T. Moore, and R. J. Joyce. "Measurement of Total Organic Halides (TOX) and Purgeable Organic Halides (POX) in Water Using Carbon Adsorption and Microcoulometric Determination," in *Water Reuse,* W. J. Cooper, Ed. (Ann Arbor, MI: Ann Arbor Science, 1981).

20. Stevens, A. A., R. C. Dressman, R. K. Sorrell, and H. Brass. "Organic Halogen Measurements: Current Uses and Future Prospects," *Journal AWWA* 77:146–154 (1985).

21. von Keller, M. "AOX-Gehalte in Fliessgewässern der Bundesrepublik Deutschland," *Deutsche gewasserkundliche Mitteilungen* 31:38–42 (1987).

22. Wigilius, B., H. Borén, A. Grimvall, G. Carlberg, I. Hagen, and A. Brögger. "Impact of Bleached Kraft Mill Effluents on Drinking Water Quality," *Sci. Total Environ.* 74:75–96 (1988).

23. Wigilius, B., B. Allard, H. Borén, and A. Grimvall. "Determination of Adsorbable Organic Halogens (AOX) and Their Molecular Weight Distribution in Surface Water Samples," *Chemosphere* 17:1985–1994 (1988).

24. Enell, M., L. Kaj, and L. Wennberg. "Long-Distance Distribution of Halogenated Organic Compounds (AOX)," in *River Basin Management,* V. H. Laikari, Ed. (Oxford: Pergamon Press, 1989).

25. Klopp, R., and K.-H. Kornatzki. "Emission und Immission von AOX im Ruhrein-zugsgebiet," *Z. Wasser-Abwasser-Forsch.* 20:160–167 (1987).

26. Borén, H., A. Grimvall, S. Jonsson, S. Karlsson, and R. Sävenhed, "Distribution and Origin of Organohalogens in the Baltic Sea," presented at the COST 641 working party on transport of organic micropollutants in estuaries, marine and brackish waters, October 4–5, 1990, Stockholm, Sweden.

27. DIN (Deutsches Institut für Normung). "Bestimmung der adsorbierbaren organisch gebundenen Halogene (AOX)," DIN 38409, Teil 14, Summarische Wirkungs-und Stoffkenngrössen, Gruppe H (Berlin: Beuth-Verlag, 1985).

28. Mason, C. P., K. R. Edwards, R. E. Carlson, J. Pignatello, F. K. Gleason, and J. M. Wood. "Isolation of Chlorine-Containing Antibiotic from the Freshwater Cyanobacterium *Scytonema hofmanni,*" *Science* 215:400–402 (1982).

29. Grimvall, A., G. Asplund, and H. Borén. "Origin of Adsorbable Organic Halogens (AOX) in Aquatic Environments," Sixth European Symposium on Organic Micropollutants in the Aquatic Environment, Lisbon, May 22–24, 1990.

30. deLeer, E. W. B. Personal communication (1989).

31. deLijser, H. J. P., C. Erkelens, A. Knol, W. Pool, and E. W. B. deLeer. "Natural Organochlorine in Humic Soils. GC and GC/MS Studies of Soil Pyrolysates," in *Lecture Notes in Earth Sciences: Humic Substances in the Aquatic and Terrestrial Environment,* B. Allard, H. Borén, and A. Grimvall, Eds. (Berlin: Springer-Verlag, 1991), pp. 485–494.

32. Nyström, A., A. Grimvall, C. Krantz-Rulcker, R. Sävenhed, and K. Åkerstrand, "Drinking Water Off-Flavour Caused by 2,4,6-Trichloroanisole," submitted to *Water Sci. Technol.*

33. Wershaw, R. L., P. J. Burcar, and M. C. Goldberg. "Interaction of Pesticides with Natural Organic Material," *Environ. Sci. Technol.* 3:271–273 (1969).

34. Sakar, J. M., R. L. Malcolm, and J.-M. Bollag. "Enzymatic Coupling of 2,4-Dichlorophenol to Stream Fulvic Acid in the Presence of Oxidoreductases," *Soil Sci. Soc. Am. J.* 52:688–694 (1988).

35. Behrens, H. "Speciation of Radioiodine in Aquatic and Terrestrial Systems under the Influence of Biochemical Processes," in *Speciation of Fission and Activation Products in the Environment,* R. A. Bulman and J. R. Cooper, Eds. (New York: Elsevier, 1985).

36. Behrens, H. "Influence of Microbial Activity on Migration of Radioiodine in Geomedia," in *The Effects of Natural Organic Compounds and of Microorganisms on Radionuclide Transport,* A. B. Muller, Ed. (Paris: OECD/NEA, RWM-6, 1986), pp. 5–18.

37. Asplund, G., H. Borén, U. Carlsson, and A. Grimvall. "Soil Peroxidase-Mediated Chlorination of Fulvic Acids," in *Lecture Notes in Earth Sciences: Humic Substances in the Aquatic and Terrestrial Environment,* B. Allard, H. Borén, and A. Grimvall, Eds. (Berlin: Springer-Verlag, 1991), pp. 475–483.

38. Christiansen, J. "The Behaviour of Iodine in the Terrestrial Environment," PhD Thesis, Risö National Laboratory, Denmark (1990).

39. Hodin, F., H. Borén, A. Grimvall, and S. Karlsson. "Formation of Chlorophenols and Related Compounds in Natural and Technical Chlorination Processes," in manuscript. Third EWPRC symposium on Forest Industry Wastewaters, June 5–8, 1990, Tampere, Finland.

CHAPTER 20

Organic Fertilizers and Humification in Soil

Paolo Sequi, Claudio Ciavatta, and Livia Vittori Antisari

INTRODUCTION

Organic fertilizers may be divided into two different groups. The first group is made up of animal and plant by-products that are rich in organic N and C, such as leather meal, ground feathers, horn and hooves, waste wool, oilseed cakes, and so on. All of these materials contain more than 5% (sometimes 10–15%) organic N, and 30–50% organic C. The second group is composed of materials comparatively poor in organic N, such as animal dung or compost from urban refuse. Their organic N content is generally about 1%, while the organic C content is variable. Such materials are often called organic amendments.

Organic fertilizers with high C and N contents do not contain humic substances. They can be applied to the soil in order to supply (1) nitrogen, which will be released more or less slowly to plant roots, and (2) carbon, which will be used by soil organisms as both a nutrient and energy source. Typical organic amendments, on the contrary, cannot be applied to the soil without first having been subjected to a period of time of organic matter fermentation or "maturation." The use of raw organic amendments, in fact, may be inappropriate for several reasons including phytotoxicity. During maturation of organic amendments, the organic carbon content generally decreases, while the proportion of humified against nonhumified carbon increases.

Two major tasks were then imperative in order to evaluate the quality and optimize the use of organic materials in agriculture. The first was to assess the actual stage of the maturation of organic amendments quantitatively, so as to be able to utilize them at the proper time. The second was to distinguish qualitatively the organic matter from different sources, so that unknown materials may be easily recognized. We approached these problems by fractionating the extractable components of organic fertilizers on a solid resin, normally used to adsorb phenolic compounds, and by characterizing the soluble components electrophoretically. The combined use of these techniques led us also to follow the course of transformation of an organic fertilizer after it had been added to the soil. The results are reported in this chapter and discussed with regard to present knowledge of humification processes in soil.

QUALITY CRITERIA FOR ORGANIC MATTER

Methods to assess the quality of organic matter in soils, manures, sludges, or composts are still not well defined. Some authors have suggested the extraction of organic matter with a solution of 0.1 M NaOH plus 0.1 M $Na_4P_2O_7$ but did not apply this procedure to organic materials other than soil.[1] Other scientists have suggested the use of selected spectral properties of humic substances (e.g., the degree of aromaticity[2]) but found that the nominal molecular weight distribution of humic acids decreases as the degree of humification increases. This is contrary to the normally accepted trend. Humification of soil organic matter has been widely studied using the ratio between humic acids (HA) and fulvic acids (FA), but the results are of uncertain interpretation because they depend on many factors, including even the geographic distribution of the soils.[3] Another commonly used spectroscopic variable is the E_4/E_6 ratio (the ratio between the absorbance at $\lambda = 465$ and $\lambda = 665$), which sometimes is considered as an index of humification.[4] However, the addition of a small amount of humic substances (i.e., humic acid from leonardite) is sufficient to change the results completely. The cation exchange capacity (CEC) of the organic fractions has also been used to evaluate the degree of humification of organic amendments,[5] but it is a very indirect criterion of evaluation.

New Parameters of Humification

Humic substances have been defined as "amorphous, polymeric, brown colored compounds, that do not belong to recognizable classes of organic compounds, such as polysaccharides, polypeptides, altered lignins, etc.".[6] Alkaline extracts from soil have been considered in the past as total humic extracts,[1] but the suspensions, especially if extracted from organic materials different from soil, contain nonhumic substances. Separation of humic acids from fulvic acids is based only on the pH value of the extracts, but the supernatant (FA) also contains many classes of organic materials which are not humic substances (e.g., polysaccharides).

Some authors[7,8] have suggested the use of polyamide columns to retain the colored fractions of FA, while others[9,10] have used Amberlite XAD-8 to selectively adsorb humic substances from freshwater. This last procedure is recommended by the International Humic Substances Society (IHSS) for the purification of humic substances from a variety of materials.

Polyvinylpyrrolidone, a cross-linked adsorbent for the chromatographic separation of aromatic acids, aldehydes, and phenols,[11,12] has been widely used to separate phenol compounds from organic extracts.[13,14] In acid media, insoluble polyvinylpyrrolidone (PVP) allows strong adsorptions and good recoveries of humic substances.[14]

Recently, we have suggested the use of selective chromatography on solid PVP to separate humified from nonhumified materials in soil, dung, compost, and sludge extracts.[15] Presently, the use of solid PVP is required by Italian law for characterization of organic amendments (peat, leonardite, and humic extracts).

The fractionation scheme of organic extracts is very simple, and identical for different matrices (soils, organic fertilizers, amendments, etc.). The separation of humified from nonhumified materials is achieved by precipitation of humic acid at a low pH value and loading of the soluble fractions on columns packed with insoluble PVP. Nonhumified materials are not retained on PVP; after washing with 0.01 N H_2SO_4, the fulvic acid fraction is eluted with 0.05 N NaOH and added to the humic acids. This procedure, reported in a recent paper,[16] has been applied with good results on organic amendments,[17,18] and also used to follow the maturation of organic materials in sludges[19] and in piles of compost from urban refuse.[20]

Three new parameters of humification have been proposed:

1. humification index (HI)[15]

$$HI = NH/(HA + FA)$$

   i.e., the ratio between nonhumified (NH) and humified (HA + FA) compounds
2. degree of humification (DH)[21]

$$DH\% = [(HA + FA)/TEC] \cdot 100$$

   i.e., the percentage of humified compounds with respect to total extracted carbon (TEC)
3. humification rate (HR)[21]

$$HR\% = [(HA + FA)/TOC] \cdot 100$$

   i.e., the percentage of humified compounds with respect to total organic carbon (TOC) in the sample

In general, the humification index (HI) is near zero (0–0.5) for humified materials (i.e., soils, organic amendments) and much higher than 1 for non-humified materials (i.e., organic fertilizers, raw composts, sewage sludges, and swine slurries). This is the parameter first proposed, and the most used up to now in the literature. The other parameters, however, are very useful on many occasions. Some relative experimental results and applications will be discussed in this chapter.

Characterization by Electrofocusing

Although the above methods have been proven to be useful for evaluating the evolution of organic matter from a quantitative point of view, they are unable qualitatively to distinguish organic matter extracted from different sources. A technique that easily serves this purpose and, in addition, produces good evidence of the organic matter evolution during humification is electrofocusing (EF), which can be considered as an extension of the electrophoretic principle. EF is carried out in a polyacrylamide gel tube where an electrophoretic carrier ampholite has been previously preblended to give a pH gradient from a lower to a higher pH value. During electrofocusing, each macromolecule moves under an electric potential in search of its isoelectric point (IEP).

Since 1972, many authors have used EF to characterize soil enzymes[22] and humic substances extracted from soil,[23] rivers,[24] and other organic materials.[25] Although the results of some authors have shown EF to be a reliable method,[26] other authors consider the EF findings as artifacts. In general they believe[27] that separation of the humic substances by the EF technique was caused by their interaction with the carrier ampholytes.

Recently, the integrity of humic substances during EF application has been demonstrated, and the possible interaction between humic substances and ampholytes has been assessed as not being responsible for the appearance of some bands as artifacts in the gels.[28] Our findings confirmed that electrofocusing is a reliable technique and can be used successfully to monitor differences in the quality of the organic matter in raw, nonpurified materials. Even using such materials, interferences are very limited, and resolution appears to be satisfactory.

Another advantage of electrofocusing is the possibility of demonstrating differences between materials that are apparently very similar from a quantitative point of view if tested by other procedures. As for other organic wastes, EF profiles have been suggested to reflect stabilization processes (i.e., humification) of the organic matter on a purely qualitative basis.[29]

Evolution of Organic Matter During Maturation of Organic Amendments

Parameters of Humification

The DH of soil organic matter, as determined by the above method, is generally higher than 60% and may reach 90% or more. Its value depends on the amount of NH materials actually present (mainly polysaccharides). Of course, lower values are always found for HR, depending, in this case, on the amount of nonextractable organic matter, the so-called humin. Such values are higher for fossil humic substances, such as leonardites, where not only DH, but also HR may sometimes be close to 100%.

Both DH and HR of nitrogen-rich organic fertilizers (dried blood, leather meal, and so on) should be equal or close to zero because humified substances are not expected to be present. Some interferences occur,[21] and we are attempting to fully eliminate them.

A very different situation occurs for fertilizer where organic matter is not stabilized. Such fertilizers include many organic amendments which need a maturation, i.e., stabilization of the organic matter. We found that stabilization of organic matter consists of processes similar to humification, and its extent may often be determined by the application of the humification parameters.

In general, the evolution of organic matter during maturation of an organic amendment is characterized by a continuous increase in humified or pseudo-humified substances in the alkali-soluble fraction, so that DH represents effectively the development of the process. Sometimes, but not always, the HR value also has a similar trend to that of DH. Figure 20.1, for instance, shows the trend of the DH during the organic matter stabilization processes in a compost from urban refuse.[30] The value increases continuously, until it reaches stable values and an asymptotic trend at the end of the stabilization processes, after 40–60 days of composting. In the case represented in the same figure, HR is not correlated to the stabilization of the amendment. It must be born in mind that in some materials there is a continuous and immediate transformation of the substances liberated from an extremely heterogeneous organic matrix, and humiclike substances do not accumulate during the process. In these cases, therefore, only DH is of practical value because humified materials are of significant importance in extracts, but only slightly relevant with respect to total organic carbon.

As previously said, generally speaking for the majority of organic substrates, both DH and HR change and reflect continuously the evolution of organic matter. For particular matrices such as pig slurries, however, only HR fits in with the actual development of the process, as shown in Figure 20.2. A better suitability of either DH or HR probably depends on the specific nature

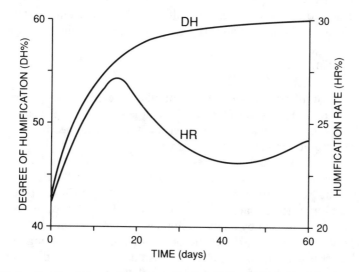

Figure 20.1. Trends of the degree of humification (DH) and of the humification rate (HR) during the thermophylic phase of the organic matter stabilization process in a compost from urban refuse. Reelaborated from Pasotti.[30]

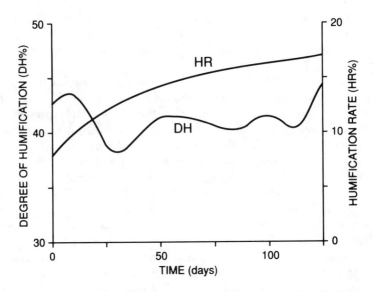

Figure 20.2. Trends of the degree of humification (DH) and of the humification rate (HR) during the stabilization process of pig slurries. Reelaborated from Govi.[29]

of the material considered, i.e., whether the bulk of organic matter in the material is evolved in the maturation process or not. If processes are effective for a small proportion of the material and involve progressively only further limited parts while the bulk remains unaltered, DH can describe the process better than HR. Especially in the case of liquid or semiliquid wastes, however, the entire mass of organic matter is simultaneously involved in the stabilization process; in such conditions, only HR can accurately describe the process.

Electrofocusing Profiles

As we mentioned above, EF patterns of alkaline extracts can give an important contribution both to the clarification of the stage of humification from a qualitative point of view and to the determination of the specific matrix from which the organic matter has been extracted.

EF profiles of alkaline soil extracts are characterized by considerable heterogeneity and a remarkable number of bands in the neutral region of the pH gradient, while EF profiles for organic nitrogen-rich fertilizers from animal residues are very differentiated. Figure 20.3 shows that EF profiles for raw pig slurries are characterized by a limited number of bands in the acidic region of the pH gradient. However, the evolution of organic matter during incubation of pig slurries is associated with a tendency to develop EF bands more similar to those of humified materials from the soil. Our results agree with those reported for sewage sludges where, during stabilization, new bands also develop in the neutral region of the pH gradient.[19]

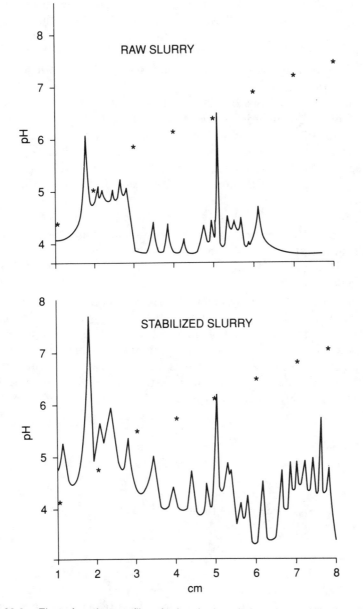

Figure 20.3. Electrofocusing profile of pig slurries during the stabilization process. Reelaborated from Govi.[29]

HUMIFICATION IN SOIL

As in the case of most nutrient cycles, soil also plays a central role in the carbon cycle. Contrary to current opinions of ecologists and agronomists, it has been estimated that animals consume only 10–20% of the carbon of photosynthates made available by plants; the remaining 80–90% is utilized by microorganisms that mainly operate in soils.[31] Such estimates have been confirmed by the evaluation of both (1) energy fluxes from sunlight to plants and soil[32] and (2) the weight ratio between organisms living above the soil surface and those inside the soil body.[33]

Although soil organisms live in conditions of starvation[34] and undergo a series of adaptations in order to face the lack of food, carbon pathways in soil are essentially of a conservative type. If a carbon source is added to the soil, for instance, complete utilization which might be expected, is never actually recorded. In fact, a substantial amount of the carbon supply — say, some 30% as a rough average — can be stored in soil, possibly in the form of new humic substances.[35]

It seems rather surprising that some of the food supply is set aside even when carbon is added to the soil as an easily available material. Mutatker and Wagner, for instance, have shown that after incubation of ^{14}C-labeled glucose in soil for 3 months, 25% of the radioactive carbon was incorporated in the soil organic matter.[36]

The incorporation was only slightly higher (34%) when the carbon supply is effected in the form of ^{14}C-labeled dextran.[37] These figures are astonishing if we consider that the maximum number of bacterial cells is generally reported to occur within the first 2 days of incubation, after which fungal growth is noted at the time the bacterial count declines; that is, a succession of different populations can be ascertained following the incorporation of glucose into soil.[38] This probably can be interpreted only on the basis of the following assumptions:

- Starving populations of living soil organisms do not utilize entirely any carbon source that has been added to the soil because a substantial percentage is stored in the soil organic matter pools.
- Biosynthesis of some pools of soil organic matter requires periods of time much shorter than foreseen.

From the point of view of chemical characteristics and agronomic behavior, humified organic matter in soil is generally considered more as a whole than as a complex of different pools. However, organic materials in soil are generally divided into pools of different stability. Although some pools have been estimated to survive in soil for thousands of years,[39] evidence has been reported in the literature that some fungi promote soil organic matter turnover. *Marasmius oreades*, for instance, has been found to rapidly degrade urease activity associated with the bulk of soil humus, as evidenced by soil discoloration.[40]

Table 20.1. Variations of Organic Carbon and Degree of Humification After Incubation of a Soil Added with Carbon and Nitrogen Sources

Treatment	Incubation	Organic Carbon (%)	Degree of Humification (%)
None	None	1.45	70.0
None	Aerobic	1.44	72.0
	Anaerobic	1.37	74.0
Glucose	Aerobic	1.50	54.5
	Anaerobic	1.45	54.5
Urea	Aerobic	1.40	50.5
	Anaerobic	1.36	67.0
Glucose +	Aerobic	1.51	58.5
urea	Anaerobic	1.44	68.0

Source: Reelaborated from Ronchi.[41]

Influence of the Addition of Organic Compounds to Soil on Quality of the Soil Organic Matter

Some results of an incubation experiment with a soil with a carbon and a nitrogen source (glucose and urea, respectively) are reported in Table 20.1. The addition of 0.3% by weight of glucose C to the soil caused an increase in total soil organic C content of 0.06–0.08%, i.e., of the order of 25% of the amount added. The addition of urea alone, on the other hand, caused some depletion of total carbon content. A situation similar to the addition of glucose C, finally, occurred if glucose C and urea N were added simultaneously to the soil prior to incubation: total organic carbon increased in an amount corresponding to about 25% of the amount added.

The relative content of humified materials, on the contrary, always decreased following any C or N addition. At least during the short term of 30 days, therefore, the supply of carbon caused an increase in organic carbon and a relative decrease in humified materials in the soil.

In another experiment, we added leather meal to the soil as a source of C and N instead of glucose and urea.[42] Actually, the initial goal of the experiment did not include the fate of the C source. Leather meal is perhaps the most important natural organic N fertilizer in Italy, but there is concern about the effect of the Cr it contains on the environment. We added leather meal to the soil in the ratio 1:10, so that the organic matter in the fertilizer prevailed over the soil organic matter, and then followed humification of the bulk mixture by the procedures described above. We were able to follow the humification process of the leather after its addition to the soil and showed that after about 300 days humification was nearly complete. Incidentally, we found that Cr was continuously released from the leather during the decomposition of organic matter and then was insolubilized; that is, the release of Cr when leather is used as fertilizer is of little agronomic or environmental significance.

An even more interesting finding with this experiment was given by the EF technique. It is clear from Figure 20.4 that the EF profiles for soil and leather meal are very different. The EF profile for the soil used was very different from that found for leather meal: soil organic matter typically showed a number of bands in the acid region, while the pattern for leather extractable organic matter was more uniformly distributed along the entire region of the pH gradient and showed many characteristic bands also at neutral pH values, and even above pH 7.

When leather meal was applied to the soil, its EF profile tended to mask the typical pattern given by the original soil bands (Figure 20.5). Such behavior was expected, because the organic matter of the leather meal added to the soil exceeded that of the native soil organic matter. However, the profile of the soil incubated for 300 days after the addition of leather meal was absolutely unexpected. About 50% of fertilizer C had evolved as CO_2 by that time, and the remaining 50% was mainly transformed to humified substances, as shown by the value of DH in Figure 20.6. The EF profile was nevertheless similar to that of the soil–leather meal mixture at the beginning of incubation, because the leather meal proteins maintained their identity even though the humification process for the applied organic fertilizer had almost reached completion. Needless to say, EF profiles of soil extracts at intermediate stages of incubation also were similar to those represented in Figure 20.6.

The findings presented in this chapter are not sufficient for a definite interpretation of the behavior of leather meal after its addition to the soil. However, it seems plausible to suggest that a part of the leather meal proteins was not utilized as food by living soil organisms, but rather transformed to humified proteins. Such proteins should act as storage proteins for the needs of successive organisms, stabilized like humified materials, though preserving their own identity.

Some Considerations about Humification in Soil

Often, perhaps too often, humification processes in soil are reduced to a matter of polymerization of phenolic substances. Aromatic compounds in general fascinate scientists, in part because of their relatively scarce involvement in the circuit of the official molecular biology research. So, properties of humus from different sources are sometimes correlated to their content of phenolics, sometimes calculated from poorly reliable NMR determinations.

Many experimental contributions have resulted from such often unexpressed convictions. Some authors have investigated so-called para-humic substances produced by pure bacterial or fungal strains in the presence of a phenolic substance as the sole carbon and energy supply (e.g., Bailly and Nkundikije-Desseaux[44]). Others have studied polymerization of specific phenolic monomers in the presence of particular enzymes such as laccase[45,46] or even inorganic constituents, such as clays or manganese oxides.[46-48]

As is well known, depolymerization of phenolic polymers, even encrusted

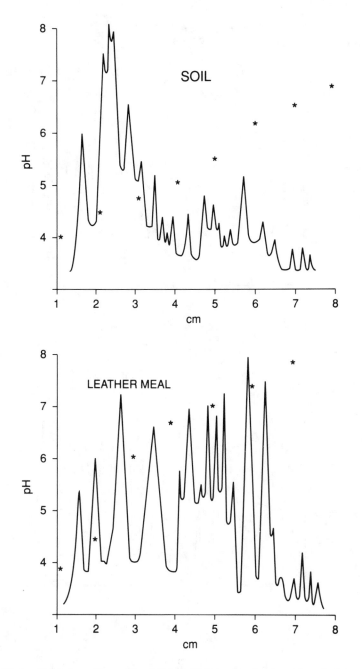

Figure 20.4. Electrofocusing profile of 0.5 M NaOH extracts of a soil and of a sample of leather meal. *Source:* Ciavatta.[43]

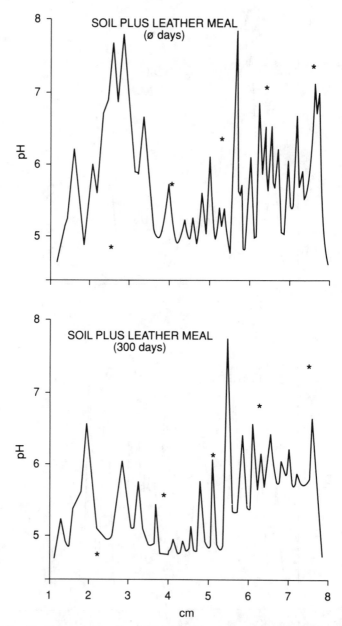

Figure 20.5. Electrofocusing profile of 0.5 M NaOH extracts of a soil incubated with leather meal for 300 days after application of leather meal at the beginning (*above*) and at the end (*below*) of the incubation period. *Source:* Ciavatta.[43]

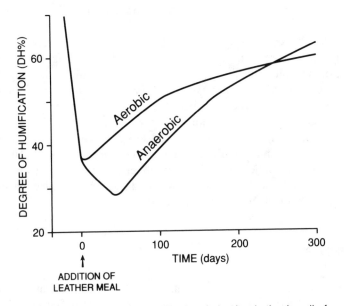

Figure 20.6. Trends of the degree of humification during incubation in soil of a sample of leather meal fertilizer under aerobic and anaerobic conditions. Reelaborated from Ciavatta and Sequi.[42]

with cellulose, and subsequent cleavage of phenolic rings have been the subject of recent research.[49,50] Experiments of polymerization of phenolic monomers are carried out at concentrations of monomers as high as 0.2–0.5 mM or 0.48–10 μg/g soil, and it is difficult to understand why such comparatively very high concentration values should occur in an overcrowded, starving world. At the same time, it should be emphasized that in the same living world substantial amounts of easily oxidizable compounds, such as glucose, are transformed in soil to stable materials.

No doubt, a puzzling problem for soil chemists arises from the heterogeneity and polydispersion of any mixture of organic molecules isolated from soil. Hayes et al., in their recent review on the structure of humic substances, concluded that, "it is possible that, at least for humic acids, no two molecules in any particular extract are the same".[51] The importance of such evidence is perhaps overestimated. Every known protein and biologically significant molecule that has been isolated and chemically characterized has been extracted from storage tissues or specifically accumulating organisms, at least for the first time. The history of molecular biology is full of complicated procedures for the purification of a definite enzyme from, say, 10 kg of a particular tuber, yielding 100 mg of a crystalline enzyme. A similar purification procedure absolutely would not be suitable, for example, for the leaves of the same plant species. We cannot pretend to obtain any isolated and purified compound from some kilograms of soil containing only some percent of organic matter, where an enormous and extremely various population of living organisms is operating.

Burns et al. have shown that extracellular soil enzymes were stabilized.[52] They suggested that resistance could be attributed to coating of soil enzymes with a net of phenolic substances, which should allow substrate and products, but not proteolytic enzymes, to pass to and from the active site. So, humification processes of soil enzymes could consist in the arrangement of a protective phenolic net against adverse actions to meet the functionality needs (e.g., improvement of the microenvironment) of soil organisms. Again, like extracellular enzymes, stabilized proteins could be a major component of soil humus. A protein applied to soil might undergo humification processes in order to be stored. Such an approach could explain the mechanisms and significance of humification processes in soil.

REFERENCES

1. Schnitzer, M., L. E. Lowe, J. F. Dormaar, and V. Martel. "Chemical Parameters for the Characterization of Soil Organic Matter," *Can. J. Soil Sci.* 61:517–519 (1981).
2. Tsutsuki, K., and S. Kuwatsuka. "Molecular Size Distribution of Humic Acids as Affected by Ionic Strength and the Degree of Humification," *Soil Sci. Plant Nutr.* 30:151–162 (1984).
3. Kononova, M. H. "Organic Matter and Soil Fertility," *Soviet Soil Sci.* 16:71–76 (1984).
4. Chen, Y., N. Senesi, and M. Schnitzer. "Information Provided on Humic Substances by E_4/E_6 Ratio," *Soil Sci. Soc. Am. J.* 41:352–358 (1977).
5. Roig, A., A. Lax, J. Cegarra, F. Costa, and M. T. Hernandez. "Cation Exchange Capacity as a Parameter for Measuring the Humification Degree of Manures," *Soil Sci.* 146:311–316 (1988).
6. Hayes, M. H. B., and R. S. Swift. "The Chemistry of Soil Organic Colloids," in *The Chemistry of Soil Constituents*, D. J. Greenland and M. H. B. Hayes, Eds. (Chichester, Eng: John Wiley and Sons, 1983) pp. 179–320.
7. Sequi, P., G. Guidi, and G. Petruzzelli. "Frazionamento e caratteristiche di solubilità degli acidi fulvici," *Agrochimica* 16:224–232 (1972).
8. Sequi, P., G. Guidi, and G. Petruzzelli. "Distribution of Amino Acid and Carbohydrates Components in Fulvic Acid Fractionated by Polyamide," *Can. J. Soil Sci.* 55:439–445 (1975).
9. Leenheer, J. A. "Comprehensive Approach to Preparative Isolation and Fractionation of Dissolved Organic Carbon from Natural Waters and Wastewaters," *Environ. Sci. Technol.* 15:578–587 (1981).
10. Thurman, E. M., and R. L. Malcolm. "Preparative Isolation of Aquatic Humic Substances," *Environ. Sci. Technol.* 15:463–466 (1981).
11. Olsson, L., and O. Samuelson. "Chromatography of Aromatic Acids and Aldehydes and Phenols on Cross-Linked Polyvinylpyrrolidone," *J. Chromatogr.* 93:188–189 (1974).
12. Newton Clifford, M. "The Use of Poly-*n*-Vinylpyrrolidone as the Adsorbent for

the Chromatographic Separation of Chlorogenic Acids and Other Phenolic Compounds," *J. Chromatogr.* 94:261–266 (1974).

13. Quarmby, C. "The Use of Polyvinylpyrrolidone in the Thin Layer Chromatography Separation of Flavonoids and Related Compounds," *J. Chromatogr.* 34:52–58 (1968).

14. Lowe, L. E. "Fractionation of Acid-Soluble Components of Soil Organic Matter Using Polyvinylpyrrolidone," *Can. J. Soil Sci.* 35:119–122 (1975).

15. Sequi, P., M. De Nobili, L. Leita, and G. Cercignani. "A New Index of Humification," *Agrochimica* 30:175–179 (1986).

16. Ciavatta, C., M. Govi, L. Vittori Antisari, and P. Sequi. "Characterization of Humified Compounds by Extraction and Fractionation on Solid Polyvinylpyrrolidone," *J. Chromatogr.* 509:141–146 (1990.)

17. Petrussi, F., M. De Nobili, M. Viotto, and P. Sequi. "Characterization of Organic Matter from Animal Manures after Digestion by Earthworms," *Plant Soil* 105:41–46 (1988).

18. Saviozzi, A., R. Levi-Minzi, and R. Riffaldi. "Maturity Evaluation of Organic Waste," *BioCycle* 198:54–56 (1988).

19. De Nobili, M., G. Cercignani, L. Leita, and P. Sequi. "Evaluation of Organic Matter Stabilization in Sewage Sludge," *Commun. Soil Sci. Plant Anal.* 17:1109–1119 (1986).

20. De Nobili, M., and F. Petrussi. "Humification Index (HI) as Evaluation of the Stabilization Degree during Composting," *J. Ferment. Technol.* 66:577–583 (1988).

21. Ciavatta, C., L. Vittori Antisari, and P. Sequi. "A First Approach to the Characterization of the Presence of Humified Materials in Organic Fertilizers," *Agrochimica* 32:510–517 (1988).

22. Cacco, G., A. Maggioni, and G. Ferrari. "Electrofocusing: A New Method for Characterization of Soil Humic Matter," *Soil Biol. Biochem.* 6:145–148 (1974).

23. Cacco, G., and A. Maggioni. "Multiple Forms of Acetylnaphthyl-Esterase Activity in Soil Organic Matter," *Soil Biol. Biochem.* 8:321–325 (1976).

24. Gjessing, E. T., and T. Gjerdhal. "Electromobility of Aquatic Humus: Fractionation by the Use of the Isoelectric Focusing Technique," in *Proceedings of the International Meeting on Humic Substances* (Wageningen, The Netherlands: Pudoc, 1972), pp. 42–51.

25. De Nobili, M., G. Cercignani, and P. Sequi. "A Comparative Study of Soil Organic Matter Evolved under Different Climatic and Pedological Conditions by Means of Isoelectric Focusing," in *Current Perspectives in Environmental Biochemistry*, G. Giovannozzi-Sermanni and P. Nannipieri, Eds. (Rome: CNR-IPRA, 1985), pp. 85–94.

26. Ceccanti, B., P. Nannipieri, and M. T. Bertolucci. "Characterization of Soil Organic Matter and Derivative Fractions by Isoelectric Focusing," *Recent Develop. Chromatogr. Electrophor.* 10:75–81 (1980).

27. Aak, O. V., V. A. Galynkin, A. P. Kaskin, and V. I. Jakovlev. "Isoelektriceskoe fokusiroyanie visokomolekulljiarnych slancevych i guminovych kislot," *Prikladnaja Biochimia i Microbiologija* 20:290–293 (1984).

28. De Nobili, M. "Electrophoretic Evidence of the Integrity of Humic Substances Separated by Means of Electrofocusing," *J. Soil Sci.* 39:437–446 (1988).

29. Govi, M. "Maturazione della sostanza organica negli effluenti degli allevamenti suinicoli," MSc Thesis, University of Bologna (1988).

30. Pasotti, L. "Valutazione della qualità del compost da R.S.U.," MSc Thesis, University of Bologna (1989).

31. Dommergues, Y., and F. Mangenot. *Ecologie microbienne du sol* (Paris: Masson, 1970).

32. Jenkinson, D. S. "The Fate of Plant and Animal Residues in Soil," in *The Chemistry of Soil Processes*, D. J. Greenland and M. H. B. Hayes, Eds. (Chichester, Eng.: John Wiley and Sons, 1981).

33. Sequi, P. "Sostanza organica e ciclo del carbonio," in *Chimica del Suolo*, P. Sequi, Ed. (Bologna, Italy: Pàtron, 1989), pp. 247–277.

34. Bakken, L. R., and R. A. Ocsen. "The Relationship between Cell Size and Viability of Soil Bacteria," *Microb. Ecol.* 13(2):103–114 (1987).

35. Lucas, R. E., J. B. Holtman, and L. J. Connor. "Soil Carbon Dynamics and Cropping Practices," in *Agriculture and Energy*, W. Lockeretz, Ed. (New York: Academic Press, 1977), pp. 505–561.

36. Mutatker, J. K., and G. H. Wagner. "Humification of Carbon-14–Labeled Glucose in Soils of Sanborn Field," *Soil Sci. Soc. Am. Proc.* 31:66–70 (1967).

37. Oades, J. M., and G. H. Wagner. "Biosynthesis of Sugars in Soils Incubated with ^{14}C Glucose and ^{14}C Dextran," *Soil Sci. Soc. Am. Proc.* 35:914–917 (1971).

38. Behera, B., and G. H. Wagner. "Microbial Growth Rate in Glucose Amended Soil," *Soil Sci. Soc. Am. Proc.* 38:591–594 (1974).

39. Jenkinson, D. S., and J. H. Rayner. "The Turnover of Soil Organic Matter in Some of the Rothamsted Classical Experiments," *Soil Sci.* 123:298–305 (1977).

40. Norstadt, F. A., C. R. Frey, and H. Sigg. "Soil Urease: Paucity in the Presence of the Fairy Ring Fungus *Marasmius oreades* (Bolt) Fr," *Soil Sci. Soc. Am. Proc.* 37:880–885 (1973).

41. Ronchi, G. "Alcuni aspetti della dinamica delle trasformazioni dell'azoto inorganico nel suolo," MSc Thesis, University of Bologna (1990).

42. Ciavatta, C., and P. Sequi. "Evaluation of Chromium Release during the Decomposition of Leather Meal Fertilizers Applied to the Soil," *Fert. Res.* 19:7–11 (1989).

43. Ciavatta, C. Unpublished results.

44. Bailly, J. R., and V. Nkundikije-Desseaux. "Sur la formation de substances noires a partir de phenols simples par des microorganismes du sol," *Plant Soil* 43:235–257 (1975).

45. Bollag, J. M. "Synthetic Reactions of Aromatic Compounds by Fungal Enzymes," in *Aquatic and Terrestrial Humic Substances*, R. F. Christman and E. T. Gjessing, Eds. (Ann Arbor, MI: Ann Arbor Science, 1983), pp. 127–141.

46. Leonowicz, A., and J. M. Bollag. "Laccases in Soil and the Feasibility of Their Extraction," *Soil Biol. Biochem.* 19(3):237–242 (1987).

47. Wang, T. S. C., M. C. Wang, Y. L. Ferng, and P. M. Huang. "Catalytic Synthesis of Humic Substances by Natural Clays, Silts, and Soils," *Soil Sci.* 135:350–360 (1983).

48. Shindo, H., and P. M. Huang. "Catalytic Effects of Manganese(IV), Iron(III), Aluminum, and Silicon Oxides on the Formation of Phenolic Polymers," *Soil Sci. Soc. Am. J.* 48:927–934 (1984).

49. Kent Kirt, T., T. Higuchi, and H.-M. Chang. *Lignin Biodegradation Microbiology, Chemistry and Potential Applications* (Cleveland, OH: CRC Press, 1980), pp. 241–255.

50. Gold, M. H., H. Wariishi, and K. Valli. "Extracellular Peroxidase Involved in

Lignin Degradation by the White Rot Basidiomycete *Phanerochaete chrysosporium*," in *Biocatalysis in Agricultural Biotechnology*, J. R. Whittaker and P. E. Sonnet, Eds. (Washington, DC: American Chemical Society, 1989), pp. 127–140.

51. Hayes, M. H. B., P. MacCarthy, R. L. Malcolm, and R. S. Swift, Eds. *Humic Substances II* (New York: John Wiley and Sons, 1989), pp. 689–733.

52. Burns, R. G., A. M. Pukite, and A. D. McLaren. "Concerning the Location and the Persistence of Soil Urease," *Soil Sci. Soc. Am. Proc.* 36:308–311 (1972).

List of Authors

Peter Adriaens, Department of Civil Engineering, Stanford University, Stanford, California 94305

George Aiken, U.S. Geological Survey, 5293 Ward Road, Arvada, Colorado 80002

Gary L. Amy, Civil, Environmental and Architectural Engineering, University of Colorado, Boulder, Colorado 80309

Livia Vittori Antisari, University of Bologna, Institute of Agricultural Chemistry, Via S. Giacomo 7, 40126 Bologna, Italy

Ronald C. Antweiler, U.S. Geological Survey, 5293 Ward Road, Arvada, Colorado 80002

G. Asplund, Department of Water and Environmental Studies, Linköping University, S-581 83 Linköping, Sweden

H. Borén, Department of Water and Environmental Studies, Linköping University, S-581 83 Linköping, Sweden

Stephen A. Boyd, Department of Crop and Soil Sciences, Michigan State University, East Lansing, Michigan 48824-1325

Christopher Cawein, Environmental Engineering Program, University of Arizona, Tucson, Arizona 85721

Yu-Ping Chin, Ralph M. Parsons Laboratory, Massachusetts Institute of Technology, Cambridge, Massachusetts 02139

Cary T. Chiou, U.S. Geological Survey, 5293 Ward Road, Arvada, Colorado 80002

Claudio Ciavatta, University of Bologna, Institute of Agricultural Chemistry, Via S. Giacomo, 40126 Bologna, Italy

Martha H. Conklin, Department of Hydrology and Water Resources, University of Arizona, Tucson, Arizona 85721

369

Harry Efraimsen, Norwegian Institute for Water Research, P.O. Box 69, Korsvoll, 0808 Oslo 8, Norway

Dennis D. Focht, Department of Soil and Environmental Sciences, University of California, Riverside, California 92521

Lewis E. Fox, 108 Pierce Hall, Harvard University, 29 Oxford Street, Cambridge, Massachusetts 02138

Egil T. Gjessing, Norwegian Institute for Water Research, P.O. Box 69, Korsvoll, 0808 Oslo 8, Norway

Magne Grande, Norwegian Institute for Water Research, P.O. Box 69, Korsvoll, 0808 Oslo 8, Norway

A. Grimvall, Department of Water and Environmental Studies, Linköping University, S-581 83 Linköping, Sweden

Richard A. Harnish, U.S. Geological Survey, Water Resources Division, P.O. Box 25046, Mail Stop 408, Denver Federal Center, Denver, Colorado 80225

Chi-Min Huang, Taiwan Sugar Research Institute, 54 Sheng Chan Road, Tainan, Taiwan

Peter R. Jaffé, Department of Civil Engineering and Operations Research, Princeton University, Princeton, New Jersey 08544

William F. Jaynes, Department of Crop and Soil Sciences, Michigan State University, East Lansing, Michigan 48824–1325

Torsten Källqvist, Norwegian Institute for Water Research, P.O. Box 69, Korsvoll, 0808 Oslo 8, Norway

A. C. Koulermos, Department of Chemical Engineering, Louisiana State University, Baton Rouge, Louisiana 70803

J. A. Leenheer, U.S. Geological Survey, 5293 Ward Road, Arvada, Colorado 80002

S. R. Lindner, National Oceanic and Atmospheric Administration, Great Lakes Environmental Research Laboratory, 2205 Commonwealth Blvd., Ann Arbor, Michigan 48105

L. W. Lion, School of Civil and Environmental Engineering, Cornell University, Ithaca, New York 14850

Houmao Liu, Civil, Environmental and Architectural Engineering, University of Colorado, Boulder, Colorado 80309

Debra J. Lonergan, Water Resources Division, U.S. Geological Survey, 430 National Center, Reston, Virginia 22092

Derek R. Lovley, Water Resources Division, U.S. Geological Survey, 430 National Center, Reston, Virginia 22092

Patrick MacCarthy, Department of Chemistry and Geochemistry, Colorado School of Mines, Golden, Colorado 80401

Mike Machesky, Penn State University, Geosciences Department, 208 Deike Building, University Park, Pennsylvania 16802

Diane McKnight, U.S. Geological Survey, 5293 Ward Road, Arvada, Colorado 80002

Laurence Miller, U.S. Geological Survey, 345 Middlefield Road, Mail Stop 465, Menlo Park, California 94025

Robert T. Mueller, Division of Science and Research, New Jersey Department of Environmental Protection, Trenton, New Jersey 08625

S. K. Ong, Battelle Memorial Institute, Environmental Technology Department, 505 King Avenue, Columbus, Ohio 43201

Joseph J. Pignatello, Department of Soil and Water, The Connecticut Agricultural Experiment Station, Box 1106, New Haven, Connecticut 06504

James F. Ranville, U.S. Geological Survey, Water Resources Division, P.O. Box 25046, Mail Stop 408, Denver Federal Center, Denver, Colorado 80225

D. D. Reible, Department of Chemical Engineering, Louisiana State University, Baton Rouge, Louisiana 70803

James A. Rice, Department of Chemistry, South Dakota State University, Brookings, South Dakota 57007-0896

Gunnhild Riise, Isotope Laboratory, P.O. Box 26, N-1432 Ås-NLH, Norway

Brenda S. Ross, Department of Crop and Soil Sciences, Michigan State University, East Lansing, Michigan 48824-1325

Paolo Sequi, University of Bologna, Institute of Agricultural Chemistry, Viale C. Berti Pichat, 10, 40127 Bologna, Italy

James A. Smith, U.S. Geological Survey, 810 Bear Tavern Road, Suite 206, West Trenton, New Jersey 08628

L. J. Thibodeaux, Department of Chemical Engineering, Louisiana State University, Baton Rouge, Louisiana 70803

G. J. Thoma, Department of Chemical Engineering, Louisiana State University, Baton Rouge, Louisiana 70803

David M. Tuck, Department of Geological and Geophysical Sciences, Princeton University, Princeton, New Jersey 08544

K. T. Valsaraj, Department of Chemical Engineering, Louisiana State University, Baton Rouge, Louisiana 70803

Walter J. Weber, Jr., Environmental and Water Resources Engineering, The University of Michigan, Ann Arbor, Michigan 48109

Robert L. Wershaw, U.S. Geological Survey, 5293 Ward Road, Arvada, Colorado 80002

Index

Abiotic hydrolysis, 14
Abiotic oxidation, 331
Abiotic processes, 14. *See also* specific
    types
Acetone, 8, 270
4-Acetoxybenzoic acid, 167
Acetylsalicylic acid (aspirin), 164, 167,
    171
Acid-base titrations, 31, 111–126
    methods in, 113–116
    results of, 116–120
Acid-extracted nitrogen, 151
Acid extraction, 131
Acid hydrolysis, 150
Acid-hydrolyzable protein, 141
Acidic functional group analysis, 39
Acidic polypeptides, 150
Acidification, 129
    H_2SO_4, 89–97
    surface water, 89–90
Acinetobacter spp., 312, 316, 318, 320
Adsorbable organic halogens (AOX),
    339, 341, 342
    net production of, 343
    occurrence of, 340
    sources of, 344
    speciation of, 344, 345
Adsorption, 302
Affinity, 29, 44, 47
Aggregation, 13, 43, 131
Agronomic behavior, 358
AHA. *See* Aldrich humic acid
Alanine, 63, 142, 155
Alcohols, 66. *See also* specific types
Aldehydes, 352
Aldrich humic acid (AHA), 256, 257,
    266, 276
Algae, 47, 48, 63, 85. *See also* specific
    types
    benthic, 76
    biomass of, 49
    blue-green, 80
    carbon fixation by, 75

    green, 80
    growth of, 49
    planktonic, 49
Aliphatic carbon, 67, 84, 141
Aliphatic hydrocarbons, 292. *See also*
    specific types
Aliphatic lipids, 69
Aliphatic structures, 14
Alkaline extracts, 352
n-Alkane, 67
Alkylammonium compounds, 189
Allochthonous material, 11, 47, 48, 52,
    75. *See also* specific types
Alumina, 101, 106, 277, 279, 285
Aluminosilicates, 25, 58
Aluminum, 8, 148, 149, 213
Aluminum oxides, 25, 70, 276
Amberlite XAD resin, 77, 352
Amicon ultrafiltration, 234
Amides, 14. *See also* specific types
Amines, 149, 150. *See also* specific
    types
Amino acid carbon, 151–152
Amino acids, 5, 8, 12, 142, 152, 155.
    See also specific types
    hydrolysis of, 175
    hydrolyzable, 151
    protonated, 32
    total, 158
Aminoglucans, 149
Amino hydrogens, 69
Amino sugars, 14. *See also* specific
    types
Amphiphiles, 24
AMW. *See* Apparent molecular weight
Anaerobic oxidation, 332
Anionic surfactants, 203
Anthropogenic inputs, 11, 327–336
AOX. *See* Adsorbable organic
    halogens
Apparent molecular weight (AMW),
    101
Aqueous activity coefficients, 257, 285

Aqueous partition coefficients, 278
Arazine, 4
Aromatic acids, 352. See also specific
    types
Aromatic amines, 149. See also
    specific types
Aromatic carbon, 84
Aromatic compounds, 256. See also
    specific types
    anthropogenic, 327–336
    heterocyclic, 331
    hydroxylated, 332
    natural, 327–336
    nitrogen-containing, 331
    structures of, 14
Aromatic hydrocarbons, 184, 292, 328
    polycyclic. See Polycyclic aromatic
        hydrocarbons (PAHs)
    polynuclear. See Polycyclic aromatic
        hydrocarbons (PAHs)
Aromaticity, 48, 69, 84
Aromatic pi-electrons, 69
Arthrobacter spp., 316
Aspartic acid, 63, 155
Aspirin, 164, 167, 171
Atomic emission spectroscopy (AES),
    56
Atrizine, 15
Autochthonous material, 47, 48, 49,
    69, 75. See also specific types

Bacteria, 52, 80, 85
Bacterial degradation of POC, 49
Batch equilibration, 204, 221
Batch experiment sorption studies,
    256–257
Benthic algae, 76
Bentonite, 213, 214
Benzene, 196, 197, 216, 218, 222, 262,
    319
Benzidine, 69
2-Benzofuran carboxylic acid, 164,
    167, 170
Benzoic acid, 317, 318
p-Benzoquinone, 327
Benzyldimethylhexadecylammonium,
    219
Benzyltributylammonium (BTBA),
    198

Benzyltriethylammonium (BTEA), 198,
    219, 219–220, 222, 223
Benzyltriethylammonium (BTEA)
    montmorillonite, 224
Benzyltrimethylammonium (BTMA),
    198, 219, 223–224
BET. See Brunauer, Emmett, Teller
Bilayer membranes, 8–9
Binding. See also specific types
    contaminant, 3–5
    equilibrium, 251
    mechanisms of, 3–5
    metal, 23, 24, 28, 111
    NOC, 252
    NOC-colloid, 257–261
    PAH-natural organic matter, 99–109
    site, 31
    solute-polymer, 256
    TCB, 263
    territorial, 31
Biodegradation, 13, 14. See also
    Degradation
    of PCBs. See Polychlorinated
        biphenyls (PCBs), biodegradation
        of
Biomass, 49
Biopolymers, 13. See also specific
    types
Biphenyls, 315, 319. See also specific
    types
Bipyridylium herbicides, 5. See also
    specific types
Bisolute effects, 292, 294–296, 300
Bitumen, 36, 38, 40–43, 44
Blue-green algae, 80
Bond number, 207, 211, 306
Brazilian fulvic acid (BFA), 113, 114,
    118, 119, 120, 122, 124
Brownian motion, 55
Brunauer, Emmett, Teller (BET)
    analysis, 282
Brunt-Vaisala frequency, 81
BTBA. See Benzyltributylammonium
BTEA. See Benzyltriethylammonium
BTMA. See Benzyltrimethyl-
    ammonium

Cadmium, 5, 23, 70, 148, 149

Calcium, 8, 58, 78, 136, 142, 212, 213, 215
  clays and, 225
  concentrations of, 112
  organo-clays and, 182
  replacement of, 216
Calcium montmorillonite, 214, 216, 219
Calorimetric acid-base titrations, 111–126
  methods in, 113–116
  results of, 116–120
Capillary forces, 207, 211
Capillary number, 206, 207, 211
Carbamate pesticides, 5. See also specific types
Carbofuran, 44
Carbohydrates, 11, 13, 14, 15, 66, 69, 141, 158. See also specific types
Carbon, 63, 66, 152. See also specific types
  activated, 291
  aliphatic, 67, 84, 141
  amino acid, 151–152
  aromatic, 84
  from bottom, 132
  dissolved. See Dissolved organic carbon (DOC)
  flux of, 49–52
  humic acid (HAC), 142, 144, 263
  humin organic, 35
  nitrogen ratio to, 11, 62
  organic. See Organic carbon
  oxygen bond to, 66
  particulate organic. See Particulate organic carbon (POC)
  precipitation of, 131
  reactive organic, 138–144
  removal of, 144
  supply of, 358
  total extracted (TEC), 353
  total organic. See Total organic carbon (TOC)
  uncharacterized, 158
  underestimates of, 140
Carbon cycling, 47, 52, 75, 357
Carbon dioxide, 163, 330, 331
Carbon dioxide evolution, 312, 313
Carbon fixation, 75

4-Carboxy-1,2-benzoquinone, 316, 317
Carboxyl groups, 8, 130
Carboxylic acids, 24, 26, 112–113, 124
Carboxylic functionalities, 256
Carboxyl peak, 84
Cascade effect, 211
Cation bridging, 5
Cation exchange capacity (CEC), 184, 187, 194, 213–214, 222
Cation exchange interactions, 4, 5, 69. See also specific types
Cationic surfactants, 202, 212
CEC. See Cation exchange capacity
Cellulose, 363
Centrifugation, 53, 79, 205, 257, 292
Cesium montmorillonite, 214
Charge density, 26
Charge reduction, 131
Charge transfer complexes, 69
Chemical bonding, 129
Chloramil (2,3,5,6-tetrachloro-p-benzoquinone), 4
α-Chlordane, 266, 268
cis-Chlordane, 256
Chlorfenvinphos, 44
Chlorinated hydrocarbons, 276, 282. See also specific types
Chlorinated hydroxybiphenyls (HBPs), 315
Chlorinated insecticides, 205–206. See also specific types
Chlorination, 345, 346
Chlorine, 78, 312
3-Chlorobanzoate, 313
3-Chlorobanzoic acid, 320
4-Chlorobanzoic acid, 316, 319
Chlorobenzene, 291, 292, 301, 319
3-Chlorobenzoic acid, 316, 319, 320
4-Chlorobenzoic acid, 316
Chlorobenzoic acids, 312, 313, 316, 318. See also specific types
Chlorobipehnyl catechol, 320
3-Chlorobiphenyl, 316, 320, 321
4-Chlorobiphenyl, 315, 316, 319
Chlorobiphenyls, 312, 318, 319. See also specific types
3-Chlorocatechol, 320
Chlorocatechols, 312. See also specific types

Chloroform, 276
Chloroform-methanol, 67
Chlorogenic acid, 164, 167, 170
3-Chloro-4-hydroxybenzoic acid, 316, 317
Chloroperoxidase, 345
Chlorophenol, 344
Chlorophyll *a*, 63
Chloropropham, 5
Chromatography, 352. *See also* specific types
   gas, 16, 38, 57, 67, 216, 236, 256, 292, 293
   high performance liquid (HPLC), 329, 330
   paper, 130
Chromium, 70
Citrate, 8
Clay colloids, 15
Clays, 212, 224, 225. *See also* specific types
   BTEA, 198, 219, 219–220, 222, 223
   BTMA, 198, 219, 223–224
   calcium, 225
   cation exchange capacity of, 184, 222
   HDTMA. *See* Hexadecyltri-methylammonium (HDTMA)
   humification and, 360
   organic matter bonding to, 5–9
   organo-. *See* Organo-clays
   organophilic, 185–190
   sodium, 225
   surfactant interaction with, 213–215
   TMA, 184, 185, 186, 195, 198, 212, 219, 221
   TMPA, 184, 185, 198, 199, 200
CMC. *See* Critical micelle concentration
Coagulation, 142
Cobalt, 148
Coculture mineralization, 316–319
Colloids, 16, 57. *See also* specific types
   aggregation of, 131
   clay, 15
   composition of, 251
   concentration of, 246
   dispersion of, 131

estuarine, 69
fraction of, 11
glycoprotein, 15
humic, 26–27
interface by, 247
mineral, 58
NOC binding to, 257–261
NOC sorption and, 262–268
pore-water. *See* Pore-water colloids
solute interactions with, 261
Combustive techniques, 56
Competition, 291, 294, 300, 302, 303, 315
Competitive exchange ions, 188
Competitive exclusion principle, 318
Competitive sorption, 220, 224, 291–305
Complexation, 148
Complexing agents, 8. *See also* specific types
Concentration gradients, 231
Conductionmetry, 158
Conductivity, 76, 207, 225
Conjugate formation, 69
Contaminant-binding mechanisms, 3–5
Contaminant-partitioning interactions, 4
Copper, 5, 48, 112, 148, 149
Cosolute properties, 299–300
Cosorbate effects, 301
Coulomb's Law, 225
Coumalic acid, 164, 167
Coumarin, 164, 171
Coumarin-3-carboxylic acid, 164, 167, 170, 174
Counterions, 26
Covalent bonding, 5
CPMAS. *See* Cross-polarization/magic angle spinning
o-Cresol, 334
p-Cresol, 328, 334
Critical micelle concentration (CMC), 201, 203, 206
Crossflow ultrafiltration, 53, 54
Cross-polarization/magic angle spinning (CPMAS)-NMR, 56, 63, 66, 67
Cryohemists, 36

Cryptophytes, 53
Crystalline enzyme, 363
Cyclic mobilization, 211
Cyclohexane, 276
Cysteine, 5

Daltons, 130
Daphnia spp., 49
DBCP, 300
DCB, 257
1,4-DCB, 262
DDT, 15, 27, 112, 203
Deflocculation, 13
Defoaming agents, 201
Degradation, 75, 85, 130
  bacterial, 49
  bio-. *See* Biodegradation
  of chlorobenzoic acid, 312
  of *o*-cresol, 334
  kinetics of, 311
  of lignin-containing plants, 83–84
  oxidative, 13
  partial, 24
  photo-, 75
  of POC, 49
  thermal, 16
Degradative plasmids, 319
Dehalogenation, 318
Demulsifiers, 201
Depolymerization, 360
Depth profiles, 76, 78, 80
Desorption, 234. *See also*
    Sorption/desorption
  kinetics of, 235
  retarded, 294
  of TOC, 269
Detergents, 201
Dextran, 358
Dialysis, 70, 102, 256
Diammonium cation, 214
Diatoms, 53
Dibenzofurans (PCDfs), 311
Dibromoethene, 216
Dicarboxylic acid, 5
3,4-Dichlorobanzoate, 317
Dichlorobanzoic acid, 316
p-Dichlorobenzene, 256
m-Dichlorobenzene, 319
1,3-Dichlorobenzene, 291

1,4-Dichlorobenzene, 291, 319
3,4-Dichlorobenzoate, 317
3,3'-Dichlorobiphenyl, 320
3,4-Dichlorobiphenyl, 317, 318
4,4'-Dichlorobiphenyl, 316, 317, 318
3,5-Dichlorocatechol, 319
1,1-Dichloromethylene, 276
2,4-Dichlorophenoxyacetate, 319
2,4-Dichlorophenoxy acetic acid, 312
Diethylammonium cations, 216
Diffusion, 86, 224–225, 246, 312
  coefficient of, 237, 239, 246
  Fickian model of, 238
  of fulvic acid, 75–86
  model of, 237
  molecular, 231, 247
  slow molecular, 294
Diffusive transport, 85
1,2-Dihydroxybenzene, 113
2,5-Dihydroxybenzoate, 335
3,4-Dihydroxybenzoate, 335
3,4-Dihydroxybenzoic acid, 316
2,3-Dihydroxybiphenyl, 316
Dimethylammonium cations, 216
N,N-Dimethylforamide, 8
Dimethylsulfoxide, 8
Diquat, 5
Direct cation-exchange interactions, 5
Disaggregation, 37, 38
Dispersants, 201
Dispersion, 131, 224–225
Displacement, 207, 296
Displacing-fluid viscosity, 211
Dissociation constants, 125
Dissolution, 282, 285, 292
Dissolved organic carbon (DOC), 49,
    133
  accumulation of in overlying water,
    236
  chemical characteristics of, 75,
    82–84
  chemistry of, 86
  composition of, 140
  concentration of, 62, 234, 236,
    237–239
  depth profiles for, 76, 80
  diffusion coefficient for, 237, 239
  diffusivities of, 239
  effect of on transport, 245–247

high levels of, 100
humic substances in, 129
increasing concentration of, 237
measurement of, 101, 132, 140
partitioning of, 235, 242
pore-water, 233–234
profile of, 81–82
reactive, 136
reactive fraction of, 141
remobilization of, 75
removal of, 136, 144
runoff and, 134
size distribution of, 236–237
size fractionation of, 234
sorption of, 242
sorption/desorption of, 234–235,
    239–242
sources of, 75, 82, 85–86
steady-state, 242
transport of, 132, 235–236, 239
uncharacterized, 158
water column stability and, 81–82
Dissolved organic compounds (DOCs),
    233
Dissolved oxygen, 11, 76
Dissolved solutes, 77
DOC. *See* Dissolved organic carbon
DOCs. *See* Dissolved organic
    compounds
Dodecyldiammonium cations, 214
Dodecyldimethyl(2-phenoxyethyl)-
    ammonium, 219
Dodecyltrimethylammonium, 219, 221
Donnan potential, 31
Drop weight method, 208
Dynamic equilibrium, 29

Easily oxidizable organic matter
    (EOOM), 233
ECD. *See* Electron capture detection
EDB, 293, 294, 295, 296, 300, 303,
    304
EDX. *See* Energy dispersive X-ray
EF. *See* Electrofocusing
EGME. *See* Ethylene Glycol
    Monoethyl Ether
Electric potential, 354
Electrofocusing, 353–354, 356, 360
Electron acceptors, 329, 331, 333

Electron capture detection (ECD), 257,
    345
Electrophoresis, 351
Electrophoretic mobility, 56, 57, 58,
    101
Electrostatic forces, 5, 225. *See also*
    specific types
Elemental analysis of humin, 38–39
Emphos CS-136, 208
Emulsifiers, 201
Energy. *See also* specific types
    fluxes in, 358
    free, 125, 253
    Gibbs free, 30
    Helmholz free, 30
    internal, 30
    linear free, 253
    reaction free, 112
Energy dispersive X-ray (EDX)
    analysis, 56, 58
Enhanced oil recovery (EOR), 206,
    211
Enthalpies, 112, 116, 119, 120, 122,
    123, 124, 291. *See also* specific
    types
    differences in, 125
    ionization, 113
Entropy, 8, 253
Enzymes, 32. *See also* specific types
    crystalline, 363
    extracellular soil, 364
    proteolytic, 364
    purification of, 363
EOOM. *See* Easily oxidizable organic
    matter
EOR. *See* Enhanced oil recovery
EPM. *See* Electrophoretic mobility
Equilibration, 204, 205, 221, 257, 276.
    See also specific types
Equilibrium, 16, 221, 252
    colloidal humic dispersions and, 145
    dynamic, 29
    ionic strength and, 111
    partial, 29, 30
    standard, 30
    thermodynamic, 31
Equilibrium binding, 251
Equilibrium constants, 29, 251
Equilibrium partitioning, 234–235

Equilibrium sorption, 242, 293, 295–296
Ester hydrolysis, 172, 173
Esterification, 15, 171, 174
Esters, 14, 15. *See also* specific types
    fatty acid methyl (FAMEs), 40, 43
    hydrolyzing, 167
    nonyl phenyl phosphate, 208
    phosphate, 14
    sulfate, 14
Estuaries, 13, 14, 16
    colloidal organic material in, 69
    colloids in, 69
    humic substances in, 129–158
        kinetics of, 144–147
        metals and, 148–152
        reactive organic carbon and, 138–144
    metals in, 148–152
Ethers, 66, 130. *See also* specific types
Ethylbenzene, 291
Ethylene Glycol Monoethyl Ether (EGME), 281–282
Exchangeable cations, 214
Extracellular organic material, 52. *See also* specific types
Extracellular soil enzymes, 364
Extracting agents, 8. *See also* specific types
Extraction, 16, 17, 48, 331. *See also* specific methods
    acid, 131
    humic acid, 141
    lipid, 56
    solutions for, 24
    solvent, 105, 311

FAMEs. *See* Fatty acid methyl esters
Fate of contaminants, 44
Fatty acid methyl esters (FAMEs), 40, 43
Ferrozine, 158
Fertilizers, 351–364
    electrofocusing and, 353–354, 356, 360
    humification and. *See* Humification
    quality criteria for organic matter in, 352–356
Ferulic acid, 329, 331

Fickian diffusion model, 238
FID. *See* Free induction decay
Filtration, 70, 89. *See also* specific types
Flocculate, 15
Flory-Huggins concept, 252, 253, 254, 255, 260, 261
Fluid-fluid interfaces, 203
Fluid-solid interfaces, 203
Fluorescence, 76, 80, 85, 158
Fluorescence quenching, 102, 103
Fluorometry, 149
Flushing experiments, 207, 210, 211
Foaming agents, 201
FQ. *See* Fluorescence quenching
Free energy, 30, 112, 125, 253
Free glycoproteins, 11
Free induction decay (FID), 78, 236
Fulvic acid, 8, 13, 24, 76–77, 130. *See also* Humic substances
    Brazilian (BFA), 113, 114, 118, 119, 120, 122, 124
    calorimetric acid-base titrations of, 111–126
        methods in, 113–116
        results of, 116–120
    characterization of, 37, 76
    chemical characterization of dissolved, 76
    defined, 129, 163
    diffusion of, 75–86
    dissolved, 76
    extracts of, 28
    fluorescence of, 158
    humic acid interaction with, 26, 28
    humic acid ratio to, 352
    isolation of, 37, 77, 174
    molecular structure of, 203
    in organic carbon, 136
    prediction of, 261
    structure of, 5, 8, 14, 16
    Suwannee River. *See* Suwannee River fulvic acid (SRFA)
Fungi, 80

Gas chromatography, 16, 38, 57, 67, 216, 236, 256, 292, 293
Gel filtration, 89
Gibbs free energy, 30

Glucuromolactone, 164
Glutamic acid, 63
Glycine, 63, 142
Glycoproteins, 11, 15. *See also* specific
    types
Glycosides, 14
Gold complexes, 148
Gravitational force, 81, 207, 211
Green algae, 80
Groundwater migration, 212
Groundwater runoff, 49

HAC. *See* Humic acid carbon
Halogenated benzene, 262
Halogenated hydrocarbons, 292. *See
    also* specific types
Halogenation, 339, 345–348
Halogens, 89. *See also* specific types
    adsorbable organic. *See* Adsorbable
        organic halogens (AOX)
    distribution of, 344
    organo-. *See* Organohalogens
Halometabolites, 339, 344
HAN. *See* Humic acid nitrogen
HBPs. *See* Chlorinated
    hydroxybiphenyls
HDTMA. *See* Hexadecyltri-
    methylammonium
Heat of vaporization, 255
Helmholz free energy, 30
Henry's constant, 280
Henry's law, 284, 285
Herbicides, 5, 16, 32, 44. *See also*
    specific types
Heterocyclic aromatic compounds,
    331
Hexachlorobiphenyl, 292
Hexadecyltrimethylammonium
    (HDTMA), 184, 185, 187, 188,
    191, 193, 194, 195, 216, 219, 221,
    224
    alkyl chains of, 217, 219
    benzene sorption to, 218
    cation exchange capacity of, 214
    high-charge vs. low-charge, 196
    in soil restoration, 199
Hexadecyltrimethylammonium
    (HDTMA) bromide, 203

Hexadecyltrimethylammonium
    (HDTMA) smectite, 185–190
Hexane, 257, 293
High performance liquid
    chromatography (HPLC), 329,
    330
HOCs. *See* Hydrophobic organic
    compounds
Hollow-fiber ultrafiltration, 70
HPLC. *See* High performance liquid
    chromatography
HS. *See* Humic substances
H_2SO_4 acidification, 89–97
HTMA. *See* Hexadecyltri-
    methylammonium
Humic acid, 8, 13, 24, 142, 252. *See
    also* Humic substances; specific
    types
    aggregation of, 43
    Aldrich (AHA), 256, 257, 266, 276
    bound, 37, 43
    characterization of, 37
    combustion of, 140
    defined, 129, 163
    displacement experiments on, 296
    extraction of, 141
    extracts of, 28
    fulvic acid interaction with, 26, 28
    fulvic acid ratio to, 352
    functionalities of, 256
    isolation of, 37, 174
    molecular structure of, 203
    organic carbon and, 136
    organic nitrogen and, 149
    partitioning into, 15
    as percentage of humic substances,
        132
    peroxidase-mediated iodination of,
        345
    precipitation of, 131
    preparation of, 292
    release of, 131
    removal of, 136
    size distribution of dispersions of,
        145
    sorption and, 262
    structure of, 14, 16
    total, 158

Humic acid carbon (HAC), 142, 144, 263
Humic acid-fulvic acid aggregates, 26, 28
Humic acid nitrogen (HAN), 142
Humic coatings on sediments, 25–26
Humic colloids, 26–27
Humic micelles, 27–28
Humic polymers, 258, 261
Humic protein, 152
Humic substance-mineral particle complexes, 23–32
    modeling of, 24–29
    thermodynamics of, 29–32
Humic substances. *See also* Fulvic acid; Humic acid; Humin; specific types
    abundance of, 129
    biological properties of, 89
    change in properties of, 89–97
    chemical properties of, 89, 129
    chlorination of, 346
    complexation of, 5
    concentration of, 163
    defined, 129, 352
    in estuaries, 129–158
        kinetics of, 144–147
        metals and, 148–152
        reactive organic carbon and, 138–144
    H_2SO_4 acidification and, 89–97
    integrity of during electrofocusing, 354
    isolation of, 163
    metal binding by, 111
    natural, 148
    natural halogenation of, 345–348
    phase concept for, 28–29
    physicochemical nature of, 89
    sulfated, 95–97
    sulfur associated with, 95
    transport of in estuaries. *See* under Transport
Humidity, 279
Humification, 352–353, 357–364
    considerations about, 360–364
    index of, 353
    organic compound addition and, 359–360

parameters of, 354–356
Humin, 24, 35–44. *See also* Humic substances
    acidic functional group analysis of, 39
    affinity of, 44
    components of, 35, 36
    defined, 35
    disaggregation of, 37, 38
    elemental analysis of, 38–39
    isolation of, 35, 37
    materials for studying, 36–38
    methods of studying, 36–38
    modeling of, 43–44
    oxidative degradation of, 13
    samples of, 36–37
Humin organic carbon, 35
HYDRAQL, 81
Hydraulic conductivity, 207, 225
Hydrocarbons, 11. *See also* specific types
    aliphatic, 292
    aromatic, 184, 292, 328
    chlorinated, 276, 282
    halogenated, 292
    mobilization of residual, 206
    partition coefficients of, 282
    polycyclic aromatic. *See* Polycyclic aromatic hydrocarbons (PAHs)
    polynuclear aromatic. *See* Polycyclic aromatic hydrocarbons (PAHs)
    saturation of, 206
    volatile, 285
Hydrochloric acid, 164
Hydrogen, 69, 113, 116, 213
Hydrogen bonding, 4, 5, 8, 69, 113
Hydrogen ion activity, 116
Hydrologic variables, 11
Hydrolysis, 14–15, 149, 151
    abiotic, 14
    acid, 150
    amino acid, 175
    ester, 172, 173
    protein, 149, 150
    of Suwannee River fulvic acid, 163–175
        equipment for, 164
        materials for, 164
        results of, 165–172

Hydrolyzing esters, 167
Hydrophobicity, 247
Hydrophobic organic compounds
    (HOCs), 233, 239
Hydrophobic partitioning, 69, 303
Hydrophobic phase partitioning, 291
Hydroquinone, 327
Hydrous iron oxides, 25
Hydroxides, 163. See also specific
    types
o-Hydroxybenzoate, 334
m-Hydroxybenzoate, 333
4-Hydroxybenzoic acid, 316
Hydroxybiphenyls, 315, 316. See also
    specific types
4-Hydroxybiphenyls, 316
Hydroxyl functionalities, 256
Hydroxyl groups, 25
Hydroxy-PCBs, 315
Hydroxy-quinone, 130
Hysteresis, 168, 172

ICP. See Inductively coupled plasma
IEP. See Isoelectric point
Illite, 185–190, 191, 224
Immunoassays, 16
Incineration, 311
Index of refraction, 255
Inductively coupled plasma-atomic
    emission spectroscopy (ICP-AES),
    56
Infrared spectral analysis, 130, 172,
    175, 256
Insecticides, 205–206. See also specific
    types
In situ dialysis, 70
Interdisciplinary research, 16
Interfacial tensions, 203, 207, 208
Internal energy, 30
In vivo fluorescence, 76, 80, 85
Ion condensation, 26
Ion exchange, 14, 15, 16, 198, 344
Ionic interactions, 31–32. See also
    specific types
Ionic strength, 111, 112, 149
Ionization, 120, 122, 125
Ionization constants, 31
Ionization enthalpies, 113
Iron, 8, 14, 48, 58, 63

dispersed, 149
ferric, 151, 152
reduction of, 327–336
size distribution of, 70
suspended, 150, 153, 158
Iron complexes, 148, 149. See also
    specific types
Iron oxides, 25, 70
Isoelectric point (IEP), 354
Isolation, 17
of fulvic acid, 37, 77, 174
of humic acid, 37, 174
of humic substances, 163
of humin, 35, 37
methyl isobutyl ketone (MIBK)
    method of, 35, 36, 37, 43, 44
peroxidase, 345

Kaolinitic sediments, 15
Kerogen, 13
KHP. See Potassium acid phthalate
Kinematic viscosity, 207
Kinetics, 16, 251, 312
degradation, 311
desorption, 235
of humic substances in estuaries,
    144–147
linear growth, 312
of metal binding, 111
Monod, 313, 318
of sorption, 235

Labile sugars, 12. See also specific
    types
Lactone, 130
Lake Fryxell, Antarctica, 75–86
Leachates, 270
Lead complexes, 148
Leaf fall, 11
Leucine, 155
Ligand exchange, 5, 8
Lignin monomers, 329
Lignins, 12, 83, 130, 329, 352
Lignoceric acid, 40
Limnological measurements, 76–77,
    78–80
Lindane, 223, 291
Linear partitioning, 275–288
Linear regression, 258, 278

Linuron, 15
Lipids, 9, 13, 14, 47, 48. *See also*
    specific types
    aggregation of, 43
    aliphatic, 69
    bound, 38, 40–43, 43, 44
    characteristics of, 38, 40–43, 66–67
    extracting agents for, 8
    extraction of, 56
Liquid lattice statistical approach, 253
Liquid-phase partition coefficients,
    277
Liquid scintillation, 205
London forces, 255
Low vapor concentrations studies,
    279–283

Macrophytes, 47. *See also* specific
    types
Magnesium, 58, 136, 142, 182, 198,
    213
Maillard reaction, 13
Maleic anhydride refractive index, 256
Manganese, 58, 63, 70, 149
Manganese oxides, 70, 360
Manganese oxyhydroxide, 14
Marinsky model, 31
Mass spectrometry, 17, 38, 57, 67, 256
MBTH. *See* 3-Methyl-2-benzo-
    thiazolinone hydrochloride
Melanoidin, 13
Membranes, 26. *See also* specific
    types
    bilayer, 8–9
    charge density of, 26
    defined, 24
    humic micelle interactions with,
        27–28
    negatively charged surfaces of, 31
Mercury, 5, 23, 148
Metabolism
    of benzene, 319
    of *p*-cresol, 334
    of methylbenzoate, 319
    of nicotinic acid, 332
    of PCBs, 315–316
Metallothioniens, 5. *See also* specific
    types
Metals, 5. *See also* specific types

binding of, 23, 24, 28, 111
complexation of, 148
in estuaries, 148–152
humic substance complexation with,
    5
oxyhydroxides of, 14
paramagentic, 63
photoreduction of, 14
sorption of, 48
thermodynamics of ionic reactions
    in, 27
Methanogenic conditions, 334
Methanol, 67, 270
Methine, 66
Methylbenzoate, 319
3-Methyl-2-benzothiazolinone
    hydrochloride (MBTH), 155
Methyl carbon, 66
Methylcyclohexane, 276
Methylcyclopentane, 276
Methylene, 66
Methylene chloride, 276
Methyl functionalities, 256
Methyl isobutyl ketone (MIBK)
    isolation method, 35, 36, 37, 43,
        44
Methyl salicylate, 256
MIBK. *See* Methyl isobutyl ketone
Micelles, 26, 28
    critical concentration of surfactants
        in, 201, 203, 206
    defined, 99–100
    humic, 27–28
    membrane interactions with, 27–28
    negatively charged surfaces of, 31
Microbial activity, 75. *See also* specific
    types
Microbial cells, 63
Microbial oxidation, 327–336
Mineral colloids, 58
Mineralization of PCBs, 312, 316–321
Mineral particle-humic substance
    complexes. *See* Humic
    substance-mineral particle
    complexes
Minerals. *See also* specific types
    modeling of, 101
    vapor sorption onto, 275–288
Mobility, 201–226

decrease in, 211–225
electrophoretic, 56, 57, 58, 101
increase in, 203–211
non-aqueous-phase liquid (NAPL),
    206–211
Models. *See also* specific types
bilayer membrane, 8–9
diffusion, 237, 238
of DOC effect on transport,
    245–247
Fickian diffusion, 238
of humic substance-mineral particle
    complexes, 24–29
of humin, 43–44
hydrophobic partitioning, 291
of minerals, 101
of PAHs, 100
of partition-interaction, 268
of polymer-NOC interaction, 252
thermodynamic partition, 251–270
Molecular diffusion, 231, 247, 294
Molecular sieving, 302, 303
Molecular weight, 89, 101, 129, 141
Monochlorobiphenyls, 315
Monochlorophenol, 344
Monod kinetics, 313, 318
Monoethylammonium cations, 216
Monomers, 28
Monomethylammonium, 214, 216, 220,
    221
Monooxygenase, 316
Montmorillonite, 212, 213, 214, 215,
    216, 219
advantages of, 224
benzyltriethylammonium, 224
monomethylammonium, 220
TMA, 222, 223
MS. *See* Mass spectrometry
Mucopolysaccharides, 13, 52
Mullica Estuary, 153–158

NADH, 315, 316
Naphthalene, 197
Naphthalene partitioning, 236, 244
NAPLs. *See* Non-aqueous-phase
    liquids
Natural aromatic compounds,
    327–336
Natural humic substances, 148

Naturally produced organohalogens,
    339–348
Natural organic matter (NOM), 218,
    327
characterization of, 100–101
PAH binding to, 99–109
solubility-enhancing effect of, 204
sources of, 100–101
Neutron activation analysis, 341
Nickel complexes, 148, 149
Nickel electron capture detection, 257
Nicotinic acid, 331, 332
Nitrogen, 62, 63, 69, 89
acid-extracted, 151
carbon ratio to, 11, 62
heterocyclic aromatic compounds
    containing, 331
humic acid (HAN), 142
organic, 149
in particulate fraction, 12
particulate organic (PON), 155
NMR. *See* Nuclear magetic resonance
NOCs. *See* Nonionic organic
    contaminants
NOM. *See* Natural organic matter
Non-aqueous-phase liquids (NAPLs),
    206–211, 285
Nonequilibrium sorption, 292,
    294–295
Nonideal behavior, 303–304
Nonionic-compound sorption by
    organo-clays, 216–224
Nonionic organic contaminants
    (NOCs)
activity coefficients of, 259
binding of, 252
colloid binding to, 251, 257–261
colloids and sorption of, 262–268
competitive sorption of, 291–305
equilibrium binding of, 251
humic polymer interactions with,
    261
partitioning of, 252
partition medium for, 194
polymer interactions with, 252
removal of, 182, 191
residual, 269
retention of, 189
solubility of, 182, 260

sorption of, 181, 262–268
transport of, 224
uptake of, 181, 189
Nonionic surfactants, 203
Nonpolar organic contaminants. *See*
    Nonionic organic contaminants
    (NOCs)
Nonpolar partitioning, 4
Nonproteinous nitrogen compounds,
    149. *See also* specific types
Nonyl phenyl phosphate esters, 208
Nonyltrimethylammonium, 221
Nuclear magnetic resonance (NMR),
    16, 28, 256
  aliphatic carbon and, 141
  carbon, 16, 48, 68
  DOC chemical charcterization and,
    84
  humin analysis and, 39–40, 43
  Lake Fryxell and, 77
  of Lake Fryxell samples, 85
  organic carbon functionality and,
    56
  cross-polarization/magic angle
    spinning (CPMAS), 56, 63, 66,
    67
  humic substance components and,
    130
  humification and, 360
  humin analysis and, 39–40, 43
  hydrogen, 13, 141
  of Lake Fryxell samples, 84, 85
  organic carbon functionality and,
    56
  Suwannee River fulvic acid and,
    168
Nutrient cycles, 357

Octanol, 251, 259
n-Octanol, 203
Octanol-water partition coefficients,
    203, 257, 259, 312
Octyldiammonium cations, 214
Oil recovery, 206, 211
Oligotrophic lake/reservoir ecosystems,
    49–52. *See also* specific types
Orbisphere probes, 76
Organic acids, 14. *See also* specific
    types

Organic amendments, 354–356
Organic carbon, 11, 218, 291
  allochthonous inputs of, 75
  analysis of, 62
  bound, 240
  combustion of, 140
  concentration of, 234
  dissolved. *See* Dissolved organic
    carbon (DOC)
  fraction, 257
  functionality of, 56
  humin, 35
  isotherms of, 216
  measurement of, 138
  particulate, 12, 47, 48, 85, 154–155
  reactive, 138–144
  in soil, 204
  sources of, 75
  total. *See* Total organic carbon
    (TOC)
  transport of, 129
Organic fertilizers, 351–364
  electrofocusing and, 353–354, 356,
    360
  humification and. *See* Humification
  quality criteria for organic matter in,
    352–356
Organic nitrogen, 149, 155
Organo-clays, 181–200, 212. *See also*
    Clays; specific types
  applications of, 198–200
  design of, 199
  expansion of, 225
  nonionic-compound sorption by,
    216–224
  preparation of, 182–184
  soil modification and, 191–195
  as surface adsorbents, 195–198
Organohalogens, 339–348
  natural production of, 340–341
  occurrence of, 340
  origin of, 341–344
Organophilic clays, 185–190
Oxalate, 8
Oxidation, 256
  abiotic, 331
  anaerobic, 332
  high-temperature, 131
  of *o*-hydroxybenzoate, 334

of *m*-hydroxybenzoate, 333
microbial, 327–336
of organic acids, 14
persulfate, 140
Oxidative degradation, 13
Oxides, 291. *See also* specific types
Oxygen, 11, 66, 76, 163
Oxyhydroxides, 14

PAHs. *See* Polycyclic aromatic
hydrocarbons
Paper chromatography, 130
Paramagnetic metals, 63
Parathion, 291
Partial degradation, 24
Partial equilibrium, 29, 30
Particle size of sediment, 9–11
Particulate organic carbon (POC), 12,
47, 48, 85, 144, 154–155
allochthonous, 52
autochthonous, 49
bacterial degradation of, 49
chemical characteristics of, 48
in Pueblo Reservoir, 69
transport and, 68
watershed, 49
Particulate organic nitrogen (PON),
155
Partition coefficients, 244, 269
aqueous, 278
of hydrocarbons, 282
liquid-phase, 277
octanol-water, 203, 257, 259, 312
saturated, 284
thermodynamic, 257
vapor-phase, 277, 282
Partitioning, 4, 15, 28, 48, 99, 220,
221, 302
decrease in, 261
DOC, 235, 242
equilibrium, 234–235
Flory-Huggins equation for, 252
hydrophobic, 69, 291, 303
linear, 275–288
naphthalene, 236, 244
of NOCs, 252
nonideal, 303
nonpolar, 4
of PCBs, 311–312

between phases, 30–31
solid-liquid, 280
solubilization reaction and, 203
Partition-interaction model, 268
Partition medium, 194, 218, 224
PCBs. *See* Polychlorinated biphenyls
PCDDs. *See* Polychlorinated
dibenzo-*p*-dioxins
PCDFs. *See* Dibenzofurans
PCE. *See* Tetrachloroethene
PCS. *See* Photon correlation
spectroscopy
Peeling process, 163
Pentachlorophenol, 292, 301
Pentyldiammonium cations, 214
Peptides, 11, 14. *See also* specific
types
Perchloroethylene, 276
Periphyton, 47, 52
Permeability, 224–225
Peronate L, 204
Peroxidase isolation, 345
Persulfate oxidation, 140
Pertronate HL, 204
Pesticides, 5, 206, 270. *See also*
specific types
Petroleum sulfonate surfactants, 203
pH, 168
changes in, 14, 149
drift in, 164
humic size distributions and, 145
humic substance interactions and,
112
metal binding and, 111
solution, 116
Phase concept for humic substances,
28–29
Phenanthrene, 99
Phenolic acids, 15, 112–113
Phenolic groups, 26
Phenolic hydrogen, 113
Phenolic hydroxyl groups, 130
Phenolic polymers, 360
Phenolic substances, 364
Phenols, 14, 328, 333, 352. *See also*
specific types
Phenylacetate, 335
Phosphate complexes, 149
Phosphate esters, 14

Photodegradation, 75
Photolysis, 14
Photon correlation spectroscopy
    (PCS), 55, 57
Photooxidation, 14
Photoreduction, 14
Phthalate, 5
o-Phthalic acid, 334, 335
p-Phthalic acid, 335
m-Phthalic acid, 335
Phthalic acids, 334. See also specific
    types
Physical-chemical interactions, 24
Physical sorption, 8
Physicochemical properties, 224, 259
Phytoplankton, 47, 53, 76
Phytotoxicity, 351
Picloram, 44
Pi-electrons, 69
Pigments, 47
Planktonic algae, 49
Plasmids, 319
Plasticizing, 302
PMA. See Polymaleic acid
POC. See Particulate organic carbon
Polyacrylic acid, 125
Polycarboxyl-phenolic aromatics, 130
2,4,4'-Polychlorinated biphenyl, 261
2,5,2'-Polychlorinated biphenyl, 256,
    265, 267, 268
Polychlorinated biphenyls (PCBs), 27,
    69, 112, 205, 270. See also specific
    types
    biodegradation of, 311-323
    applications of, 321-323
    coculture mineralization and,
        316-319
    metabolism and, 315-316
    products of in soil, 311-315
    strain construction strategies
        and, 319-321
    coculture mineralization of,
        316-319
    cometabolism of, 311, 313
    congeners of, 292, 316-319
    degradation of, 311, 317
    isomers of, 4
    kinetics of degradation of, 311
    metabolism of, 311, 315-316

mineralization of, 312, 316-321
partitioning of, 311-312
strain construction strategies for
    mineralization of, 319-321
Polychlorinated dibenzo-p-dioxins
    (PCDDs), 292, 301, 311
Polycondensation reactions, 13
Polycyclic aromatic hydrocarbons
    (PAHs), 16, 69, 99-109, 206, 270.
    See also specific types
    modeling of, 100
    NOM sources and, 100-101
    subsurface transport of, 108
Polyelectrolytes, 118, 129
Polyions, 26
Polymaleic acid (PMA), 256, 258, 266
Polymeric gels, 31
Polymerization, 363
Polymers, 13, 270. See also specific
    types
    humic, 258, 261
    NOC interaction with, 252
    nonlinear, 254
    phenolic, 360
    polar, 255
    solute binding to, 256
Polynuclear aromatic hydrocarbons.
    See Polycyclic aromatic
    hydrocarbons (PAHs)
Polypeptides, 150, 151, 352
Polysaccharides, 47, 163, 352. See also
    specific types
Polyvinylpyrrolidone (PVP), 352, 353
PON. See Particulate organic nitrogen
Pore-water colloids, 231-248
    DOC concentration profile and,
        237-239
    DOC size distribution and, 236-237
    DOC sorption/desorption and,
        234-235, 239-242
    DOC transport and, 235-236
    experiments with, 233-236
    modeling of, 245-247
    naphthalene partitioning and, 236,
        244
    sediment physical properties and,
        233-234
Potassium, 58
Potassium acid phthalate (KHP), 164

Precipitation, 131, 292
Primary amines, 149. *See also* specific types
Primary productivity, 11, 14
Productivity, 11, 14
Protein hydrolysis, 149, 150
Proteins, 5, 8, 11, 13, 14, 47, 363. *See also* specific types
  acid-hydrolyzable, 141
  humic, 152
  solubility of, 15
  stabilized, 364
Proteolytic enzymes, 364
Protocatechuate, 316
Protonated amino acids, 32
Protonation, 119, 122, 123, 124
Proton consumption, 116
Proton release, 111
Pseudomonas spp., 315, 316, 319, 320
Pueblo Reservoir, 47-70
  carbon flux in, 49-52
  chemical characteristics of, 55-58, 62-66
  description of, 52-53
  lipid characterization in, 66-67
  physical characteristics of, 55-58
  sample collection in, 52-53
  sample fractionation from, 53-55
PVP. *See* Polyvinylpyrrolidone
Pyridine, 8
Pyrolysate, 344
Pyrolysis, 256
Pyronate 40, 204
Pytophosphate, 8

Quaternary ammonium cations, 212, 214, 215, 222, 224
Quaternary ammonium cations (QUATS), 184, 185, 186, 187, 195, 198
Quaternary ammonium compounds, 216, 224
QUATS. *See* Quaternary ammonium cations
Quinone, 130

Raoult's law, 251, 252, 253
Reaction free energies, 112
Reactive DOC, 136

Reactive organic carbon, 138-144
Reduction of sulfate, 334
Refractive index, 255, 256
Regression analysis, 260
Relative humidity, 279
Research needs, 16-17
Residual saturations, 207
Residue on evaporation (ROE), 81
Resins, 77, 352. *See also* specific types
Reverse-phase separation method, 102
Rewetting agents, 201
ROE. *See* Residue on evaporation
Rotifers, 49
Runoff, 49, 134

Salicylate, 5
Salicylic acid, 113
Salinity, 14, 134, 144, 149, 158. *See also* entries beginning with Salt
  changes in, 11
  coagulation and, 142
  constant, 155
  gradients of, 76
  humic size distributions and, 145
  increasing, 16
  increasing concentrations of, 15
Salt extracts, 149, 158. *See also* Salinity
Salt filter residue (SFR), 141, 144
Salting in, 14, 15, 16
Salting out, 14, 15, 16
Saturated partition coefficients, 284
Saturation constant, 313
Saturations, 207
Scanning electron microscopy (SEM), 56, 57, 58
Scatchard-Hildebrand equation, 255
Seasonal changes, 11
Secondary productivity, 11
Sediment
  defined, 23
  DOC transport in, 235-236
  humic coatings on, 25-26
  kaolinitic, 15
  NOC sorption by, 262-268
  particle size of, 9-11
  physical properties of, 233-234
  suspended, 239-242
Sedimentation, 13, 182

SEM. *See* Scanning electron microscopy
Semiinfinite slab relationship, 247
Sensitivity analysis, 263
Separation, 102, 269. *See also* specific types
Sep-Pak method, 102
Serum albumin, 151, 155
Settling, 53
SFR. *See* Salt filter residue
Sieving, 302, 303
Silt, 11, 57
Silver, 5
SiO₄, 213
Site binding, 31
Smectites, 185–190, 195–198, 199, 200
Snowmelt, 11
Sodium, 78, 182, 212, 213, 215, 225
Sodium azide, 257
Sodium dodecyl sulfate, 203
Sodium hydroxide, 164
Sodium montmorillonite, 213, 214, 216, 219
Soil. *See also* specific types
    competition in, 291
    competitive sorption of NOCs by, 291–305
    degradation products of PCBs in, 311–315
    geographic distribution of, 352
    humification of. *See* Humification
    modification of, 191–195
    PCBs in, 311–315
    restoration of, 199
    stabilization of, 199
    surface adsorption in, 302
    vapor sorption onto, 275–288
Solid-liquid partitioning, 280
"Solids effect", 269
Solids-liquid separation, 269
Solubility, 15, 182, 201, 203–206, 260, 301
Solubilization, 28, 203
Solutes. *See also* specific types
    activity coefficients for, 260
    colloid interaction with, 261
    dissolved, 77
    polymer binding to, 256
    uptake of, 220, 291
Solvents, 8, 105, 311. *See also* specific types
Sorbate-organocation systems, 302
Sorbates, 216, 291, 302. *See also* specific types
Sorbents, 276, 291, 301–302. *See also* specific types
Sorption, 148
    batch experiment studies of, 256–257
    of benzene, 218
    competitive, 220, 224, 291–305
    of DOC, 242
    equilibrium, 242, 293, 295–296
    isotherms of, 184, 285
    kinetics of, 235
    of metals, 48
    of NOCs, 181, 262–268
    nonequilibrium, 292, 294–295
    nonionic-compound, 216–224
    of PCE, 291
    physical, 8
    reduced, 203–206
    of TCE, 275, 279
    of tetrachloromethane, 219, 221
    van der Waals, 8
    of vapors onto soil, 275–288
Sorption/desorption, 231, 234–235, 239–242
Spectral analysis, 130. *See also* specific types
Spectrometry. *See* Spectroscopy
Spectrophotofluorimetry, 158. *See also* Spectroscopy
Spectrophotometry, 131. *See also* Spectroscopy
Spectroscopy, 48. *See also* specific types
    atomic emission (AES), 56
    infrared, 130, 175, 256
    mass, 17, 38, 57, 67, 256
    nuclear magnetic resonance. *See* Nuclear magnetic resonance (NMR)
    photon correlation (PCS), 55, 57
SRFA. *See* Suwannee River fulvic acid
Standard equilibrium, 30
Steady-state DOC, 242
Steric effects, 299

Stoichiometry, 330
Stokes-Einstein equation, 237
Stokes settling, 53
Substitution reaction, 215
Succinic acid, 125
Sugars, 12, 14. *See also* specific types
Sulfate, 331, 334
Sulfated humic substances, 95–97
Sulfate esters, 14
Sulfolane, 8
Sulfur, 89, 95
Surface-active agents. *See* Surfactants
Surface adsorbents, 195–198
Surface adsorption, 302
Surface charge, 225
Surface tension, 201
Surfactants, 201–226, 270. *See also*
    specific types
  anionic, 203
  applications of, 201
  cationic, 202, 212
  chemical properties of, 203–211
  classification of, 201
  clay interaction with, 213–215
  commercial, 201, 203, 204
  concentration of, 207, 208
  critical micelle concentration of, 201,
    203
  defined, 201
  mobility decrease and, 211–225
  mobility increase and, 203–211
  molecular structure of, 201, 203
  non-aqueous-phase liquid mobility
    and, 206–211
  nonionic, 203
  petroleum sulfonate, 203
  reduced sorption and, 203–206
  solubility enhancement and,
    203–206
Suwannee River fulvic acid (SRFA),
    112, 114, 116, 117, 118, 119, 120,
    122, 124, 142
  hydrolysis of, 163–175
    equipment for, 164
    materials for, 164
    results of, 165–172
Syringic acid, 329, 331

Tannins, 15

TCB. *See* Trichlorobenzene
TCE. *See* Trichloroethylene
TEC. *See* Total extracted carbon
Temperature, 11
Territorial binding, 31
Tertiary oil-recovery techniques. *See*
    Enhanced oil recovery (EOR)
2,3,5,6-Tetrachloro-*p*-benzoquinone
    (chloramil), 4
3,3'-4,4'-Tetrachlorobiphenyls, 317
2,3,7,8-Tetrachlorodibenzodioxane, 4
Tetrachloroethene (PCE), 208, 209,
    210, 211, 291
Tetrachloromethane, 205, 219, 221
Tetradecyltrimethylammonium, 219
Tetraethylammonium, 214, 216, 219,
    223
Tetramethylammonium (TMA), 184,
    185, 186, 195, 198, 212, 216, 219,
    221
Tetramethylammonium (TMA) illite,
    224
Tetramethylammonium (TMA)
    montmorillonite, 222, 223
Tetramethylammonium (TMA)
    smectite, 196, 197
Thermal-degradation methods, 16
Thermal destruction, 311
Thermodynamic equilibrium, 31
Thermodynamic partition coefficient,
    257
Thermodynamic partition model,
    251–270
Thermodynamics, 27, 29–32
Tin complexes, 148
Titration calorimetry. *See* Calorimetric
    acid-base titrations
Titrations, 31, 130, 164
  acid-base. *See* Acid-base titrations
  forward, 170
  manual, 165
TMA. *See* Tetramethylammonium
TMPA. *See* Trimethylphenyl-
    ammonium
TOC. *See* Total organic carbon
Toluene, 276, 277, 283, 328
Toluidine, 69
Total extracted carbon (TEC), 353

Total organic carbon (TOC), 233, 256, 257, 267, 353
AOX and, 344
concentration of, 265
desorption of, 269
high, 263
humic acid and, 262
humin in, 35
increasing levels of, 268
low levels of, 270
Transformations of organic substances, 14–16
Transport, 75, 224–225. *See also* specific types
diffusive, 85
of DOC, 132
DOC effect on, 245–247
of HOC, 239
humic substance-mineral particle complexes and, 23–32
modeling of, 24–29
thermodynamics of, 29–32
of humic substances in estuaries, 129–158
kinetics of, 144–147
metals and, 148–152
reactive organic carbon and, 138–144
of NOCs, 224
of organic carbon, 129
of PAHs, 108
physicochemical factors affecting, 224
pore-water colloids and, 231–248
DOC concentration profile and, 237–239
DOC size distribution and, 236–237
DOC sorption/desorption and, 234–235, 239–242
DOC transport and, 235–236
experiments with, 233–236
modeling of, 245–247
naphthalene partitioning and, 236, 244
sediment physical properties and, 233–234
in Pueblo Reservoir, 67–70
subsurface, 108

unidirectional, 246
of VOCs, 275
Triazine herbicides, 16, 32, 44. *See also* specific types
2,4,6-Trichloroanisole, 344
Trichlorobenzene (TCB), 203–204, 223, 256, 257, 262, 263, 291
1,1,1-Trichloroethane, 276
Trichloroethylene (TCE), 276, 278, 283, 293, 294, 299, 300
dissolution of, 280, 282
EDB and, 304
equilibrium sorption and, 295
negligible, 281
sorption of, 275, 279
sorption isotherms for, 285, 296
vapor partition coefficients and, 282
vapor pressure and, 279
2,4,6-Trichlorophenol, 344, 345, 346
4-(2,4,6-Trichlorophenoxy)-2,6-trichlorophenol, 346
Triethylammonium cations, 216
2,3,4'-Trihydroxybiphenyl, 316
Trimethylammonium cations, 216
1,3,5-Trimethylbenzene, 276, 277, 283, 284
Trimethylphenylammonium (TMPA) clays, 184, 185, 198, 199, 200
Trimethylphenylammonium (TMPA) smectites, 195–198
Triton X-100, 203, 205
Tyrosine, 335

UF. *See* Ultrafiltration
UFR. *See* Ultrafilter residue
Ultrafilter residue (UFR), 141
Ultrafiltration, 53, 54, 70, 101, 234, 236, 344
Ultramembrane filtration, 89
Ultrasonic dispersion, 9
Uncharacterized carbon, 158
UNIFAC, 253
Uranium complexes, 148

Valine, 142
Van der Waals forces, 8, 131, 299
Vaporization, 255

Vapor-phase partition coefficients,
    277, 282
Vapor pressure, 279
Vapors. *See also* specific types
    dissolution of, 285
    high concentration of, 285
    low concentration of, 279–283
    mixtures of, 283–285
    sorption of onto soil, 275–288
Vermiculite, 185–190, 191, 193
Vertical displacement, 207
Viscosity, 207, 211
VOCS. *See* Volatile organic
    compounds

WALTZ method, 78

Water-air interface, 281
Water column stability, 81–82
Watershed POC, 49
Wet sedimentation, 182
Wetting agents, 201
Whatman GFC, 141
Wyoming bentonite, 213, 214

X-ray analysis, 56, 58
Xylene, 277
p-Xylene, 276, 283

Yeasts, 80
YSI probes, 76

Zinc, 5, 58, 70, 148, 149, 213
Zooplankton, 49